Advanced Sciences and Technologies for Security Applications

The series Advanced Sciences and Technologies for Security Applications comprises interdisciplinary research covering the theory, foundations and domain-specific topics pertaining to security. Publications within the series are peer-reviewed monographs and edited works in the areas of:

- biological and chemical threat recognition and detection (e.g., biosensors, aerosols, forensics)
- crisis and disaster management
- terrorism
- cyber security and secure information systems (e.g., encryption, optical and photonic systems)
- traditional and non-traditional security
- energy, food and resource security
- economic security and securitization (including associated infrastructures)
- transnational crime
- human security and health security
- social, political and psychological aspects of security
- recognition and identification (e.g., optical imaging, biometrics, authentication and verification)
- smart surveillance systems
- applications of theoretical frameworks and methodologies (e.g., grounded theory, complexity, network sciences, modelling and simulation)

Together, the high-quality contributions to this series provide a cross-disciplinary overview of forefront research endeavours aiming to make the world a safer place.

The editors encourage prospective authors to correspond with them in advance of submitting a manuscript. Submission of manuscripts should be made to the Editor-in-Chief or one of the Editors.

More information about this series at http://www.springer.com/series/5540

Dimitris Gritzalis • Marianthi Theocharidou
George Stergiopoulos
Editors

Critical Infrastructure Security and Resilience

Theories, Methods, Tools and Technologies

Springer

Editors
Dimitris Gritzalis
Department of Informatics
Athens University of Economics
and Business
Athens, Greece

George Stergiopoulos
Department of Informatics
Athens University of Economics
and Business
Athens, Greece

Marianthi Theocharidou
Directorate E. Space, Security and Migration
European Commission – Joint Research
Centre
Ispra, Italy

ISSN 1613-5113 ISSN 2363-9466 (electronic)
Advanced Sciences and Technologies for Security Applications
ISBN 978-3-030-00023-3 ISBN 978-3-030-00024-0 (eBook)
https://doi.org/10.1007/978-3-030-00024-0

Library of Congress Control Number: 2018961423

This Springer imprint is published by the registered company Springer Nature Switzerland AG
The registered company address is: Gewerbestrasse 11, 6330 Cham, Switzerland

Contents

Part I
Governance & Risk Management

Resilience Approach to Critical Information Infrastructures

Eric Luiijf and Marieke Klaver

Abstract This chapter discusses new societal risk due to the fast information and communication as well as operational technology changes which are not yet fully taken into account by governmental policymakers and regulators. Internet-of-things, cloud computing, mass consumer markets and embedded operational technologies are some of the areas outlined in this chapter which may be the cause for serious disruptions of critical infrastructures, critical information infrastructures, essential services, and the undisturbed functioning of the society. Current national protection approaches mainly focus on the classical telecommunication sector and the stove-piped critical sectors such as energy, health, transport, etcetera. This chapter argues that a change of mind and actions are needed to properly govern the new cyber risk before serious incidents occur and that such a new approach is urgently needed to make the societies at large more resilient.

Keywords Policy and management · Policy analysis · Critical information infrastructure · Critical infrastructure protection · Operational technology · Internet of Things · Essential services

E. Luiijf (✉)
Luiijf Consultancy, Zoetermeer, The Netherlands
e-mail: luiijfconsultancy@ziggo.nl

M. Klaver
Netherlands Organisation for Applied Scientific Research TNO, The Netherlands
e-mail: marieke.klaver@tno.nl

D. Gritzalis et al. (eds.), *Critical Infrastructure Security and Resilience*, Advanced Sciences and Technologies for Security Applications,
https://doi.org/10.1007/978-3-030-00024-0_1

1 Introduction

The fast-changing world of information and communication technologies (ICT) and the increasing use of Operational Technology (OT)[1] introduces new cyber security-related risk to critical infrastructures (CI), critical information infrastructures (CII), essential services, and societies at large. In an attempt to mitigate and manage this cyber risk, nations have created or are creating CI protection (CIP) and cyber security related laws and regulations. Most nations solely focus on the well-known classical telecommunication sector and the ICT in their stove-piped vertical critical sectors such as energy, health, transport, etcetera. Only recently, some additional cyber-related essential services such as cloud, certificate and root/Domain Name Services (DNS) services are for instance recognized as part of the United States CI [6] and by the European Union in the so-called network and information security (NIS) directive [8]. By May 2018, the latter directive had to be transposed by the EU Member States in national legislation. At the end of November 2018, the EU Member States should have designed the operators of essential services (OES) and digital service providers (DSP). In this chapter, we will debate that national governments and regulators overlooked major areas of ICT and OT services critical to nations. Both unexpected massive scale disruptions of those services or cyberattacks stemming from such ICT and OT may cause serious effects to CI, CII and societies at large.

Therefore, we will analyze the full spectrum of the cyber risk elements that stems from the omnipresent use of ICT and OT in all aspects of our modern societies. We will show the pitfalls of the current approach to dealing with this risk. Key elements of the cyber risk to society are currently largely overlooked by governmental policymakers and regulators. Such ICT and OT elements are hidden in plain sight being key services to current CI and CII services as well as widely used ICT-services on the one hand. On the other hand, new ICT developments either may pose a new threat to CI, CII, and society, or soon will need to be recognized as CII by nations.

Last but not least, nations push CI operators to put a lot of efforts (and costs) to secure and protect certain critical (tele)communication services that already may be considered overrated as critical to society and/or for which the criticality for society is diminishing rapidly due to modal shifts to internet-based services.

Following this introduction, we provide some key definitions for this chapter. To lay some groundwork, we summarize an analysis of an extensive set of CI and NIS policies by nations with respect to what nations consider as their critical and essential information technology-based services.

[1] Operational technology (OT) according to [3] is the technology commonly found in cyber-physical systems that is used to manage physical processes and actuation through the direct sensing, monitoring and or control of physical devices. OT generally monitors and controls physical processes with sensors and actuators such as motors, valves, and pumps.

In the next section of this chapter, we propose and outline a model comprising six areas of ICT-based services which contain possible critical or essential ICT (and OT) services to a nation.

Using this model, we show the gaps with the current national approaches to critical and essential ICT-based services, and the risk they pose to society. Not all nations are on the same pace or use the same (sub)set of ICT and OT as another nation. Therefore, there is not a single recipe for nations to apply a new approach to their identification of, and legislation and regulation for such services. We will, however, pose a set of recommendations which national policy-makers may use to derive a resilient CI, CII and network & information system security-policy that flexibly adapts to the ever and fast-changing digital technologies critical to one's nation. The approach includes recommendations which may help to increase the nation's resilience against the full set of threats to such services in all hazards approach.

2 Definitions

Critical Infrastructure (CI) is defined as "*an asset, system or part thereof located in Member States which is essential for the maintenance of vital societal functions, health, safety, security, economic or social well-being of people, and the disruption or destruction of which would have a significant impact in a Member State as a result of the failure to maintain those functions.*" (derived from [17]).

Critical Information Infrastructure (CII) is defined as "*Those interconnected information and communication infrastructures which are essential for the maintenance of vital societal functions, (health, safety, security, economic or social well-being of people) – the disruption or destruction of which would have serious consequence.*" [13].

Digital Service Provider (DSP) means "*any legal person that provides a digital service within the meaning of point (b) of Article 1 of Directive (EU) 2015/1535 of the European Parliament and of the Council which is of a type listed in Annex III of* [8]" [8]. is also known as the European Union NIS directive. The list in the Annex comprises online marketplaces, online search engines and cloud computing services.

Governance is "*all of the processes of governing, whether undertaken by a government, a market or a network, over a social system (family, tribe, formal or informal organization, a territory or across territories) and whether through the laws, norms, power or language of an organized society*" [1].

Operator of Essential Services (OES) means "*a public or private entity of a type referred to in Annex II, which meets the criteria laid down in Article 5(2) of* [8]" [8].

3 Analysis of CI, CII and NIS Policies and CII Elements

Using [4] as a source for pointing to a long list of national definitions of Critical (Information) Infrastructure Sectors and services as well as National Cyber Security Strategies, national policy-level approaches identified ICT-related CI in various ways:

- Nations that define ICT-related critical sectors and implicitly their critical processes:
  - (Tele)communication sector, e.g. Australia, Bangladesh, Chile, India, Indonesia, Jersey, Republic of Korea, Trinidad and Tobago, United Kingdom;
  - ICT sector at large, e.g. Austria, Canada, Czech Republic, France, Germany, Ghana, Japan, Malaysia, Poland, Qatar, Slovakia, Slovenia, Spain, Sweden, Switzerland;
  - Electronic communications, e.g. Belgium, Norway, Turkey.
- Nations that define and outline their critical ICT in terms of critical products, services, and objects:
  - Product/service-oriented approach.
    For example, national approaches by Croatia (electronic communication, data transmission, information systems, providing audio and audio-visual media services), El Salvador (networks and telephone plants, radio and television stations), Estonia (telecommunications, transmission and notification systems, software, hardware and networks, including the infrastructure of the Internet);
  - Critical objects-oriented approach.
    The Virgin Islands, for example, regards facilities and distribution systems for critical utilities such as telecommunications as CI;
  - A split "C(ommunications)" and "IT" approach.
    The United States, as example, distinguishes critical telecommunications with the classical phone/fax/SMS and mobile services, and information technology (IT) comprising critical control systems and critical IT-services such as life-critical embedded systems, physical architecture, and Internet infrastructure.
- Nations that have specified specific critical ICT-services; two approaches can be distinguished:
  - Using a limited services list approach.
    For example, the Netherlands: "internet and data services, voice services and text messaging, internet access and data traffic";
  - Using an open-ended, flexible services list approach.
    An example is Denmark: "phone, internet, information networks, processing and transmission of data, navigation, satellite/radio/TV transmission, post and courier services, *etcetera*."

Most of the CI-related efforts stem from the national homeland security, anti-terrorism and all hazard disaster approaches. In addition, the economic pillar responsible ministry in nations often covers the digital domain and cyber security policies. Within the EU, the Directive on security of network and information systems [8] defines the following critical and essential information service areas as a minimum set for each of the EU Member States:

- Operator of Essential Services, being (a) the traditional CI operators in the energy (electricity, oil, gas), transport (air, rail, water, road), banking and financial market infrastructures, health, drinking water supply and distribution sectors, and the (b) the operators of digital infrastructure comprising of Internet Exchange Points (IXP), Domain Name Service (DNS) providers, and Top-Level Domain (TLD) registries.
- Digital Service Providers of on-line market places, on-line search engines, and cloud computing services (i.e. application services).

These two policy areas show the two main streams for identifying critical ICT-elements. Part of the elements identified are ICT/OT-related services that are critical on their own value, and the other ICT elements are critical because of the importance for the more traditional CI sectors such as energy and transport. In each of these more traditional CI, the importance of embedded ICT/OT services increases rapidly. It is notoriously difficult to identify all of these CII elements and services as CII elements tend to be more interwoven and tend to hide themselves in CI, in cyber-physical processes, and in stacks of information-based services. The speed of innovation and uptake of new digital technologies by societies in processes that evolve into critical processes is high. As a result, new critical ICT- and OT-based functions and services appear seemingly out of the blue in the telecommunications and IT CI sector(s), the classical sector-specific CI (as shown in Fig. 1), and also beyond these established domains.

National strategies for proper governance including legislation and regulation regarding C(I)I and essential services should cover two aspects:

- Guaranteeing the adequateness of the level of protection and resilience of C(I)I and essential ICT- and OT-based services;
- Protecting ICT and OT against vulnerabilities and malicious use, e.g. as part of a distributed denial of service network which may seriously impact the functioning of C(I)I.

Based upon this view, we analyzed the current and future critical and essential services which need to be covered. As a result, a conceptual model for the whole cyber domain is proposed comprising six areas to be covered by governance at the national level.

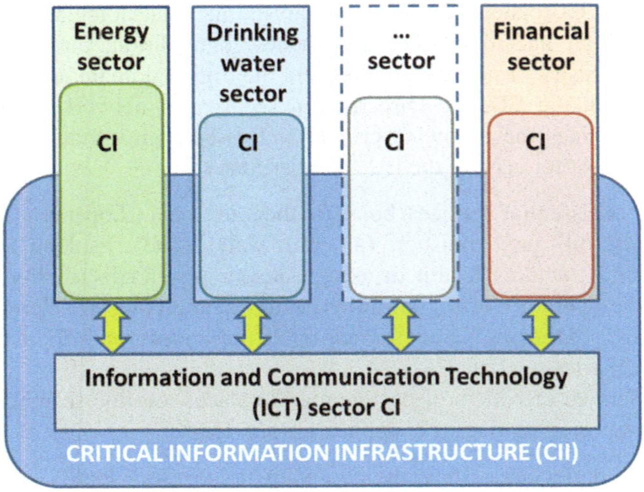

Fig. 1 Critical information infrastructure [13]

4 Conceptual Model

In support of the identification of CII elements, a conceptual model was developed that represents six different ICT/OT areas that need to be monitored for the need of governance at the national level or even at the international level. These six areas (see Fig. 2) and their cross-area delivered products and services (arrows in Fig. 2) are:

1. Key manufacturers.

   A relatively small set of extremely large, globally operating manufacturers of hardware and software produce key components for ICT- and OT-systems. Their components are used at a large scale in C(I)I and/or the mass consumer market, e.g. processor chips by Intel and AMD, operating systems and business applications by Microsoft (e.g. Windows, Office) and Google (Android), networking by CISCO systems, and OT by for instance Wonderware, Rockwell Automation, Siemens, and Honeywell.

   Not all subcomponents to their products are homebred. Multiple key manufacturers may use the same embedded software libraries produced by very specialised companies or open source software providers.

   When a serious security flaw in such a product is found and published, easily many hundreds of million systems both in CII, essential services, and consumer systems may be vulnerable overnight. Examples are the Branchscope, Meltdown, Spectre vulnerabilities in Intel processors [15], the vulnerabilities used by the Stuxnet malware, and the WPA2 crack vulnerability making WiFi protocol implementations insecure [20].

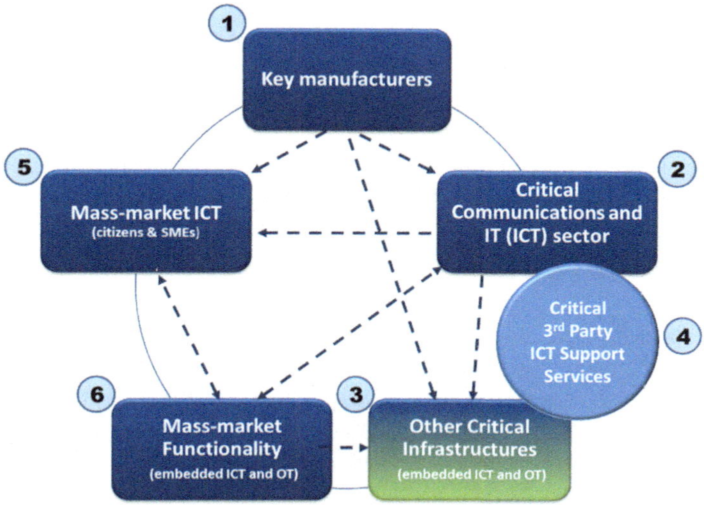

Fig. 2 The six CII and essential ICT/OT services elements and cross-area delivered products and services

Such vulnerabilities may actively be attacked and exploited within hours after they become public knowledge. Hundreds of thousands OT systems or millions of ICT systems worldwide may be affected. Innocent end-users, small and medium enterprises (SME), and organisations may be too late to take mitigation actions, if they exist at all. The result of a major vulnerability in such products may be disruptive to society.

An example was the Heartbleed vulnerability which affected the privacy of all registered users of amongst others Blogger/Blogspot, Dropbox, Facebook, Electronic Frontier Foundation, Etsy, Google, Imgur, Instagram, Netflix, OKCupid, Pinterest, Stack Overflow, Wikipedia, Woot, Wordpress.com/Wordpress.org, and YouTube [21]. It was estimated that millions of end-users using services from some 600,000 flawed servers worldwide were at risk.

Fortunately, most key manufacturers administer their products in a paternal way; even after products are beyond end of support, critical patches may appear to mitigate very serious cyber security risk. In case of embedded libraries, however, key manufacturers may be slow or even refrain from resolving major vulnerabilities causing cyber risk to many.

Nations have to continuously monitor such risk to their nation and population and take coordinated international action in case manufacturers react slowly or not at all.

2. Critical communications and IT (ICT) sector.

This is the critical CII element which provides the national critical core services and functions of the classical communications sector (wireline and cable infrastructure, mobile telecommunications, navigation systems, ground and space segments for satellite communications, and broadcast). An area with fast technological changes and organizational changes due to liberalisation, privatization, mergers, and acquisitions. Over the last decade, "internet access" services were added to their set of CI services by an increasing number of nations, The Netherlands being the first in 2001 [16]. Only recently, nations recognise other Internet-related services as CII or as essential services, e.g. in the NIS directive [8]: key digital infrastructure service providers Internet Exchange Points, Top-Level Domain registrars and root/Domain Name Service providers, and DSP.

- Despite the fact that the Internet and its services are critical to modern societies, many nations do not pursue much the governance of this CII area. Nations take the stance that private industry has the lead. It is only when market failure occurs, that regulators and government may reluctantly step in. Earlier occasions have not led to a wake-up call. We just mention to (a) the bankruptcy filing by KPNQwest affecting 67 country code top-level domains in May 2002 [19], (b) hostile takeover of a CII operator by a foreign company, and (c) cut undersea cables depriving multiple nations from internet services at the same time. Telecommunication backbone operators are recognized as CII in the USA but have not been identified as critical or essential by nations in Europe.
- An issue in this second area is the risk of foreign influence on a national CII or essential services by mergers and acquisitions. Several nations pro-actively have developed or are developing legislation to block foreign takeovers of CII operators, e.g. Australia, the United Kingdom and Canada; the European Commission started to develop regulation to protect essential assets to the Union and its Member States [9]. Other nations may find themselves in a position where a foreign acquisition happens and they have to try to remedy this risk to national security after the fact.

3. ICT and OT embedded in other CI sectors.

   Major technological changes in (embedded) ICT and OT in 'traditional CI' services such as the energy and financial sectors may cause the need to add new critical services to the national set of CII and essential services. As a result, the criticality of one or more CII and essential services may fade overtime. An example of possible new CII is the blockchain infrastructure including cryptocurrency services, whereas the use of national beeper infrastructures diminishes fast as their functionality is replaced by the internet and mobile telephone technologies.

   Moreover, all nations that have created of list with CI sectors, have put power on their list of critical services. However, currently roll-outs occur of smart meters, solar power panels and other distributed energy resources at a massive scale. All equipment which has embedded ICT and sometimes OT.

   Is the risk of a nation-wide power blackout to be left to the market or do national authorities have a say? An example was the recent need for a firmware upgrade of almost 230,000 solar power plants (millions of solar panels) in Germany as the exploitation of a software vulnerability could cause such a blackout [2]. How can a nation ensure that the majority of the plant owners update their system?

4. Critical services by third parties to ICT service providers.

   Certain services provided by third parties to the ICT sector such as name and address services can be crucial for the operations of CII and essential services, and implicitly CI. Both technological and organizational changes in this area may (silently) cause shifts in the set of CII.

   Such services that support the functioning of critical and essential ICT-services are often provided out of plain sight and are overlooked by policymakers. The services may be offered by relatively small businesses. These businesses may even fail to meet the number of employees and turnover criteria set by governments to identify their set of CII and essential operators, e.g. as part of the implementation of the NIS directive.

   Only recently, services in this fourth area are recognized by a number of nations as critical or essential. Both the USA and the EU Member States (by means of the NIS directive) recognize top-level domain name registrars, DNS service providers, and Internet Exchange Points as critical respectively essential to their nations.

   Unfortunately, this set of services is incomplete. An essential service in the Internet domain comprises the trust and security infrastructure. High-trust certificate, Public Key Infrastructure (PKI) and Trusted Third Party (TTP) services, especially those used in e-government and banking, are not yet recognized by the USA and EU as essential, despite the lessons identified by The Netherlands in 2011. The certificate infrastructure used by the Dutch government and the municipalities was considered compromised. National crisis management response actions were required which included the nationalization of DigiNotar, the certificate operator, and postponing of the certificate invalidation by Windows Update for the .nl domain [14]. The latter action provided time for other certificate providers to generate a new certificate root with some 3500 new strong certificates in a trusted way. Moreover, installation of these certificates in e-government and trusted machine-to-machine information services of 400–500 municipalities and governmental agencies required a lot of effort.

   More of this type of essential third-party services critical to the functioning of internet and telecommunication exist but are hidden to the plain sight of the public and governmental policymakers. Hidden, that is, until a major incident occurs and an unexpected crisis occurs.

   The risk of a large-scale impact is high. Therefore, some form of governmental oversight is essential for this type of essential services where a higher level of assurance and business continuity is required. One might consider regulatory measures similar to those that apply to the CII and essential operators in the telecommunication and IT sector(s), e.g. by applying the same breach reporting requirements as applied to the set of digital service providers referred to by the NIS directive.

5. Mass market ICT.

   Although citizens, small and medium enterprises (SME) and organizations use a lot of ICT embedded in functions and for accessing information services, we are still on the verge of a mass market uptake of ICT and OT in our daily lives. Service disruptions of Facebook and WhatsApp already cause people to dial the 1-1-2 (or 9-1-1) emergency number as people perceive that their social live breaks down as they 'do not live in cyberspace anymore' [11].

   - When larger numbers of citizens and SME distrust their ICT, cannot access their social networks or electronic banking, etcetera for more than a couple of days, the socio-psychological impact may come close or exceed the national criteria used to define C(I)I.

     EU's NIS directive takes into account digital service providers as market places and search engines but lacks the societal impact of disrupted mass-scale end-user ICT-based services and social media. Moreover, Google started in 1998, Facebook in 2004, and Whatsapp in 2009; can you pinpoint today's start-up that will be considered as an essential service in about 5 years by your nation?

6. Mass market functionalities with *embedded and connected* IT and OT.

   Psychologically, people are addicted to anything that make one's live more convenient, provides us happiness, or provides delight. That drives the mass markets of (a) Internet of Things, and (b) enhanced functionalities by embedded IT and OT which people do not recognize anymore as IT/OT. Increasingly, consumer and professional product functions are based on IT and OT embedded in the product. Most often, these products connect to and interact with the internet, and with ICT that is part of the fifth area above: 'Mass market ICT'.

   - Market analysts expect an exponential uptake of Internet of Things (IoT) and smart appliances with estimates that range between 28 and 50 billion IoT devices worldwide by 2021 (see e.g. [5]).

   - The development of the IoT products is predominantly by other manufacturers than the traditional IT manufacturers. Most of those manufacturers have a reputation in producing functional equipment, e.g. kitchen appliances, for the consumer market. The new functionality is based on the embedded use of IT and OT. The manufacturers, however, lack long-term knowledge and experience in cyber security.

   - The other set of manufacturers are start-ups who invented new functionality (with embedded IT and OT) which make people happy and one's live easier. Think about manufacturers of washing machines, smart fridges, digital TVs, thermostats, carbon monoxide and smoke detector sets (e.g. Google Nest), smart toasters, home lighting, smart doorbells, and home automation equipment.

   - Such mass-market IoT devices provide major cyberattack opportunities as the IoT devices are spread in manifold across multiple nations and are connected to a wide array of telecommunication and internet service providers. The first glimpses of IoT-based cyberattacks to C(I)I have already been shown by e.g. the Mirai, Hajime and Bricker botnets [10].

- To reduce this societal risk by insecure IoT, *manufacturers* on the one hand should properly implement security standards and take adequate preventive and response measures such as timely provision of patches in case of a vulnerability. On the other hand, the EU and a set of its Member States currently discuss governance measures especially for IoT such as certification of hardware and software, and ICT-product liability [7]. How such an approach will work for securing mass-market products such as smart dish washing machines, smart doorbells, and smart BBQs is currently unclear. If this governance by nations is not arranged fast and too a full extend, nations need to be prepared to deal with major incidents with (inter)national impact.
- Another threat arises from the massive roll-out of enhanced functionalities in cyber physical systems including autonomous vehicles and robots. Embedded ICT and OT, but hidden to the unconscious end-users, support new functionalities to our daily life. Consider the amount of ICT- and OT-supported functions in a modern car. In a split moment software decides to hit the brakes, explode the air bag, correct the drive by a lane departing system, or automatically park the car. And then we are not yet discussing autonomous driving by for instance Teslas and Google cars. Soon we will see platooning 'trains' of trucks on the road [18].
- An exploited vulnerability, e.g. through Ecall, mobile connectivity or WiFi, may result in a safety risk to persons in or near that cyber-physical system. Just consider all cars of one brand exploding airbags while driving on the highway due to malware.
- Nations, their policy-makers, legislation, regulators, and crisis management currently do not consider how deal with for example:
- millions of smart TVs affected by malware or acting as a denial-of-service attack platform to C(I)I,
- millions of smart fridges, smart washing and dish-washing machines with a serious exploited vulnerability which causes instabilities in the smart power grid,
- an exploited cyber security flaw or software failure affecting the safety of millions of vehicles taking part in collaborative driving or autonomous driving.
- In a first reaction, authorities will state that the consumer is responsible for the cyber security problem and that he/she should discuss the problem with the manufacturer. But when power grids blackout and hospitals shut down operations as a result of such an attack?
- Proper governance of such threats and part of C(I)I should start exploring scenarios about the 'unthinkable' moment one has to act.

5 Conclusion and Recommendations

The (new) stakeholders in these six areas that can be considered as (potential) key players in CII and the provision of essential IT- and OT-based services need to be involved in the governance and the protection thereof. The first challenge for nations is to identify sufficiently early that such new services are part of their set of CII or set of essential services. This means that nations need to keep track of technological developments as well as changes in the ownership of organizations, e.g. through mergers and acquisitions, as part of this identification process.

Secondly, the obligations of CII stakeholders and essential service operators need to be applied to new stakeholders as well. On the one hand this can be difficult to pursue as new entrants might be reluctant to become part of the existing CIIP or essential service protection community. They may be unwilling to bear the costs of increased protection for society. On the other hand, the long existing trust-based operator communities may be reluctant to admit operators of fast rising essential services to their inner circle of sharing cyber security information.

Moreover, some traditional CII operators may fall below the threshold of criticality criteria as their systems are no longer deemed critical or essential at the national level. They may be reluctant to give up their position to the inner information sharing circle.

The analysis above showed that the current risk and approaches by governments to protect the CII and essential services in their nation are too much focused on the classical telecommunications and IT-sector. CII and essential IT- and OT-based services increasingly appear in other, less or even none governed areas of the current ICT landscape. New globally communicating IT- and OT-based products and services appear fast on the market. Such products and services may be used at a massive scale, e.g. IoT. In case their level of cyber security is insufficient, citizens may expect that authorities step in and take action such as creating legislation or more oversight by regulators. Moreover, governments need to be alert to reduce the risk of mass-scale cyberattacks through mass market equipment on the one hand, and on the other hand of disruptions in new CII and essential services which suddenly disrupt the perceived undisturbed way of living of tens of million people.

Based upon the analysis above, we recommend governmental policy-makers to:

- Use the list of identified ICT areas (Fig. 2) as a checklist for checking your policies and identification of possible new CII and essential IT- and OT-based services as well as CII services that are not critical or essential anymore.
- Collaborate internationally in order to define policies for international ICT products and services that are identified as CII or essential service in one's nation.
- Use horizon scanning and technology watching to identify both emerging technologies, and IT- and OT-based products and services that may become critical or essential rapidly. After identification, proper governance activities need to be started.

- Consider how to govern the risk of cyberattacks stemming from mass-market products and services. How can nations and key stakeholders prevent, prepare for and manage response to mass-scale cyberattacks that exploit IoT vulnerabilities? And how to guarantee the safety of citizens when cyber-physical systems in vehicles, robots, etcetera are vulnerable and attacked?

Acknowledgments This chapter is a follow-up on earlier work by the authors in the domain of legal risk regulation which was published in [12].

References

1. Bevir M (2012) Governance: a very short introduction. Oxford University Press, Oxford
2. Boemer JC et al (2011) Overview of German grid issues and retrofit of photovoltaic power plants in Germany for the prevention of frequency stability problems in abnormal system conditions of the ENTSO-E region continental Europe. In: 1st international workshop on integration of solar power into power systems, p 6
3. Boyes H, Isbell R (2017) Code of practice cyber security for ships. London, United Kingdom
4. CIPedia(c) (n.d.). Available at: http://www.cipedia.eu. Accessed 18 June 2018
5. CISCO (n.d.) Internet of Things (IoT). Available at: https://www.cisco.com/c/en/us/solutions/internet-of-things/overview.html. Accessed 16 June 2018
6. DHS (2017) Critical infrastructure sectors. Available at: https://www.dhs.gov/critical-infrastructure-sectors
7. ENISA (2017) Considerations on ICT security certification in EU Survey Report. Heraklion, Greece https://doi.org/10.2824/090677
8. European Commission (2016) Directive (EU) 2016/1148 of the European Parliament and of the Council of 6 July 2016 concerning measures for a high common level of security of network and information systems across the Union. Brussels, Belgium. Available at: http://data.europa.eu/eli/dir/2016/1148/oj
9. European Commission (2017) Proposal for a Regulation of the European Parliament and of the Council establishing a framework for screening of foreign direct investments into the European Union. Brussels, Belgium. Available at: https://ec.europa.eu/transparency/regdoc/rep/1/2017/EN/COM-2017-487-F1-EN-MAIN-PART-1.PDF
10. Fisher D (2017) BrickerBot, Mirai and the IoT Malware Knife Fight. Digital Guardian blog. 26 April. Available at: https://digitalguardian.com/blog/brickerbot-mirai-and-iot-malware-knife-fight.
11. Justicenews (2015) Facebook outage sparks calls to 911. Justice Bews Flash. 27 January. Available at: http://www.justicenewsflash.com/2015/02/02/facebook-outage-sparks-calls-to-911_20150202133988.html.
12. Luiijf E, Klaver M (2015) Governing critical ICT: elements that require attention. Eur J Risk Regul 6(2):263–270. https://doi.org/10.1017/S1867299X00004566
13. Luiijf E, Van Schie T, Van Ruijven T (2017) Companion document to the GFCE-MERIDIAN good practice guide on critical information infrastructure protection for governmental policy-makers. The Hague, Netherlands. Available at: https://www.thegfce.com/good-practices/documents/publications/2017/10/30/companion-document-to-the-gfce-meridian-good-practice-guide-on-ciip
14. Ministry of Security and Justice (2011) Dossier DigiNotar webpage, Dossier DigiNotar. Available at: https://www.ncsc.nl/english/current-topics/Files/dossier-diginotar.html.
15. Olenick D (2018) Researchers uncover BranchScope, a new Intel processor vulnerability. SC Magazine, 27 March. Available at: https://www.scmagazine.com/researchers-uncover-branchscope-a-new-intel-processor-vulnerability/article/754159/

16. StasV&W (2001) Nota Kwetsbaarheid op internet (KWINT). The Hague, The Netherlands: Tweede Kamer der Staten Generaal. Available at: https://zoek.officielebekendmakingen.nl/dossier/26643/kst-26643-30
17. The Council of the European Union (2008) Council Directive 2008/114/EC of 8 December 2008 on the indentification and designation of European critical infrastructures and the assessment of the need to improve their protection. Off J Eur Union 75–82
18. TNO (2017) Truck platooning technology ready for the public roads. The Hague, The Netherlands. Available at: https://www.tno.nl/en/about-tno/news/2017/10/truck-platooning-technology-ready-for-the-public-roads/. Accessed 18 June 2018
19. Touton L (2002) IANA handling of root-zone changes. Available at: http://www.dnso.org/clubpublic/council/Arc11/msg00123.html.
20. Vanhoef M, Piessens F (2017) Key reinstallation attacks: forcing nonce reuse in WPA2. In: Proceedings of the 2017 ACM SIGSAC Conference on Computer and Communications Security (CCS 2017). ACM, pp. 1313–1328. Available at: https://lirias.kuleuven.be/handle/123456789/620017
21. Wagenseil P (2014) Heartbleed: who was affected, what to do now. Tom's Guident, April. Available at: https://www.tomsguide.com/us/heartbleed-bug-to-do-list,news-18588.html.

Methodologies and Strategies for Critical Infrastructure Protection

Nikolaos Petrakos and Panayiotis Kotzanikolaou

Abstract The protection of critical infrastructures at a national level is not a trivial task. In involves various steps such as the indentation, the prioritization and the protection of those infrastructures and services that are vital for the wellbeing of the society. Although some sectors, subsectors and services seem to be very important for all countries, others may differ in their significance based on the specific economic, environmental and social characteristics of each country. In this chapter we review existing methodologies and national strategies for critical infrastructure protection. We examine methodologies for identifying and assessing critical sectors and services, relying on top-down and bottom-up administrative approaches. We examine common practices that have been applied in various countries to identify critical infrastructures and to establish national protection plans. Finally, we describe a set of goals that are commonly found in different methodologies and best practices for critical infrastructure protection.

Keywords Critical Infrastructure (CI) · CI identification · CI assessment criteria · CI protection strategies

1 Introduction

According to the European Council [4] the term Critical Infrastructures (CI) means an asset, system or part of a system that is essential for the proper operation of vital societal functions, related with health, safety, security, economic or social well-being of people. Failure to maintain a tolerable operation level, or worse the complete collapse of a CI would have a significant impact on human, social, economic and national levels. Business and industry from both the private and the public sectors, largely depend on CIs for their vital functions. Typical examples

N. Petrakos · P. Kotzanikolaou (✉)
Department of Informatics, University of Piraeus, Piraeus, Greece
e-mail: npetrakos@unipi.gr; pkotzani@unipi.gr

D. Gritzalis et al. (eds.), *Critical Infrastructure Security and Resilience*,
Advanced Sciences and Technologies for Security Applications,
https://doi.org/10.1007/978-3-030-00024-0_2

of CIs include the Information and Communication Technology (ICT), Energy, Transport, Economy (Financial Sector), Health, Defense, Food, Water Supply and Governmental services.

Most CIs can be modelled as cyber-physical systems, where the information (cyber) part, controls the physical components and the underlying structures, to manage, control and optimize the goals and functionality of the CIs. For example, the control of the physical components of a power plant (*e.g.* generators or distribution elements) can be performed via interactive sensors and actuators, which periodically check the status, transmit information to central information control systems and accept network commands to modify the status of the physical components accordingly.

The identification and evaluation of CIs is a challenging task, while the process and methodologies used to identify and assess critical infrastructures, especially those that may affect people and systems at a national level, may differ from one country to another. In this chapter we review methodologies for the identification and evaluation of CIs. We examine various criteria that have been utilized in existing methodologies and we describe relevant best practices from various countries. Finally, we describe some goals that are commonly found in different methodologies and best practices for critical infrastructure protection.

2 Methodologies for Identifying and Evaluating Critical Infrastructures (CIs)

The identification of national critical infrastructures is mandated by various national and international regulation. Within the framework of a national protection programme, each EU member state has to identify the "National Critical Areas/Sectors", record and evaluate their systems or components [4], as well as to record and evaluate (potential) interdependencies between the identified CIs. Also, to develop and/or update an *Operator Security Plan* and an *Emergency Plan* to protect their national CIs.

Since, in most cases, CI owners and/or operators are private entities, any process to identify national CIs, and any other process under a national protection programme, requires the exchange of information between the parties involved, in accordance with the principle of stakeholder cooperation, in particular, *public-private partnership* (PPP) [18]. The above framework for the implementation of each national protection programme is depicted in Fig. 1.

EU Council [4] provides further direction in identifying and designating national CIs for each member state, as it indicates the obligation for member states to identify any potential European Critical Infrastructure (ECI) in their territory, *i.e.*, those CIs, the disruption or destruction of which would have significant cross-border impacts. Indeed, often, the damage or loss of a CI in one member-state may have negative effects on several others [8]. The Directive concentrates on the energy and transport

Fig. 1 Components of a national protection program, as identified in [12]

sectors, while it leaves open the inclusion of other sectors within its scope, giving a priority to the information and communication technology (ICT) sector.

The necessary condition for the implementation of the Directive in each member state is, therefore, the identification of the CIs in each member state so that any potential ECI can be drawn from this list [6]. For each national CI, identified as an ECI, each member state is responsible, among other things, for gathering information on security risks, threats and vulnerabilities per ECI. The ultimate goal of the Directive is to implement, in all designated ECI, Operator Security Plans (OSP) in order to protect them at an operator level, within the framework of a common European strategy.

2.1 Critical Infrastructure Identification

At a primary level, the exact definition of a CI varies from country to country. In general, Critical Infrastructures are infrastructures whose disruption, failure or destruction would have a significant impact on public health, public and civil matters, the environment, security, social and economic well-being. Besides the differences at a definition level, there are often differentiations at a more substantial level, with an impact on the identification of CI. In Germany, for example, CI is divided into vital technical infrastructures, and vital socio-economic service infrastructures [16]. In another example, in Great Britain [15], infrastructures are divided into critical national infrastructures and other critical infrastructures.

Based on the relevant literature (*e.g* [11]) and widely used international practices, a formal process for identifying and designating national CIs can be implemented in four consecutive stages:

1. *Identification of critical sectors/subsectors.* At this stage, the sectors and/or subsectors that are considered important for national interests are identified.

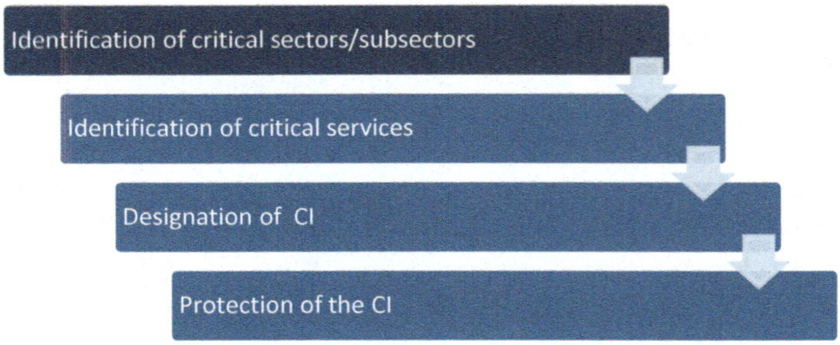

Fig. 2 Typical stages of a CI protection program

2. *Identification of critical services.* Critical (or vital) services of the sector/subsector are identified and designated for each critical sector.
3. *Designation of CI.* For each critical service, the critical assets/components that comprise the CI are identified and designated.
4. *Protection of the CI.* Procedures for protection and security are implemented for each CI.

The above stages and their connections are depicted in Fig. 2.

2.1.1 Identification of Critical Sectors/Subsectors

Atthe stage of identifying critical sectors, each member state draws up an initial list of its national critical sectors, that is, the sectors that exist within the geographical boundaries of its territory and which include potential CI. The process of selecting the national critical sectors and subsectors is not always straightforward. All national sectors are not equally vital/critical in each country. Some sectors can be classified as critical and some as less critical or less important. Moreover, not all services of a sector/subsector are equally critical, which makes it difficult to identify the initial list of critical sectors at a strategic level. The European Commission [8] recognizes the diversity of the national critical sectors in each member state, *i.e.* the difficulty of identifying the critical sectors/sub-sectors in each country and of identifying, designating and prioritizing CI within every critical sector.

However, in view of creating a common framework for the European Programme for Critical Infrastructure Protection (EPCIP), a common list of critical sectors/subsectors is encouraged. Such an indicative list is presented in Table 1 (European Commission [9]). Several Member States have adopted this list to identify their national critical domains. In line with the widely used CI identification methodology by sector, ENISA (Rossella and Cédric, Methodologies for the identification of Critical Information Infrastructure assets and services, [20]) proposes a variation of the indicative table of the Green Paper [9], which is depicted in Table 1 and incorporates the concept of *service* per sub-sector. The concept of service is often used as a synecdoche for the term infrastructure, as it integrates on

Table 1 Indicative critical sectors and their related critical services

Sector	Subsector	Service
1. Energy	Electricity	Generation (All Forms)
		Transport/Distribution
		Electricity market
	Petroleum	Extraction
		Refinement
		Transport
		Storage
	Natural gas	Extraction
		Transport/Distribution
		Storage
2. Information & Communication Technologies (ICT)	Information Technologies	Web services
		Datacentre/Cloud services
		Software as a Service (SaaS)
	Communications	Voice/Data communication
		Internet
3. Water	Drinking water	Water storage
		Water distribution
		Water quality assurance
	Wastewater	Wastewater collection and treatment
4. Food		Agriculture/Food production
		Food supply
		Food distribution
		Food quality/safety
5. Health		Emergency healthcare
		Hospital care
		Supply of pharmaceuticals, vaccines, blood, medical supplies
		Infection/epidemic control
6. Financial services		Banking
		Payment transactions
		Stock exchange
7. Public order & safety		Maintenance of public order
		Judiciary and penal systems
8. Transport	Aviation	Air navigation services
		Airports operation
	Road transport	Bus/Tram services
		Maintenance of the road network
	Train transport	Management of public railway
		Rail transport services
	Maritime transport	Monitoring and management of shipping traffic
		Ice-breaking operations
	Postal/Shipping	

Table 1 (continued)

Sector	Subsector	Service
9. Industry	Critical industries	Employment
	Chemical/nuclear industry	Storage and disposal of hazardous materials
		Safety of high-risk industrial units
10. Civil administration		Government functions
11. Space		Protection of space-based systems
12. Civil protection		Emergency and rescue services
13. Environment		Air pollution monitoring and early warning
		Meteorological monitoring and early warning
		Groundwater monitoring and early warning
		Marine pollution monitoring and early warning
14. Defense		National defense

Rossella and Cédric, Methodologies for the identification of Critical Information Infrastructure assets and services. Guidelines for charting electronic data communication networks, [19]

a level sufficiently abstract and sufficiently descriptive the concept of a set of assets/products and processes that (ultimately) need protection.

2.1.2 Identification of Critical Services

Two main approaches can be found in the literature for identifying national critical services by sector [12]; (Rossella and Cédric, Methodologies for the identification of Critical Information Infrastructure assets and services. Guidelines for charting electronic data communication networks, [17, 19]: the State-driven (top-down) and the Operator-driven (bottom-up) approaches, briefly described below.

State-Driven Approach (Top-Down)

In this approach (Fig. 3), the government has a leading role in defining and prioritizing critical services – through a coordinating body – or a competent Authority for the protection of CI. At the central level, a list of indicative national critical sector services is compiled. Alternatively, a list of national critical services is drawn up at a cross-sectoral (or transverse) level. Critical services are then evaluated, with specific criteria, and prioritized to give the final list of national critical services. For each critical service, a list of stakeholders-operators is drawn up, from which (or in cooperation with) a list of the most critical goods, products and systems supporting this service is extracted [12]. This approach has been used in countries such as Switzerland, Estonia, the Netherlands and the Czech Republic.

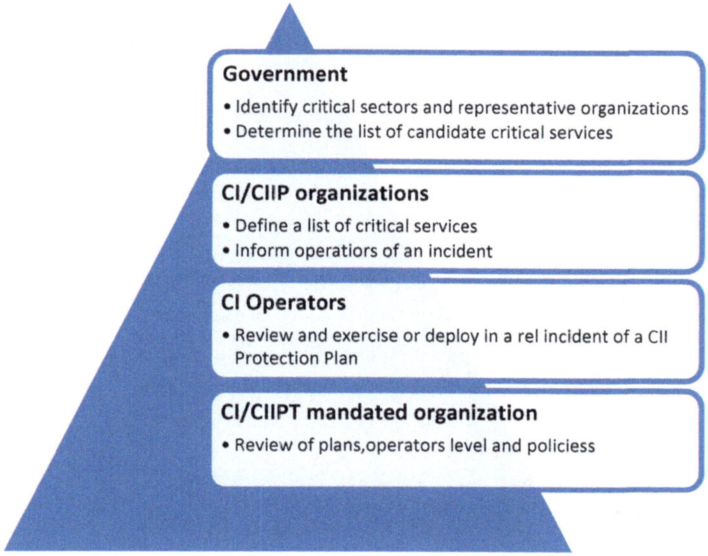

Fig. 3 A Top-down approach for the Identification and Protection of Critical Services. (Rossella and Cédric, Methodologies for the identification of Critical Information Infrastructure assets and services. Guidelines for charting electronic data communication networks, [19])

Operator-Driven Approach (Bottom-Up)

In the operator-driven approach (Fig. 4), the leading role is assigned to the operators of critical infrastructures. In particular, after national critical sectors/subsectors have been identified at the central level (based on Public-Private Partnerships – PPP), a list of stakeholders-operators of CIs, also known as *Vital Operators* (VO), is drawn up, who are requested to identify and evaluate the critical services and the most critical assets/systems that they comprise. In several European countries, this responsibility is assigned to the relevant VO by the body responsible for the relevant critical sector (e.g. the relevant ministry). It should be stressed that in this approach, the VO goes into the foreground, as opposed to the concept of critical service that dominates the State-driven approach. In a way, the VO is itself a crucial asset that needs protection [12]. This approach has been used in countries such as France and the United Kingdom. Each approach requires the implementation of a metrics to assess and prioritize the criticality of each asset (e.g. services, goods, systems, etc.) of a critical sector. This process, despite its importance, is not obvious, especially since, almost always, the criteria for the criticality of the data involved vary from country to country. Moreover, an infrastructure may be critical because of the interdependencies between the service that the infrastructure supports and other services within the same or a different sector.

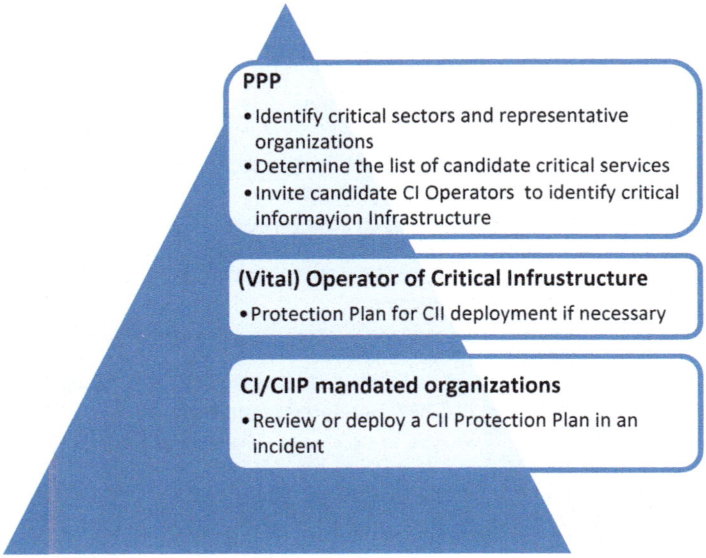

Fig. 4 Operator-driven approach. (Rossella and Cédric, Methodologies for the identification of Critical Information Infrastructure assets and services. Guidelines for charting electronic data communication networks, [19])

2.2 *Dependencies and Interdependencies*

An asset (e.g., infrastructure, service or subsector) may be critical not only due to the direct impact that may cause due to its malfunction or loss thereof (also known as first-order effect) but also due to the impact on other critical assets (also known as second-order or more generally multi-order effects) [12]. The first-order effects reflect the direct vitality of a critical asset for the society [14], *e.g.* according to the cross-cutting criteria mentioned above, while the multi-order effects reflect the indirect vitality of the asset on other critical assets.

Typically, second-order effects are understood either as dependencies, where a critical element depends on another element or as interdependencies where two critical elements are mutually affected at national or even transnational levels (Rossella and Cédric, Methodologies for the identification of Critical Information Infrastructure assets and services. Guidelines for charting electronic data communication networks, [19]). It is noted that dependencies or interdependencies may exist either within the sector/subsector in which a service operates or between two or more sectors/subsectors at the national level or between two sectors/subsectors operating in different member states.

2.3 *Criteria for Critical Service/Infrastructure Assessment*

In the literature, two families of criteria that can be used to assess the criticality (and then the prioritization) of a potential critical service or infrastructure are found, as described below.

2.3.1 Sectorial Criteria

Sectorial criteria are technical or functional criteria by which potential CIs can be identified and prioritized. For example, sectorial criteria may relate to (usually quantifiable) specific properties or features of an infrastructure that supports this sector service. These features may either be technical (e.g. minimum diameter of oil or gas pipeline, minimum capacity, electric power in Megawatt etc.) or not (e.g. repair time or cost) and vary depending on the sector. For example, in the case of an Information CI, the sectoral criteria could be: the data transit speed, the information system recovery time, the number of personal data records maintained or processed by the system, etc.

2.3.2 Cross-Cutting Criteria

The cross-cutting criteria assess ex-ante the gravity of the impacts that the malfunction or disruption or the destruction of a potential CI would have. The designation reflects the impact at the national level of an unexpected incident affecting that infrastructure in a worst-case scenario in the critical service (*e.g.* of a sector/sub-sector or at a cross-sectoral level) provided through the affected infrastructure. A potential CI is understood to be a CI when the impact of an incident affecting the infrastructure meets at least one or more quantitative and/or qualitative criteria. Criticality criteria may include [4, 8]:

1. *The extent of the geographic area.* An infrastructure is rated for the minimum extent of the geographic area that could be affected by an incident affecting the infrastructure.
2. *Casualties.* The criterion is the minimum number of victims and/or injuries that an incident affecting the infrastructure can cause.
3. *Economic effects.* This criterion is the macroeconomic impact (e.g. loss of Gross National Product, losses due to dependencies, land loss, population relocation costs and pollution costs) and/or macro-social impact, including potential environmental impacts.

4. *Public effects.* The criterion assesses how a (potential) infrastructure impact can affect a large proportion of people who enjoy the critical service that depends on this infrastructure. The evaluation is carried out in two steps [4]:
   - The incident category is identified as follows: (A1) Damage to the citizens' health, (A2) Loss of public confidence and (A3) Disruption of the citizens' daily lives.
   - For each category, the following are identified: (B1) The number of people (potentially) affected, (B2) the gravity of impacts and (B3) the duration of the consequences.

2.3.3 Combination of Sectoral and Cross-Cutting Criteria

Typically, sectorial and cross-cutting criteria can be used in combination when determining and prioritizing a CI. Thus, in line with the approach adopted under the EPCIP [4]:

1. For each national critical sector, and for each sector infrastructure under consideration, sectorial criteria are applied to designate the infrastructure as a potential CI in a sector.
2. For each potential CI, it is checked whether the infrastructure meets the criteria of the CI definition.
3. The potential CI should meet at least one cross-cutting criterion of criticality from the list of criteria in Sect. 2.3.2.

3 Best Practices in European Countries

In recent years, most EU member states have, or are in the process of designing, consistent cyber security policies for CIs [7], (Rossella and Cédric, Methodologies for the identification of Critical Information Infrastructure assets and services, [20]). The EU highlights the need to use best practices and methods to protect CI. Based on public sources and the related literature, we review some best practices implemented in EU countries, for identifying and protecting European CIs.

Figure 5 describes the maturity levels for assessing the progress made by a member state in protecting its national CI. Due to the importance of the protection of the Information CI for most national critical sectors today, the (interpolated) Level 2 has been added to cover most national critical sectors to include countries which, although they do not have or have not completed their critical infrastructure protection strategy, have already drawn up a cybersecurity strategy to protect their critical information and communication infrastructure.

The following are some interesting cases of full or partial application of the CI identification approach methodologies mentioned in the previous sections. In particular, we present aspects of the methodological approaches applied by member

Fig. 5 Maturity evaluation of national CI protection in the EU [7]

Criticality Scale	Description
CAT 5	This is infrastructure the loss of which would have a catastrophic impact on the UK. These assets will be of unique national importance whose loss would have national long-term effects and may impact across a number of sectors. Relatively few are expected to meet the Cat 5 criteria
CAT 4	Infrastructure of the highest importance to the sectors should fall within this category. The impact of loss of these assets on essential services would be severe and may impact provision of essential services across the UK or to millions of citizens
CAT 3	Infrastructure of substantial importance to the sectors and the delivery of essential services, the loss of which could affect a large geographic region or many hundreds of thousands of people
CAT 2	Infrastructure whose loss would have a significant impact on the delivery of essential services leading to loss, or disruption, of service to tens of thousands of people or affecting whole counties or equivalents
CAT 1	Infrastructure whose loss could cause moderate disruption to service delivery, most likely on a localised basis and affecting thousands of citizens
CAT 0	Infrastructure the impact of the loss of which would be minor (on national scale).

states such as France, Germany and European Countries like Great Britain[1]. Some of these countries are either pioneers in the methodological design of CI identification and designation (*e.g.* France), and/or are referenced in the literature as good practices (*e.g.* UK) ([12, 17], (Rossella and Cédric, Methodologies for the identification of Critical Information Infrastructure assets and services, [20]).

[1] UK under BREXIT process.

3.1 France

France is one of the first European countries that designed and implemented a policy for the identification and protection of CIs. France has defined 12 vital sectors [22] which are divided in 3 main areas as follows:

1. State sectors (Public Services, Military Operations, Judicial Functions, Space and Research).
2. Civil protection sectors (Health, Water Management, Food).
3. Areas of economic and social life of the nation (Energy, Electronic communications, Audiovisual and Information Systems, Transport, Economy, Industry).

As part of the national security strategy ([12, 23], an Operator-driven national approach to identifying the national CI was followed. In particular, the government, through the establishment of relevant mandates defines a list of *Vital Operators*, where each operator is related to one critical sector. The administrator of each Vital Operator is obliged to:

- Appoint security officers, both centrally and locally.
- Carry out a risk assessment to identify the critical assets/systems in its area of responsibility, and set up an Operator Security Plan (OSP) to protect it.
- Identify the assets/systems that will be the subject of the OSP, to be implemented under the operator's responsibility, as well as an *external protection plan*, to be implemented by the responsible public body.

Within the framework of the same approach to the protection of national Information CI [10], decrees 2015-351 (27 March 2015), 2015-350 (27 March 2015) and 2015-349 (27 March 2015) establish obligations for the 200 vital operators in relation to the security of their Information Systems.

3.2 Germany

According to the German Constitution, it is the state's task to guarantee public safety and to ensure the provision of the essential goods and services. The Federal Office of Civil Protection and Disaster Assistance (BBK) is responsible to stimulate the operators of Critical Infrastructures to proactively secure CIs and prepare effective crisis management plans. According to [2] the following infrastructures and sectors that have been defined as critical in Germany: Energy (Electricity, gas, mineral oil), Water (supply, disposal), Food (Food retail industry, food industry), Culture and media (Broadcasting, print media, cultural assets), Information (Telecommunication, information technology), Finance (Financial service providers, insurances, banks, stock exchanges), Health (health care, laboratories, drugs, vaccines), Transport (aviation, shipping, rail, roads, logistics) and State/Administration (Parliament, government, judiciary, emergency services).

Since 2009 Germany has adopted a National Strategy for Critical Infrastructure Protection (CIP Strategy) [16]. The strategy engages the Federal and the local governments to enhance and implement CI protection in their respective areas of responsibility. It comprises the following work packages, which in part are implemented in parallel, and is based on the co-operative approach adopted by the Federal Administration with the involvement of the other major players, *i.e.* operators and the relevant associations:

1. Definition of general protection targets.
2. Analysis of threats, vulnerabilities, and management capabilities.
3. Assessment of the threats involved.
4. Prioritization of protection targets, taking into account any existing measures.
5. Implementation of goal attainment measures primarily by means of: association-specific solutions and internal regulations; self-commitment agreements by business and industry; development of protection concepts by companies.
6. Continuous, intensive risk communication process.

3.3 Great Britain

In the United Kingdom, national infrastructure is defined as facilities, systems, locations or networks needed to operate the country and provide basic services on which people's daily lives rely on in the UK [15]. Some assets of the national infrastructure are called critical in the sense that their loss would lead to significant economic or social consequences or casualties in the UK. These critical assets make up the National Critical Infrastructure (CNI).

CNI in Great Britain is made up of 13 national infrastructure sectors: Chemicals, Civil Nuclear Communications, Defense, Emergency Services, Energy, Finance, Food, Government, Health, Space, Transport and Water. Several sectors have defined 'sub-sectors'; Emergency Services for example can be split into Police, Ambulance, Fire Services and Coast Guard.

Each sector has one or more Lead Government Department(s) (LGD) responsible for the sector, and ensuring protective security is in place for critical assets. Methodologically, the identification and designation of CI in the United Kingdom have been characterized as a hybrid [12], *i.e.* having elements from both approaches described in Sect. 2. In fact, each ministry, responsible for its sector/subsector of competence, defines a list of Critical Operators (CO) of the services that make up the sector/subsector, and each CO, in its turn, identifies the most critical assets/systems that make up the services.

At the same time, the Centre for the Protection of National Infrastructures (CPNI), as the responsible public body, independently conducts its own study on the assessment of potential CI, according to cross-cutting assessment criteria for the impact on the British society. In particular, a rating scale of 6 levels, CAT 0 to CAT 6 (Fig. 6) was applied for the assessment of the criticality of the impact, where

Level 0	Absence of any activity relevant to CIP	Croacia, Irland, Portugal
Level 1	Embodiment of Directive 2008/114 (EU Council, 2008) at national level. Nothing further	Greece, Bulgaria, Denmark, Malta
Level 2	Establishment of a national Strategy on Protection of CI	Belgium, Cyprus, Letonia, Lithuania, Hungary, Slovenia, Sweden
Level 3	Implementation of a National Protection Plan on CI	Austria, France, Germany, Switzerland, Estonia, Spain, G. Britain, Holland, Poland, Romania, Slovakia, Czech, Finland

Fig. 6 Scale of evaluation of CI in the UK. (Ref. [21])

a national CI is defined as an asset having a CAT 3 or CAT 4 or CAT 5 level, while any asset of lower level is understood as critical, although its criticality is more local or regional. Any CAT0-CAT2 graded infrastructure is called Wider National Infrastructure (WNI).

A risk-based method is followed for the final ranking of a critical asset in Great Britain [21]. Specifically, for each asset, in addition to the CAT rating, the assessment of the probability of successful completion of a threat to that asset is taken into account. This probability is assessed based on the one hand on the vulnerabilities or weaknesses of the asset being evaluated, and on the other, on the likelihood of a threat materialising (e.g. human or physical).

4 United States of America (USA)

The USA is the first country to develop initiatives to protect CI. According to the National Infrastructure Protection Plan [1], CI is defined as the systems and means, both physical and virtual, which are so vital for the USA that any unavailability or destruction would have a significant impact on safety, economy, public health, or any combination thereof.

Presidential Policy Directive 21 [5] defines the Department of Homeland Security as the responsible body for critical infrastructure security through the Office of Infrastructure Protection. The US Department of Homeland Security has identified 17 critical sectors. For each sector, a Sector-Specific Agency was appointed as a coordinator. These sectors are: (1) Chemical Sector, (2) Commercial Facilities, (3) Communications, (4) Critical Manufacturing, (5) Dams, (6) Emergency Services, (7) Information Technology, (8) Nuclear Reactors, (9) Materials & Waste, (10) Food and Agriculture, (11) Defence Industrial Base, (12) Energy, (13) Healthcare and Public Health, (14) Financial Services Sector, (15) Water and Wastewater Systems, (16) Government Facilities and (17) Transportation Systems.

For better cooperation and coordination of the Public-Private Partnership (PPP), bodies such as Government Coordinating Councils (GCC) and Sector Coordinating

Councils (SCG) are established for each critical sector. Governance councils involve government representatives at the local, regional and federal level, while sectoral councils involve representatives of the Owners/Operators of CI. The competencies of key stakeholders are summarized as follows:

- Sector Coordinating Councils (SCC): Independent and self-managed councils, which are actively involved in policy-making and sector-specific strategies, involving, inter alia, representatives of the owners and operators of CI sector. SCC act as key links for communication between the Government and the Private Operator of the CI.
- Government Coordinating Councils (GCC): Councils, by CI sector, composed of representatives of the Government and the States with the primary responsibility for coordination between private and public bodies.
- Critical Infrastructure Cross Sector Councils (CICSC): These councils include representatives of all SCC sectoral councils for cross-sectoral coordination and cross-sectoral issues (e.g. dependencies and interdependencies).
- Federal Senior Leadership Councils (FSLC): Councils linking sector councils with the rest of the state institutions regarding the security and resilience of CI.

The above structures are endorsed by the Critical Infrastructure Partnership Advisory Council (CIPAC). The Department of Homeland Security implemented CIPAC [3] as a mechanism to directly engage private and public interests in CI issues with a view to creating high-level policies to mitigate the risks and consequences of external threats.

5 Critical Infrastructure Protection Strategy

From the description of the national protection strategies described above, it is shown that despite their differences, in all strategies there can be found some key priority areas contributing to an effective holistic protection strategy for CIs. These areas involve:

1. ***Vision/Goals***: The development of any CI protection strategy involves the formulation of the strategic objectives (vision) that are broken down into individual quantifiable goals.
2. ***CI Security Administration***: The administration/governance structure concerns the designation of competent and mandated bodies for the protection of CI, the definition of roles and responsibilities per body, as well as the framework for cooperation between public and private bodies.
3. ***Public-Private Partnerships***: Each national protection programme involves the cooperation of the stakeholders (EU Commission 2005), in particular through Public-Private Partnership (PPP), including the public bodies and the owners/operators of CI.

4. ***Exchange of Information***: The exchange of information refers to the awareness of threats/vulnerabilities, to ensuring early warning to stakeholders and, more generally, to sharing information and adequate knowledge of risks and threats.
5. ***Legislative/Regulatory Framework***: The adoption of laws is an important tool to ensure, inter alia, that public and private bodies respond to their roles and responsibilities, as well as complying with specific safety standards.
6. ***Identification and assessment of national CI***: The identification and assessment of national critical assets (sectors, subsectors, services and specific subsystems) is a prerequisite for the implementation of the national policies for CI protection. An important criterion for classifying CI is, among others, the extent and importance of interconnections and interdependencies among CI.
7. ***Risk Assessment***: A key element of the CI protection strategy is the methodical evaluation of threats and the assessment of the resulting security risks of national CI.
8. ***Risks and Crisis Management***: Response measures to emergencies ensure the continued operation or rapid recovery of the critical asset.

6 Conclusions

The identification and assessment of Critical Infrastructures at a national level is one of the top priorities, as evidenced by a number of national and international initiatives. Both international literature and reality have shown that due to the interdependencies among CIs, the occurrence of a threat or failure in a CI very often leads to cascading impacts on other interconnected CI, leading to cumulative large-scale effects [13]. Such phenomena can be systematically modelled only through a holistic approach for the protection of CIs that is based on well-defined methodologies and tested best practices. Regardless of whether a state-driven or operator-driven approach is followed, the initial identification of national CIs is a prerequisite for their subsequent systematic assessment according to sectorial and/or cross-cutting criteria.

References

1. Austin Smith (2017) Presidential Policy Directive 21:Implementation: an interagency security committee white paper. Interagency Security Committee
2. BKK Annual Report (2015) 10–13. Germany: federal office of civil protection and disaster assistance (BKK)
3. Chris Boyer – AT&T (2017) Critical infrastructure partnership overview. AVP – global public policy. Retrieved from https://www.oecd.org/going-digital/digital-security-in-critical-infrastructure/digital-security-workshop-february-2018-%20Boyer.pdf
4. Council Directive 2008/114/EC (2008). Official J Eur Union 51: 75

5. Critical Infrastructure Security and Resilience – PPD-21 (2013) Washington DC. Retrieved from https://www.dhs.gov/sites/default/files/publications/PPD-21-Critical-Infrastructure-and-Resilience-508.pdf
6. EU Commission (2012) Review of the European Programme for Critical Infrastructure Protection (EPCIP)
7. EU Cybersecurity Dashboard: A Path to a Secure European Cyberspace (2015) BSA – the software alliance. Retrieved from http://cybersecurity.bsa.org/assets/PDFs/study_eucybersecurity_en.pdf
8. European Commission (2007) European programme for critical infrastructure protection. Off J
9. European Commission (2005) Green paper on a European programme for critical infrastructure protection, Brussels, COM, pp 576. Retrieved from https://eur-lex.europa.eu/legal-content/EN/TXT/?uri=CELEX:52005DC0576
10. French National Digital Security Strategy (2015) Secrétariat général de la défense et de la sécurité nationale (SGDSN). Retrieved from https://www.enisa.europa.eu/topics/national-cyber-security-strategies/ncss-map/France_Cyber_Security_Strategy.pdf
11. Gritzalis D, Stergiopoulos G, Kotzanikolaou P, Magos E, Lykou G (2016) Critical infrastructure protection: a holistic methodology for Greece. Conference on security of industrial control and cyber physical systems (CyberICPS). Springer, 19–34
12. Klaver M (2011) Good practices manual for CIP policies, for policy makers in Europe. Brussels: RECIPE. Retrieved from http://www.oiip.ac.at/fileadmin/Unterlagen/Dateien/Publikationen/FINAL_RECIPE_manual.pdf
13. Kotzanikolaou P, Theoharidou M, Gritzalis D (2013) Assessing n-order dependencies between critical infrastructures. Int J Crit Infrastruct 9(1–2):93–110
14. Luiijf E, Burger H, Klaver M, Marieke H (2003) Critical infrastructure protection in the Netherlands: a Quick-scan. EICAR Denmark, Copenhagen
15. National Security Strategy and Strategic Defence and Security Review (2015) UK Government. Retrieved from https://assets.publishing.service.gov.uk/government/uploads/system/uploads/attachment_data/file/478933/52309_Cm_9161_NSS_SD_Review_web_only.pdf
16. National Strategy for CIP (2009) Federal Republic of Germany
17. Novotný P, Rostek P (2014) Perspective of cross-cutting criteria as a major instrument to determination of critical infrastructure in the Czech Republic. (Vol. 2). Research papers faculty of materials science and technology Slovak University of technology
18. Public Private Partnerships (PPP) – Cooperative models (2017) ENISA. Retrieved from ENISA: https://www.enisa.europa.eu/publications/public-private-partnerships-ppp-cooperative-models
19. Rossella M, Cédric L-B (2014) Methodologies for the identification of critical information infrastructure assets and services. Guidelines for charting electronic data communication networks, European union agency for network and information security. ENISA, Heraklion
20. Rossella M, Cédric L-B (2015) Methodologies for the identification of critical information infrastructure assets and services. ENISA. Brussels: European Union Agency for Network and Information Security (ENISA)
21. Strategic Framework and Policy Statement on Improving the Resilience of Critical Infrastructure to Disruption from Natural Hazards (2010) London: UK Cabinet Office. Retrieved from https://www.gov.uk/government/publications/strategic-framework-and-policy-statement-on-improving-the-resilience-of-critical-infrastructure-to-disruption-from-natural-hazards
22. The Critical Infrastructure Protection in France. (2017). Paris: Secrétariat général de la défense et de la sécurité nationale (SGDSN). Retrieved from Secrétariat général de la défense et de la sécurité nationale: http://www.sgdsn.gouv.fr/uploads/2017/03/plaquette-saiv-anglais.pdf
23. The French White Paper on defence and national security (2013) Paris: permanent representation of France to NATO

Risk Analysis for Critical Infrastructure Protection

Richard White

Abstract Until recently, infrastructure owners and operators only had to worry about local acts of nature and the occasional vandal to maintain their services to a prescribed standard. All that changed with the 1995 Tokyo Subway Attacks and 9/11 which ushered in the unprecedented threat of domestic catastrophic destruction by non-state actors. Now infrastructure owners and operators find themselves under almost constant global cyber attack, the consequences of which could be catastrophic. Critical infrastructure protection has been a core mission of the Department of Homeland Security since its foundation in 2002. This chapter examines the work of the Department to protect the nation's critical infrastructure, and efforts to develop a uniform risk analysis to guide its strategic planning and facilitate cost-benefit-analysis of mitigation measures on the part of infrastructure owners and operators.

Keywords Risk analysis · Critical infrastructure · Cyber attack · Homeland security · NIPP · RMF · RAMCAP · LIRA

1 Introduction

Homeland security encompasses actions designed to safeguard a nation from domestic catastrophic destruction. For most of history, only nature and nations could inflict domestic catastrophic destruction, either in the form of disaster or warfare. That changed with the 1995 Tokyo Subway Attacks and 9/11. The first incident demonstrated the ability of non-state actors to deploy a weapon of mass destruction (WMD). The second incident demonstrated the ability of non-state actors to achieve WMD effects by subverting critical infrastructure. The first incident ushered in the concept of homeland security to deal with the unprecedented new threat of

R. White (✉)
University of Colorado, Colorado Springs, CO, USA
e-mail: rwhite2@uccs.edu

D. Gritzalis et al. (eds.), *Critical Infrastructure Security and Resilience*, Advanced Sciences and Technologies for Security Applications,
https://doi.org/10.1007/978-3-030-00024-0_3

domestic catastrophic destruction wielded by non-state actors. The second incident propelled homeland security to the forefront of United States policy concerns resulting in the largest reorganization of Federal government since the end of World War II.[1] Keeping WMD out of the hands of non-state actors and protecting the nation's critical infrastructure were core missions assigned to the new Department of Homeland Security by the 2002 Homeland Security Act, and remain central to its mission today.

Whereas international efforts have made WMD and their agents even more difficult to obtain, critical infrastructure have conversely become increasingly more vulnerable to attack. This unfortunate trend is attributable to the inescapable allure of computer automation as it becomes increasingly more capable and affordable due to the continued capacity of chip manufacturers to double processing power approximately every two years in accordance with Moore's Law.[2] Computer automation in the form of Industrial Control Systems (ICS) and Supervisory Control and Data Acquisition (SCADA) systems have reduced costs and introduced efficiencies that were once unthinkable. They have also created vulnerabilities in the form of entry points for unauthorized agents to gain access and control of critical infrastructure, either directly or indirectly through the Internet. Infrastructure owners and operators who previously only had to contend with local acts of nature and the occasional vandal now find themselves under nearly constant global cyber attack. Because there is no cure for cyber attack, critical infrastructure owners and operators will have to continue fending off this siege for the foreseeable future. Needless to say the stakes are high. A successful cyber attack shutting down the North American electric grid, undermining the US Federal Reserve, or initiating simultaneous meltdowns in two or more nuclear power plants would dwarf any past national disaster. The prospects of such an attack are what keep critical infrastructure protection a priority homeland security mission. Perhaps lesser known, though, is the central role risk analysis plays in protecting the nation's critical infrastructure.

Mitigation measures needed to reduce vulnerabilities in today's critical infrastructure are many and varied, and in a lot of cases, beyond the reasonable ability of infrastructure owners and operators to afford, or rate payers to bear. Whenever given a problem with more tasks than resources it becomes necessary to prioritize. Both infrastructure owners and operators and the Department of Homeland Security seek objective measures for prioritizing their mitigation efforts. Typically these involve measures of risk, however that may be defined, in order to facilitate cost-benefit-analysis and ultimately apply scarce resources where they provide the greatest return on investment. This chapter examines the problem of critical infrastructure

[1]More precisely, the 2002 Homeland Security Act was the largest reorganization of Federal government since the National Security Act of 1947 formalized the structural changes that occurred during World War II creating a new Department of Defense and Central Intelligence Agency.

[2]Observation made by Intel founder Gordon Moore in 1965 that the number of transistors per silicon chip doubles about every 18 months.

protection and work done by the Department of Homeland Security to formulate an objective risk analysis capable of guiding national investments in homeland security.

2 What Is Critical Infrastructure?

In the United States, critical infrastructure is currently defined according to 2013 Presidential Policy Directive No. 21 (PPD-21) as "systems and assets, whether physical or virtual, so vital to the United States that the incapacity or destruction of such systems and assets would have a debilitating impact on security, national economic security, national public health or safety, or any combination of those matters." [21] PPD-21 further identifies 16 critical infrastructure sectors as listed in Table 1.[3]

Among the sixteen critical infrastructure sectors, four in particular are deemed "lifeline" sectors (Table 1 bold accent). According to the 2013 National Infrastructure Protection Plan, communications, energy, transportation, and water constitute "lifeline" sectors because they are essential to the operation of most other critical infrastructure sectors, and to each other (US Department of Homeland Security [22], p. 17). If we take a closer look at the four lifeline sectors, we can see that they, in turn, are comprised of at least twelve subsectors as listed in Table 2.

If we look even closer, we can see that these lifeline subsectors are themselves comprised of critical assets as indicated in Table 3.

Critical infrastructure owners and operators depend upon their critical assets functioning within specified parameters to reliably deliver the goods and services

Table 1 PPD-21 Critical infrastructure sectors

1. Chemical	7. Emergency services	13. Information technology
2. Commercial facilities	**8. Energy**	14. Nuclear reactors, materials, & waste
3. Communications	9. Financial services	**15. Transportation systems**
4. Critical manufacturing	10. Food & Agriculture	**16. Water & wastewater systems**
5. Dams	11. Government facilities	
6. Defense industrial base	12. Healthcare & Public health	

Refs. [17, 18]

[3]PPD-21 released in 2013 by the Obama administration was only the most recent executive order to define critical infrastructure. Critical infrastructure was originally defined in PDD-68 released in 1998 by the Clinton administration. PDD-68 identified twelve infrastructure sectors. PDD-68 was superseded by HSPD-7 released in 2003 by the Bush administration identifying eighteen infrastructure sectors. Although the number of critical infrastructure sectors changed in each iteration, the definition of critical infrastructure remained relatively unchanged. It is not inconceivable that a future executive order might again change the number of critical infrastructure sectors.

Table 2 Lifeline infrastructure subsectors (Partial List)

Communications	Energy	Transportation	Water
1. Terrestrial	3. Electricity	6. Aviation	11. Drinking Water
2. Satellite	4. Natural Gas	7. Highway	12. Wastewater
	5. Oil	8. Railroad	
		9. Maritime	
		10. Pipeline	

Table 3 Lifeline Subsector Critical Assets (Partial List)

Subsector	Critical assets			
1. Terrestrial Comms.	Connections	Landlines	Exchanges	Control centers
2. Satellite Comms.	Connections	Satellites	Ground stations	Control centers
3. Electricity	Generators	Transmission lines	Distribution networks	Control centers
4. Natural gas	Storage facilities	Pumping stations	Distribution networks	Control centers
5. Oil	Storage facilities	Pumping stations	Distribution networks	Refineries
6. Aviation	Airports	Aircraft	Maintenance facilities	Air traffic control
7. Highway	Stations	Cars/Trucks/Buses	Roads/Bridges/Tunnels	Traffic control
8. Railroad	Stations	Trains	Rails/Bridges/Tunnels	Traffic control
9. Maritime	Ports	Ships	Maintenance facilities	Traffic control
10. Pipeline	Storage facilities	Pumping stations	Pipe networks	Control centers
11. Drinking water	Collection facilities	Treatment plants	Distribution networks	
12. Wastewater	Collection facilities	Treatment plants	Distribution networks	

which generally benefit them and the greater economy. If critical assets cease to function as required, either due to internal or external fault, then both the owners and operators and the greater economy may suffer. Depending on the extent of the fault, they may all suffer greatly.

On August 14, 2003, a cascading power failure left 50 million people in the northeastern United States and eastern Canada in the dark. It was the largest blackout in American history. The nearly week-long blackout caused an estimated \$4–\$10 billion in economic losses and resulted in a 0.7% drop in Canada's gross domestic product [31]. A John Hopkins study determined that New York City experienced a 122% increase in accidental deaths and 25% increase in disease-related deaths, and that ninety people died as a direct result of the power outage [3].

3 Critical Infrastructure Protection

The 2003 Blackout was caused by cascading failure stemming from a high-voltage power line brushing against some overgrown trees in northern Ohio [10]. The blackout may be termed a "technical disaster" as it was an accident resulting from neither the actions of nature nor man. Because it is a public utility subject to regulation, the electricity subsector maintains high reliability standards established by the North American Electric Reliability Corporation (NERC) and overseen by the Federal Energy Regulatory Commission (FERC) [24, 25].

Historically, NERC standards maintained high service reliability by requiring electricity companies to anticipate potential outages due not only to normal wear but also transient natural phenomenon, primarily those induced by weather. Planning for weather disruptions is readily accommodated because meteorological data collected over hundreds of years supports probability projections on the likelihood of a particular natural transient occurring within a given region. Thus, every year planners may expect to cope with tornadoes and flooding in the Midwest, snowstorms in the Northeast, hurricanes in the Southeast, and wildfires in the West and Southwest. On the other hand, planning for potential service disruptions stemming from malicious acts of man are not so readily predictable as there is no similar body of data to support forecasting, and perhaps never will be. Unlike the deterministic behavior of nature governed solely by the laws of physics which may be rendered mathematically within a calculable range of precision, the behavior of individuals is indeterministic due to willful actions that induce so many assumptions as to render precise mathematical calculation impractical. Nor was such calculation necessary as for most of the utilities' history manmade destruction was mostly contained to local acts of vandalism and did not pose a threat of propagating across the larger enterprise. That view, however, began to change as a result of the 1995 Tokyo Subway Attacks.

In March 1995, Aum Shinrikyo, a quasi-religious cult, attempted to overthrow Japanese government and initiate apocalypse by releasing the deadly nerve agent Sarin on the Tokyo subway system during morning rush hour. Tragically, twelve people lost their lives, but experts believe it was only luck that prevented thousands more from being killed [13]. It was the first use of a weapon of mass destruction by a non-state actor.

Prior to this incident, WMD were thought to require the resources of a nation state to acquire. This lent some comfort in that nations had developed sophisticated national security apparatus and treaties to keep each other in-check. The fact that WMD had fallen into the hands of non-state actors sent shock waves through national security establishments everywhere. They were unprepared to deal with WMD threats from non-state actors. The same means and methods that deterred WMD use among nation states were useless against non-state actors.

The profound implications stemming from the 1995 Tokyo Subway Attacks prompted many nations to re-examine their domestic security arrangements, especially in the United States. Both Congress and the President chartered various commissions to assess the nation's ability to withstand WMD attack by non-state actors. Among these charters was the President's Commission on Critical Infrastructure Protection. Although the commission found no immediate threat to US critical infrastructure, its 1997 report warned of a growing vulnerability to cyber attack, and the risk it presented to "national security, global economic competitiveness, and domestic well being." [15] In response to the commission report, President Clinton in May 1998 issued Presidential Decision Directive No. 63 (PDD-63) establishing an organizational and procedural framework for US government to work with industry in reducing critical infrastructure vulnerabilities to both cyber and physical attack [19]. Unfortunately, PDD-63 proved too little too late to protect America's aviation infrastructure and spare the nation from the devastating attacks of 9/11.

On September 11th, 2001, nineteen hijackers gained control of four passenger jets and flew them into structures representing the economic and military strength of the United States. In a span of only two hours, they utterly destroyed the Twin Towers in New York City and severely damaged the Pentagon outside Washington DC. Alerted to these suicide attacks, passengers aboard the fourth aircraft rose up against the hijackers, forcing them to abort their mission against the nation's capital and crash instead into an empty field outside Shanksville Pennsylvania. Altogether the attacks left 3000 dead and caused $40 billion in direct damages. The attacks were noted for their "surpassing disproportion". The hijackers had achieved WMD effects without using WMD by subverting the nation's aviation infrastructure and turning passenger jets into guided missiles. The devastation wrought by non-state actors propelled homeland security to the forefront of US policy concerns [1].

Less than a month after 9/11, President Bush issued Executive Order 13228 creating the Office of Homeland Security within the Executive Office of the President [20]. The newly appointed Homeland Security Advisor, former Pennsylvania Governor Tom Ridge, quickly set about formulating homeland security strategy. In June 2002, the Office of Homeland Security released the first National Strategy for Homeland Security clearly framing the homeland security mission:

> Homeland security is a concerted national effort to prevent terrorist attacks within the United States and reduce America's vulnerability to terrorism, and minimize the damage and recover from attacks that do occur [14]

The 2002 National Strategy for Homeland Security defined homeland security in terms of terrorism. This would prove to be an unfortunate choice of terms. In the many studies commissioned in the wake of the 1995 Tokyo Subway Attacks, the word "terrorism" became shorthand for "WMD attack by non-state actors". This was an unfortunate conflation because the word "terrorism" already had its own definition under Title 18 Section 2331 United States Code. Under this definition, terrorism is a crime distinguished by motive, that is to say, violent acts calculated to coerce government. Both the 1995 Tokyo Subway Attacks and 9/11 were, by definition, terrorist acts. Their distinction, however, as acts of domestic catastrophic destruction perpetrated by non-state actors, have been overshadowed by the original definition. Thus, most people confuse homeland security with terrorist motive rather than the means by which non-state actors may inflict domestic catastrophic destruction. Fortunately, that distinction did not escape Tom Ridge and his staff at OHS.

Within the 2002 National Strategy for Homeland Security, OHS identified six critical mission areas essential to accomplishing the homeland security mission. The two mission areas most essential to reducing the nation's vulnerability to domestic catastrophic destruction were 4) Protecting Critical Infrastructure and Key Assets, and 5) Defending Against Catastrophic Threats in the form of chemical, biological, radiological, and nuclear agents [14]. That these were essential homeland security missions was affirmed in November 2002 when Congress passed and the President signed the Homeland Security Act creating the Department of Homeland Security. It was the largest reorganization of US government since the end of World War II combining twenty-two Federal agencies and 230,000 personnel into a single Executive department [16]. The 2002 Homeland Security Act made these two critical missions inherent functions of the new Department of Homeland Security [6]. Despite many reorganizations to perceived and emerging threats, critical infrastructure protection remains an essential mission and function of the Department of Homeland Security.

4 NIPP & RMF

Among its many provisions, the 2002 Homeland Security Act required the new Department of Homeland Security to "develop a comprehensive national plan for securing the key resources and critical infrastructure of the United States ... " [6] Although the Department of Homeland Security was activated in January 2003, it did not produce its first National Infrastructure Protection Plan until February 2005. Despite its delayed appearance, the 2005 plan was only an interim one, and was quickly replaced by an approved one in 2006. The 2006 National Infrastructure Protection Plan was itself replaced in 2009 by the new Obama administration, and updated once more in 2013. The Trump administration has not issued its own plan, making the 2013 National Infrastructure Protection Plan the prevailing guidance for national efforts to protect critical infrastructure.

Despite the many changes, the National Infrastructure Protection Plan remains predicated on the organizational and procedural framework established in President Clinton's PDD-63. Organizationally, the whole plan is based on voluntary cooperation between government and industry. This is primarily necessitated by the fact that most critical infrastructure is privately owned[4] and the Federal government doesn't

[4]From the outset, the US government has claimed that 85% of critical infrastructure is privately owned. Despite this claim, nobody knows the true percentage of private versus public infrastructure.

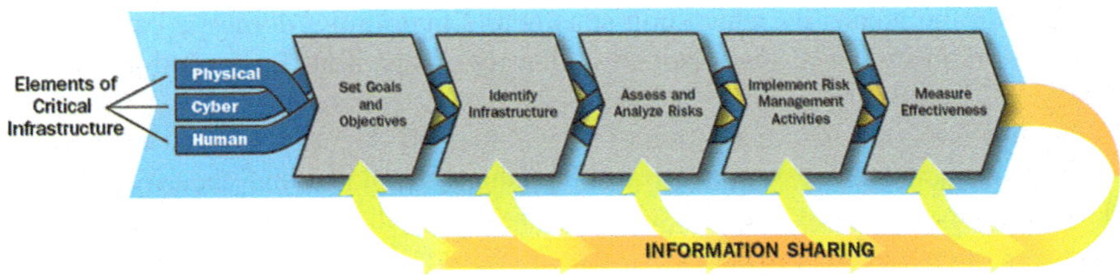

Fig. 1 2013 NIPP risk management framework [22]

have specific regulatory authority over industry security practices.[5] Accordingly, the National Infrastructure Protection Plan is built upon a Public/Private Partnership that works in voluntary cooperation with industry to promote improved security practices.

The Public/Private Partnership is managed by DHS. The Department, in turn, appoints Sector Specific Agencies, other Federal agencies with functional or regulatory ties to a specific industry, to work with industry representatives participating in Sector Coordinating Councils. There are sixteen Sector Coordinating Councils, one for each of the infrastructure sectors currently identified in PDD-21. Each of the Sector Coordinating Councils meets periodically to discuss practices and update their corresponding Sector-Specific Plans. The Sector-Specific Plan summarizes the current state of the industry and recommends practices for improving overall security. Although the Sector-Specific Plans are nominally updated every four years, the first were issued in 2007, the second update completed in 2010, and the most recent updates released in 2016. The analysis conducted in the Sector-Specific Plans generally conform to the five steps of the Risk Management Framework [22] (Fig. 1).

Procedurally, the National Infrastructure Protection Plan advocates the Risk Management Framework (RMF) for incrementally enhancing security across all critical infrastructure sectors. In its current form, the RMF is a continuous improvement process comprised of five basic steps: (1) Set Goals and Objectives, (2) Identify Infrastructure, (3) Assess and Analyze Risks, (4) Implement Risk Management Activities, and (5) Measure Effectiveness [22]. The RMF was established to meet objectives set out in the 2002 National Strategy for Homeland Security to "ensure that the taxpayers' money is spent only in a manner that achieves specific objectives with clear performance-based measures of effectiveness." (Office of Homeland Security [14], p. xiii) This requirement for objective and measurable

[5]Although US law may grant regulatory control over many facets of critical infrastructure, those same laws may not necessarily authorize regulatory authority over industry security measures. Thus, for example, although the 1970 Clean Air Act, 1972 Clean Water Act, and 1974 Safe Drinking Water Act give the Environmental Protection Agency authority to regulate drinking water and waste treatment utilities, those same laws do not give EPA authorization to regulate security measures for those utilities.

performance stems from the 1993 Government Performance Results Act,[6] a law requiring all Federal agencies to objectively and measurably account for their taxpayer funded expenditures [4]. More important than serving as the outline for the Sector-Specific Plans, the RMF also provided the blueprint for the Department of Homeland Security's overall critical infrastructure protection efforts.

While the Department of Homeland Security has undergone many changes since it was founded in 2003, critical infrastructure protection remains a primary mission as stipulated in the 2014 Quadrennial Homeland Security Review [26, 27]. Critical infrastructure protection is the responsibility of the DHS National Protection and Programs Directorate (NPPD). This Directorate works directly with infrastructure owners/operators, and State and Local governments to implement the steps of the Risk Management Framework. The NPPD Office of Infrastructure Protection manages the National Critical Infrastructure Prioritization Program which conducts an annual census of critical infrastructure assets in coordination with State and Local governments. At the request of infrastructure owners/operators, DHS Protective Security Advisors are prepared to conduct voluntary Site Assistance Visits and Security Surveys [26, 27]. Moreover, NPPD also staffs the National Infrastructure Coordinating Center maintaining 24-hour watch over the nation's infrastructure and stands ready to coordinate Federal support in a national emergency.

Despite these efforts, the Risk Management Framework is fraught with problems at every step: (1) Reluctance by infrastructure owners/operators to share data; (2) Multiple infrastructure databases with questionable and incomplete listings; (3) No uniform analysis for comparing risks across assets and sectors; (4) No direct funding for security improvements to private industry; and (5) No established metric for guiding national strategy [33]. Perhaps the most critical shortcoming is the absence of a uniform cross-sector risk analysis formulation. Absent this capability, the Department of Homeland Security cannot attain its original goals of (1) informing near-term action, and (2) rationally guiding long-term resource investments (Office of Homeland Security [14], p. 33). In short, the Department lacks the means for strategic planning. Strategic planning depends upon a standard metric. Without a standard metric, it is impossible to determine where you are and where you need to go. The strategic significance of such a measure is not lost on DHS and was the reason it invested early in a cross-sector risk analysis formulation called RAMCAP.

5 RAMCAP

The Risk Analysis and Management for Critical Asset Protection was developed by the American Society of Mechanical Engineers (ASME) at the request of the White House shortly after 9/11 to create a means for prioritizing protection of the nation's critical infrastructure and support resource allocation decisions for risk-reduction

[6]GPRA was amended in 2011 by the GPRA Modernization Act of 2010.

Risk = **T**hreat x **V**ulnerability x **C**onsequence

Fig. 2 RAMCAP risk formulation

initiatives. Convening a team of distinguished risk analysis experts from industry and academia, ASME defined a seven-step methodology enabling asset owners to perform analyses of their risks and risk-reduction options relative to specific malevolent attacks. Risk was defined as a function of the likelihood of a specific attack, the asset's vulnerability to these attacks, and the consequences of the attack. With this information, alternative risk-reduction actions could be evaluated for their ability to reduce the vulnerability, likelihood, and/or consequences of attack. Reductions in risks could be used in estimating net benefits (benefits less costs) and benefit-cost ratios that would support informed decisions to allocate resources to specific risk-reduction actions [2] (Fig. 2).

The initial version of RAMCAP was the 2004 draft Risk Analysis and Management for Critical Asset Protection: General Guidance, a detailed description of the general process. The General Guidance was widely circulated in draft and reviewed extensively by panels of applied risk management and security experts. It was seen as a highly competent and comprehensive synthesis of the best available methods appropriate for both academic and risk professionals. It did not prove, however, as useful to infrastructure owners and operators. RAMCAP was consequently redesigned with a key criterion (among others) to better facilitate self-assessment by on-site staff in a relative short period of time (typically less than a week of work by a team of 3–6 people, after assembly of the necessary documents). In response to this change, the General Guidance, which was never published, was streamlined and simplified into two documents in 2005, the semi-technical Introduction to Risk Analysis and Management for Critical Asset Protection and a nontechnical Risk Analysis and Management for Critical Asset Protection Applied to Terrorism and Homeland Security [2].

The methodology described in those three initial RAMCAP documents was deemed as meeting requirements in early drafts of the National Infrastructure Protection Plan for a simple and efficient process to support consistent, quantitative risk analysis and with results that could be systematically and directly compared. In 2006, the earlier documents were updated and republished as RAMCAP: The Framework, Version 2.0, which was still primarily oriented towards terrorism. Following Hurricane Katrina in 2005, RAMCAP underwent further refinement to accommodate all-hazard risks from both natural and manmade threats. In 2009, All-Hazards Risk and Resilience: Prioritizing Critical Infrastructure Using the RAMCAP Plus Approach was published, updating RAMCAP Framework 2.0 and providing the basis for a generic, all-sector standard by ASME Codes and Standards [2]. Unfortunately, these changes did nothing to save RAMCAP from its fate.

RAMCAP was the recommended risk methodology of choice advocated in the first official release of the National Infrastructure Protection Plan in 2006. According to the 2006 NIPP, RAMCAP satisfied baseline criteria for risk assessment supporting "national-level, comparative risk assessment, planning, and resource prioritization." (US Department of Homeland Security [23], p. 36) But something went wrong. Despite the tremendous efforts by ASME to build stakeholder consensus in a uniform risk analysis methodology, the stakeholders rejected RAMCAP. Although no official explanation was given, speculative accounts suggest two main reasons why RAMCAP fell flat: (1) it wasn't easy, and (2) it wasn't "invented here". Whatever the reason why RAMCAP failed to gain universal acceptance among infrastructure owners and operators, it was never mentioned again in subsequent releases of the National Infrastructure Protection Plan. DHS abandoned RAMCAP. Even so, all was not lost. RAMCAP became the basis for the J100–10 American Water Works Association (AWWA) National Standard for Water and Wastewater Vulnerability Assessments. It was because of this association with the water infrastructure that RAMCAP resurfaced for another look by DHS in 2014.

6 DWRP

In 2014, the Department of Homeland Security Science and Technology Directorate (DHS S&T) launched the Drinking Water Resilience Project (DWRP) to address a mounting crisis with the US drinking water infrastructure. In the US, about 156,000 public water systems provide drinking water to about 320 million people through more than 700,000 miles of pipes. Unfortunately, much of the system is starting to come to the end of its useful life, with many of the pipes over 100 years old. As a consequence, there are an estimated 240,000 water main breaks per year contributing to an estimated 1.7 trillion gallons of water lost to broken and leaky pipes. The cost to fix the system is estimated somewhere between $650 billion and $1 trillion [11]. Most water utilities are unprepared to take on this expense. The Environmental Protection Agency (EPA) doesn't have the money, [28, 29] nor does Congress, having allocated only $17.3 billion to the Drinking Water State Revolving Fund (DWSRF) over the past 20 years; [30] less than 3% needed to fix the problem based on the lowest estimate.

Leaky pipes are not the only concern. Climate change also poses a threat to the nation's drinking water infrastructure. Higher air and water temperatures promote increased growth of algae and microbes, increasing the need for drinking water treatment. Higher air and water temperatures also melt the polar ice caps causing global sea levels to rise. Sea-level rise increases the salinity of both surface and ground water, resulting in salt-water intrusion into coastal drinking water supplies. Reduced annual precipitation and extended drought threaten in-land water supplies. Climate change presents yet another challenge for which water utilities are unprepared to pay the bill [28].

As if these concerns weren't enough, since 9/11, water utilities have had to contend with the possibility of malicious attack. The fact that 15% of utilities provide service to more than 75% of the US population make drinking water infrastructure a potential high-value target (US Department of Homeland Security [22], p. 17). A carefully coordinated cyber attack could conceivably disrupt the distribution systems to major metropolitan areas.

Again, faced with more tasks than resources, DWRP sought to find a way to prioritize national infrastructure investments, but not just for water. DWRP sought to prioritize risk across all lifeline infrastructure sectors. The obvious place to start was RAMCAP. First, DHS S&T wanted to know if the RAMCAP risk formulation did indeed produce comparable results across infrastructure sectors. Next, they wanted to know if RAMCAP took into account emerging threats from aging infrastructure, climate change, and cyber attack. The short answer to both questions was "no".

Detailed analysis including modeling and simulation determined that RAMCAP did not account for emerging threats from aging infrastructure, climate change, or cyber attack. Nor could RAMCAP account for mobile assets, leaving out the entire aviation subsector of the transportation sector. Most significantly, RAMCAP allowed wide variability in its calculations, making the results incomparable across assets or sectors. Overcoming these shortfalls would require a major overhaul of RAMCAP [7].

7 LIRA

The Lifeline Infrastructure Risk Analysis (LIRA) methodology was specifically designed to overcome RAMCAP's shortfalls and produce a uniform risk analysis formulation whose results could be compared across both infrastructure assets and sectors. LIRA was predicated on the same risk formulation as RAMCAP; that is to say, Risk = Threat × Vulnerability × Consequence. LIRA also employs the same set of Reference Scenarios as RAMCAP, expanded, however, to account for emerging threats from aging infrastructure, climate change, and cyber attack. Here the similarities end, though, and RAMCAP and LIRA differ significantly from each other [7].

Among the major differences between RAMCAP and LIRA is that RAMCAP risk analysis is bottom-up, and LIRA risk analysis is top-down. The basic difference between these two approaches is that RAMCAP risk analysis is conducted at the component level, while LIRA risk analysis is conducted at the system level. As with other aspects of the RAMCAP risk analysis formulation, the focus on components introduces a wide range of variability to the risk results depending which components are chosen to analyze. LIRA eliminates this variability by examining the system as a whole and basing risk analysis on the ability of that system to perform its assigned function [7].

To further reduce variability in risk results, LIRA provides default data values whereas RAMCAP does not. This simple enhancement is significant in a number of different ways. First, it substantially reduces the amount research and expertise required to perform risk analysis. RAMCAP left it to the team performing risk analysis to individually ascribe Threat and Vulnerability values to each of the 41

Table 4 RAMCAP vs. LIRA risk analysis methods

RAMCAP			LIRA		
#	Step	Description	#	Step	Description
1	1.	Asset characterization	1	1.	System asset identification
2	2.	Threat characterization	2	2.	System failure mode definition
3	3.	Consequence analysis	3	3a.	First order consequences
4	4.	Vulnerability analysis	4	3b.	Second order consequences
5	5.	Threat analysis	5	3c.	Third order consequences
6	6a.	Risk calculation	6	3d.	Consequences assessment
7	6b.	Resilience calculation	7	4.	Vulnerability analysis
8	7a.	Applied countermeasures	8	5.	Probability analysis
9	7b.	Calculate net benefit	9	6.	Risk assessment
10	7c.	Compare countermeasures	10	7.	Mitigation optimization
11	7d.	Risk management	11	8.	Resilience optimization
			12	9.	Risk management
			13	10.	Input data analysis

reference scenarios addressing a wide range of natural and manmade hazards. LIRA provides default Threat and Vulnerability values for each its 54 Reference Scenarios, allowing the team or individual performing risk analysis to change any value with which they have more insight or expertise; in other words, you don't have to be an expert in everything. Second, the added benefit to this approach is that an initial risk analysis with LIRA can be completed in a matter of hours, compared to RAMCAP which requires days if not weeks, even though RAMCAP boasts fewer steps than LIRA (See Table 4). In short, the default data values not only make risk results more comparable across infrastructure assets and sectors, they also make LIRA easier to use than RAMCAP [7].

Another significant difference is that LIRA accommodates mobile infrastructure assets; RAMCAP does not. Mobile infrastructure assets in the form of passenger jets proved particularly lethal on 9/11. Not including them in any comparative risk analysis of lifeline assets is a major oversight. LIRA not only includes passenger and cargo jets, but also accommodates risk analysis for rail freight service, major transportation corridors (rail or highway), and cruise ships (See Table 5) [7].

Arguably, the most critical component of LIRA is the default database. Where possible, data was pulled from nationally available sources such as the National Oceanic and Atmospheric Agency (NOAA), US Geological Survey (USGS), and the National Aeronautics and Space Administration (NASA). These sources sufficed for probability data pertaining to weather and geological phenomena. The data was geo-coded so it could be simply referenced by US postal ZIP code. Similar data for technical and manmade disasters were not correspondingly available. To compensate, the initial LIRA database was seeded with approximate data corresponding to estimates for whether the given scenarios were unlikely, likely, or very likely to occur. Similar estimates were provided for vulnerability. Obviously it would have been preferable to use real data for the LIRA risk calculations. It

Table 5 LIRA subsector system assets

#	Sector	Subsector	System asset	Concerning party
1.	Water/Wastewater	Water	Water treatment & Distribution utility	Utility owner/operator
2.		Wastewater	Sewer treatment & Collection utility	Utility owner/operator
3.	Energy	Electricity	Electrical utility	Utility owner/operator
4.		Natural gas	Gas utility	Utility owner/operator
5.		Oil	Oil refinery	Refinery owner/operator
6.	Transportation	Aviation	Passenger/Cargo jet	Air service owner/operator
7.		Highway	Major transportation bridge	State DOT
8.		Rail freight	Rail freight service	Rail owner/operator
9.		Mass transit	Major transportation corridor	Route owner/operator
10.		Pipeline	Oil pipeline	Pipeline owner/operator
11.		Maritime	Shipping port	Port owner/operator
12.		Maritime	Cruise ship	Cruise line owner/operator
13.	Information	Internet	Internet exchange point	Internet service provider
14.		Internet	Domain name servers	Root server administrator

is for this reason that a data feedback mechanism was incorporated into Step 10 of the LIRA process (See Table 4). Every time LIRA is used, it collects real-world data. The data is kept anonymous to protect the user's privacy. Otherwise, new input data (i.e., default overrides) is collected in an archive database for later analysis and incorporation back into the default LIRA database.

To demonstrate the feasibility of the LIRA methodology, and to begin building the LIRA database, a prototype application was deployed called the LIRA Database Validation Tool (LIRA-DVT). LIRA-DVT incorporates the full functionality of LIRA. To distinguish between real and estimated data, LIRA-DVT color codes all displayed values. Real data is color coded green, including user inputs. Estimated data is color coded black. Data input is protected by user login. The login is anonymous and requires no personally identifiable information. Data created under the user login is only accessible by that user login. This feature not only provides security for the user data, it also allows them to save any intermediate work or return to previous results. LIRA-DVT is accessible from a web browser at https://lira.uccs.edu/app/ (Fig. 3).

8 Are We There Yet?

Does LIRA solve the problem of uniform risk analysis for DHS and the National Infrastructure Protection Plan Risk Management Framework? The answer is "partly". LIRA still faces the second major challenge that confronted RAMCAP, and

Fig. 3 LIRA-DVT

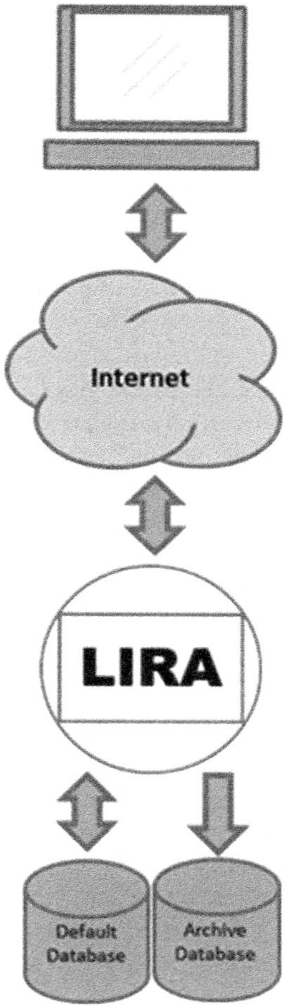

that's the "not invented here" syndrome. For good reasons, infrastructure owners and operators are dubious about outside solutions to their problems. Of course, there's concern about potentially leaking sensitive business information. For similar reasons, there is also concern about exposing themselves to lawsuits. And then there are those like former President Reagan who said that the most terrifying words in the English language are "I'm from the government and I'm here to help."

Perhaps the biggest impediment to accepting an outside solution is the perception that "one size can't fit all". Again, this observation is not unfounded. By one estimate there are more than 250 critical infrastructure risk analysis methods [9]. One reason for this vast proliferation is that each method is specifically tailored to address a different purpose, and all risk analysis entails a set of tradeoffs. RAMCAP, upon which LIRA was based, is no exception.

Both RAMCAP and LIRA are the product of a conscious series of tradeoffs in choosing the type and terms of risk analysis in order to achieve their desired purpose: a uniform method for comparing risk across infrastructure assets and sectors. The first tradeoff involved the question of completeness: do you analyze the asset or the network? Lifeline infrastructure assets are highly interrelated and dependent on each other and other infrastructure sectors. Many consider risk analysis incomplete unless it takes into account the effects of these interdependencies. Network analysis, however, is much more complicated than individual asset analysis, and doesn't facilitate "self-assessment" which was a key design criteria for RAMCAP. Moreover, network analysis is sensitive to the fidelity of the analysis conducted on individual nodes, so an argument can be made that good network analysis must necessarily begin with good asset analysis [34].

In analyzing an asset, the next tradeoff pertains to the desired level of confidence or resolution of the results. In this regard, the choice is between qualitative versus quantitative risk analysis. Qualitative risk analysis simplifies risk assessments by reducing inputs to a manageable set of judgments. A general criticism of qualitative methods, though, is that the poor resolution of input data can lead to erroneous or misleading output results. The general requirement for "objective and measurable performance" driven by the Government Performance Results Act steered RAMCAP towards a quantitative approach [34] (Fig. 4).

The choice of quantitative risk analysis is subsequently tempered by complimentary requirements for precision and accuracy. The question of precision hinges on the choice between relative or absolute results. Ideally, an absolute risk measure would be the preferred choice. Unfortunately, since there can be no concept of absolute safety or security, the choice of measure must also be relative, as is the case with RAMCAP. Despite this selection, it does not preclude the accuracy of results. Accuracy in this regard refers to the choice of formal or informal methods for guiding risk analysis. Bayesian Networks, Conditional Linear Gaussian Networks, Stochastic Models and other formal quantitative methods have proven records of performance in diverse fields such as engineering, finance, healthcare, and meteorology. What trips them up with critical infrastructure is the dearth of data for statistical analysis of catastrophic incidents, particularly involving malicious human intent. Various attempts to work around this obstacle often lead to formulations that are neither transparent nor repeatable. The need for consistency has fueled the development of informal quantitative methods, including the RAMCAP formulation which calculates Risk as the product of Threat, Vulnerability, and Consequence [34].

What terms you use and how you define them are equally important to determining the purpose and functionality of a risk formulation. In the absence of specific guidance, it should not be unexpected that two different assets might apply the same risk formulation in two different ways. For the purpose of consistency which underpins the ultimate goal of comparison, it is better if two assets apply the same formulation in the same systematic way. By extension, it would seem equally desirable to apply rigorous methods of estimation such as Delphi, Fault Trees, Event Trees, and Reliability Block Diagrams. Such rigorous methods, though, require substantial investments in time and resources, making them impractical for

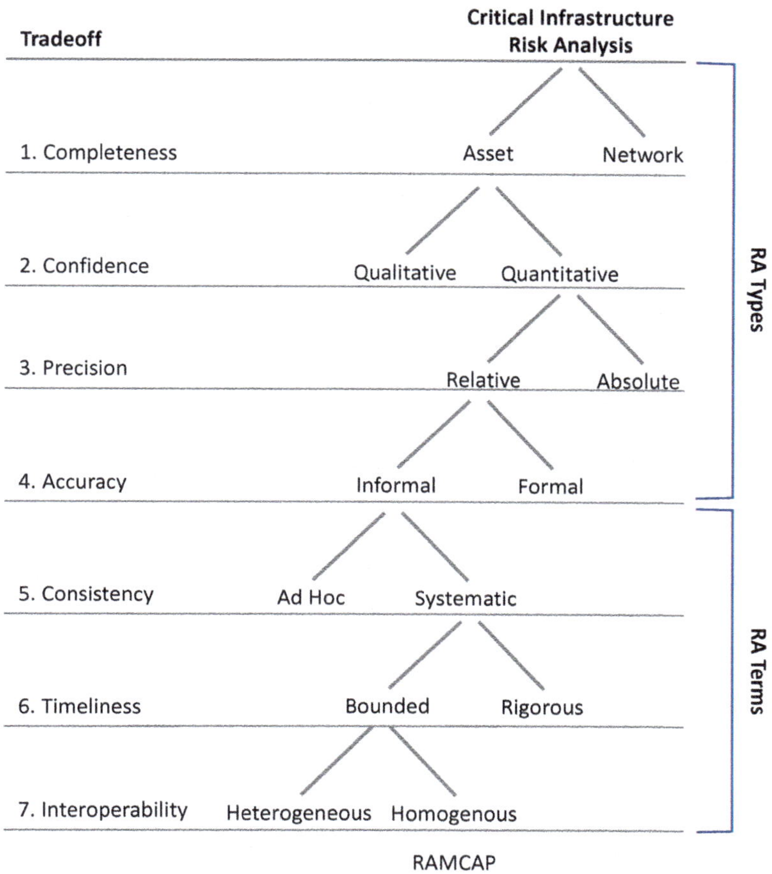

Fig. 4 Risk analysis design tradeoffs

large-scale application across a diverse population. Alternatively, a less rigorous but bounded system can support both requirements for consistency and timeliness. RAMCAP is bounded by its Reference Scenarios. These perform another important function by providing a homogenous base for comparison that ultimately support uniform risk analysis across assets and sectors [34].

As we can see, both LIRA and RAMCAP are products of design tradeoffs selected to fulfill specific program objectives. By the same token, alternative tradeoff selections will produce different risk formulations addressing different program objectives. And these are certainly not the only tradeoffs that may be considered. Accordingly, it is easy to see why there are so many different types of critical infrastructure risk formulations. Similarly, it is also easy to see understand why infrastructure owners and operators may seek alternative risk formulations to address their own specific program objectives.

9 Conclusion

I think it is important to conclude by returning to the threat of cyber attack that makes critical infrastructure protection a priority national policy concern. The 1997 report by the President's Commission on Critical Infrastructure Protection correctly predicted the increasing vulnerability of critical infrastructure to cyber attack [15]. Although the report resulted in PDD-63 forming the foundation for today's National Infrastructure Protection Plan, concern about cyber attack was subordinated to greater concerns from physical attack as demonstrated by the absence of cyber threats in the RAMCAP Reference Scenarios. This is understandable from the standpoint that 9/11 was a physical attack and the resulting wave of homeland security measures were primarily aimed at preventing future such attacks [14]. Perhaps it was the cyber attack on Georgia that preceded Russia's invasion in August 2008 that caused a re-evaluation of priorities [5]. Whether it was because of that or STUXNET, the threat of cyber attack against the nation's critical infrastructure received renewed attention and elevated priority in the 2010 Quadrennial Homeland Security Review [24, 25]. In February 2013, President Obama issued Executive Order 13636 directing the National Institute of Standards and Technology (NIST) to develop a Cybersecurity Framework to form the basis of a critical infrastructure cybersecurity program [17, 18]. A year later NIST released its Cybersecurity Framework (NIST 800–53), a process maturity model for incrementally improving cybersecurity practices across an organization.[7] [12] LIRA includes Reference Scenarios that assign threat and vulnerability values based on a process maturity-like assessment. On December 23rd, 2015, a cyber attack on the Ukrain power grid succeeded in shutting down 30 substations and cutting off power for six hours to 225,000 customers. On March 15th, 2018, in an unprecedented announcement the US Government publicly accused Russia of trying to hack into the US electricity grid [32]. In point of fact, it is understood that the US electric grid is vulnerable to cyber attack and its loss could precipitate the worst disaster in US history [8].[8] By the same token, risk analysis can help mitigate this concern by concentrating resources where they're most needed. In short, risk analysis is essential to critical infrastructure protection, which, in turn, is essential to homeland security, which is about safeguarding the US from domestic catastrophic destruction.

[7]The NIST Cybersecurity Framework is but one of a number of process maturing models for improving critical infrastructure cybersecurity. The NIST Cybersecurity Framework itself was based upon the 2012 Electricity Subsector Cybersecurity Capability Maturity Model (ES-C2M2) developed with support from the Department of Energy. In 2012 the Department of Transportation released its Roadmap to Secure Control Systems in the Transportation Sector. And in May 2013, DHS reported it was employing the Cyber Assessment Risk Management Approach (CARMA) to assess cybersecurity in the Information Technology Sector (i.e., "Internet").

[8]To date, the worst disaster in US history outside the Civil War was the 1900 Galveston Hurricane in which an estimated 6000–12,000 people perished.

References

1. 9/11 Commission (2004) A failure of imagination: the 9/11 commission report. US Government Printing Office, Washington, DC
2. American Water Works Association (2010) Risk analysis and management for critical asset protection (RAMCAP) standard for risk and resilience management of water and wastewater systems. American Water Works Association, Washington, DC
3. Anderson GB, Bell ML (2012) Lights out: impact of the August 2003 power outage on mortality in New York, NY. Epidemiology 23(2):189–193
4. Brass CT (2012) Changes to the government performance and results act (GPRA): overview of the new framework of products and processes. Congressional Research Service, Washington, DC
5. Bucci S (2009) A most dangerous link. US Naval Institute, Annapolis
6. Congress US (2002) Homeland security act of 2002. US Government Printing Office, Washington, DC
7. George R, White R, Chow CE, Boult T (2017) Apples-to-Apples: LIRA vs. RAMCAP. Homeland Security Affairs, Volume November, p. Article 17071
8. Idaho National Laboratory (2016) Cyber threat and vulnerabilty analysis of the US electric sector. Idaho National Laboratory, Idaho Falls
9. Lewis TG, Darken RP, Mackin T, Dudenhoeffer D (2012) Model-based risk analysis for critical infrastructures. In: Critidal infrastructure security: assessment, prevention, detection, response. WIT Press, Ashurst/Southampton, pp 3–19
10. Minkel J (2008) The 2003 Northeast blackout – five years later. [Online] Available at: https://www.scientificamerican.com/article/2003-blackout-five-years-later/. Accessed 7 Mar 2018
11. Morrow M (2016) America's water infrastructure is in need of a major overhaul. [Online] Available at http://www.foxbusiness.com/features/2016/01/28/america-s-water-infrastructure-is-in-need-major-overhaul.html. Accessed 6 Feb 2016
12. National Institute of Standards and Technology (2014) Framework for improving critical infrastructure cybersecurity. National Institute of Standards and Technology, Washington, DC
13. Neifert A (1999) Case study: sarin poisoning of subway passengers in Tokyo, Japan, in March, 1995. Camber Corporation, Huntsville
14. Office of Homeland Security (2002) National strategy for homeland security. The Whitehouse, Washington, DC
15. President's Commission on Critical Infrastructure Protection (1997) Critical foundations: protecting America's infrastructures. US Government Printing Office, Washington, DC
16. The President of the United States (2002) A reorganization plan for the department of homeland security. US Government Printing Office, Washington, DC
17. The White House (2013a) Executive order 13636, improving critical infrastructure cybersecurity. The Federal Register, Washington, DC
18. The White House (2013b) PPD-21, critical infrastructure security and resilience. The White House, Washington, DC
19. The Whitehouse (1998) PDD-63, critical infrastructure protection. The Whitehouse, Washington, DC
20. The Whitehouse (2001) EO 13228, establishing the office of homeland security and the homeland security council. The Whitehouse, Washington, DC
21. The Whitehouse (2013) Presidential policy directive – critical infrastructure security and resilience. Office of the Press Secretary, Washington, DC
22. US Department of Homeland Security (2013) National infrastructure protection plan. US Department of Homeland Security, Washington, DC
23. US Department of Homeland Security (2006) National infrastructure protection plan. US Department of Homeland Security, Washington, DC
24. US Department of Homeland Security (2010a) 2010 quadrennial homeland security Review. US Department of Homeland Security, Washington, DC

25. US Department of Homeland Security (2010b) Energy sector-specific plan. Department of Homeland Security, Washington, DC
26. US Department of Homeland Security (2014a) 2014 quadrennial homeland security review. US Department of Homeland Security, Washington, DC
27. US Department of Homeland Security (2014b) National protection and programs directorate (NPPD) office of infrastructure protection (IP). US Department of Homeland Security, Washington, DC
28. US Environmental Protection Agency (2014a) Climate change adaptation plan. US Environmental Protection Agency, Washington, DC
29. US Environmental Protection Agency (2014b) EPA response to EO13636, improving critical infrastructure cybersecurity. US Environmental Protection Agency, Washington, DC
30. US Environmental Protection Agency (n.d.) How the drinking water state revolving fund works. [Online] Available at: http://www.epa.gov/drinkingwatersrf/how-drinking-water-state-revolving-fund-works#tab-1. Accessed 6 Feb 2016
31. US-Canada Power System Outage Task Force (2006) Final report on the implementtion of task force recommendations, s.l.: s.n
32. Volz D, Gardner T (2018) In a first, US blames Russia for cyber attacks on energy grid. [Online] Available at: https://www.reuters.com/article/us-usa-russia-sanctions-energygrid/in-a-first-u-s-blames-russia-for-cyber-attacks-on-energy-grid-idUSKCN1GR2G3. Accessed 3 Apr 2018
33. White R (2014) Towards a unified homeland security strategy: an asset vulnerability model. Homeland Security Affairs 10:Article 1
34. White Ricahrd, Burkhard A, Boult T, Chow CE (2016) Towards a comparabgle cross-sector risk analysis: a re-examiniation of the risk analysis and management for critical asset protection (RAMCAP) methodology. s.l., s.n., pp 28–40

Risk-Based Analysis of the Vulnerability of Urban Infrastructure to the Consequences of Climate Change

Erich Rome, Manfred Bogen, Daniel Lückerath, Oliver Ullrich, Rainer Worst, Eva Streberová, Margaux Dumonteil, Maddalen Mendizabal, Beñat Abajo, Efrén Feliu, Peter Bosch, Angela Connelly, and Jeremy Carter

Abstract This chapter gives an introduction to risk-based vulnerability assessment of urban infrastructure regarding the consequences of climate change, by describing an approach developed as part of the EU-funded research and innovation project *Climate Resilient Cities and Infrastructures*. The approach is modular, widely applicable, and supported by a suite of software tools. It guides practitioners and end-users through the process of risk-based vulnerability assessment of urban systems, including built-up areas and (critical) infrastructure. How the approach can be adapted to and applied in a local context is demonstrated via its exemplary application in case studies with the four European cities Bilbao (Spain), Bratislava (Slovakia), Greater Manchester (United Kingdom), and Paris (France). Essential

E. Rome (✉) · M. Bogen · D. Lückerath · O. Ullrich · R. Worst
Fraunhofer IAIS, Sankt Augustin, Germany
e-mail: erich.rome@iais.fraunhofer.de; manfred.bogen@iais.fraunhofer.de; daniel.lueckerath@iais.fraunhofer.de; oliver.ullrich@iais.fraunhofer.de

E. Streberová
Hlavné mesto SR Bratislava, Bratislava the Capital of the Slovak Republic, Bratislava, Slovakia
e-mail: eva.streberova@bratislava.sk

M. Dumonteil
Ecole des Ingénieurs de la Ville de Paris, Paris, France
e-mail: margaux.dumonteil@eivp.fr

M. Mendizabal · B. Abajo · E. Feliu
Tecnalia, Parque Tecnológico de Bizkaia, Derio, Bizkaia, Spain
e-mail: maddalen.mendizabal@tecnalia.com; benat.abajo@tecnalia.com; efren.feliu@tecnalia.com

P. Bosch
TNO, Utrecht, The Netherlands
e-mail: peter.bosch@tno.nl

A. Connelly · J. Carter
The University of Manchester, School of Environment Education and Development, Manchester, UK
e-mail: Angela.Connelly-2@manchester.ac.uk; Jeremy.Carter@manchester.ac.uk

D. Gritzalis et al. (eds.), *Critical Infrastructure Security and Resilience*, Advanced Sciences and Technologies for Security Applications, https://doi.org/10.1007/978-3-030-00024-0_4

concepts for risk and vulnerability assessments and the current state of the art from related research projects are discussed before a detailed description of the developed approach and its supporting tools is given.

Keywords Climate change adaptation · Risk assessment · Urban systems · Infrastructure · Vulnerability · Risk analysis

1 Introduction

Urbanisation is an immensely influential trend in human history. The United Nations' annually updated World Urbanisation Prospects reported in 2014: *"In 2007, for the first time in history, the global urban population exceeded the global rural population, and the world population has remained predominantly urban thereafter."* [31]. The mostly developed and densely populated Europe reached this turning point even earlier. In 2014, a majority of the world's population has been living in cities and urban areas (world: 54%, EU: 72.5%) and projections for 2050 predict even larger shares (world: 66 + %, EU: 80 + %; [12, 33, 34]). According to the United Nations Environment Program (UNEP)[1], cities generate up to 80% of a country's GDP, but also consume 75% of the natural resources and account for 60–80% of greenhouse gas emissions. That is, urbanization and economic growth happening in cities are the biggest contributors to climate change.

Cities are heavily affected by consequences of climate change (CC), but also have the largest potential for mitigation and adaptation. Adapting to urbanisation, climate change, social, economic and security trends is a challenging endeavour for cities and prone to potential conflicts of interest. It requires managing tasks like accommodating a growing – and ageing – population, providing the required services, fostering economic sustainability, and keeping the city liveable and attractive. Implementing climate change adaptation is still a comparably new responsibility that comes on top of these 'traditional' tasks of cities. It is a given that cities are short of funds, personnel and expertise for this new task.

There are, of course, potential synergies that cities could mobilise in order to use their scarce resources in an efficient way. Climate change adaptation measures could also have benefits for civil protection (CP) as well as critical infrastructure protection (CIP) and resilience (CIR). In this respect, the Fifth Assessment Report of the Intergovernmental Panel on Climate Change [17] has introduced an important change of paradigm, namely to perform risk assessment rather than solely the indicator-based vulnerability assessment suggested in earlier Assessment Reports. Risk assessment is standard in the domains of CP and CIP/CIR, and thus it is now possible to use the same or similar concepts in all three related domains, which is a prerequisite for using synergies.

[1] Source: UNEP brochure "Global Initiative for Resource Efficient Cities"

The European Union's H2020 research and innovation project *Climate Resilient Cities and Infrastructures* [25] investigated how cities and urban infrastructure can be made more resilient against consequences of climate change. For this purpose, RESIN has developed a set of methods and tools in close cooperation with city stakeholders. In this chapter, we will introduce one of them – namely the risk-based vulnerability assessment method *Impact and Vulnerability Assessment of Vital Infrastructures and Built-up Areas* (IVAVIA). It enables municipal decision makers to consider, analyse, and evaluate risks and vulnerabilities of urban areas and their infrastructures under specific extreme weather events and climate change-related scenarios. In the domains of CIP and CIR, an all hazards approach is standard, i.e. extreme weather events and consequences of climate change are a subset of all the hazards that need to be considered. CI risk assessment for weather and climate related hazards could benefit from the specialised methods developed for the climate change domain. In particular, it could help the actors in CIP/CIR to familiarise with the medium to long term changes in the frequency and/or severity of the CC related hazards and foster the coordination of measures for enhancing CIP/CIR and CC adaptation measures.

The remainder of this chapter is structured as follows: In the next section, we will briefly introduce the most essential concepts underlying risk and vulnerability assessments, before characterising the state-of-the-art and how IVAVIA is related to it. We continue with a short overview of the RESIN project. Thereafter, we will present the three stages of IVAVIA more in-depth. IVAVIA, like other RESIN methods and tools, has been developed, tested and assessed in a close 'co-creation' process together with four European cities. We will briefly describe the city case studies that RESIN has performed for IVAVIA and present the main assessment results that the city partners Bilbao, Bratislava, and Greater Manchester achieved by applying IVAVIA. We conclude with summarising the main results and insights gained.

2 Risk and Vulnerability Assessments – Essential Concepts

Up to their Fourth Assessment Report, the Intergovernmental Panel on Climate Change promoted scenario-based vulnerability assessments that took account of the bio-physical and socio-economic characteristics of a given system [18]. Yet, whilst this vulnerability-based concept shared affinities with those employed in allied disciplines, such as disaster risk management, they deployed the concepts in different ways [32]. This meant that close working between the two communities, desirable as it was in terms of implementing climate change adaptation in practice, was challenged.

The conceptual framing of climate change adaptation needed to evolve, and in 2012, the IPCC switched to 'risk' as the organising concept to understand climate change [16]. Hence, the climate change research community was advised to undertake risk assessments, which encompassed impact and vulnerability assessments [10]. The Fifth Assessment Report [17] sets out the concepts explicitly (Fig. 1).

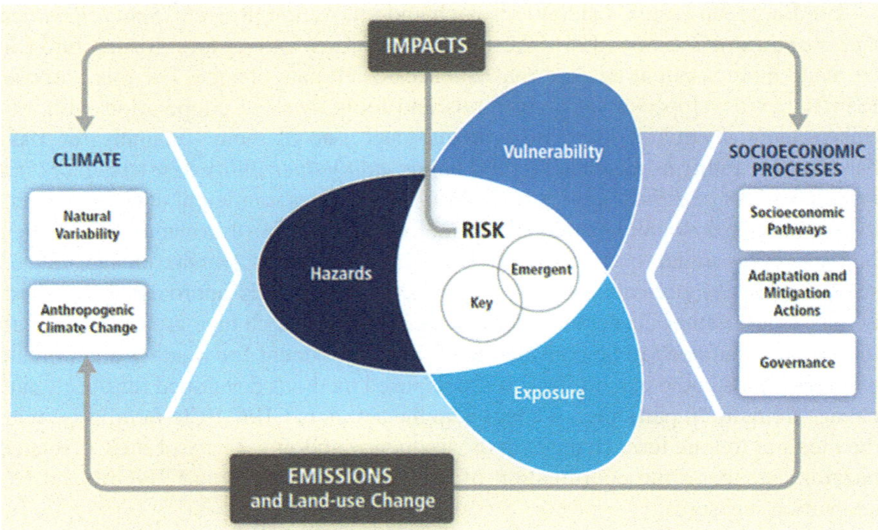

Fig. 1 Risk as compositions of hazards, exposure, and vulnerability. (Source: Ref. [17])

Compared to the Fourth Assessment Report [15], where exposure to climate change hazards was once considered to be part of vulnerability – alongside sensitivity to hazards and capacity to adapt – the move to risk has separated out exposure. In the IPCC's language, risk is therefore a function of hazard, exposure and vulnerability.

A *hazard* is "*. . . the potential occurrence of a natural or human-induced physical event or trend, or physical impact that may cause loss of life, injury, or other health impacts, as well as damage and loss to property, infrastructure, livelihoods, service provision, and environmental resources*" [17]. A climate-related hazard is a special case that is (at least partially) caused by climatic *drivers*. Examples for climate-related hazards include flooding, heatwave, and drought [7], while examples for related climatic drivers include sea-level rise, increased temperatures, and lack of precipitation.

Exposure refers to the objects or systems that might potentially be exposed: The presence of people, livelihoods, species or ecosystems, environmental services and resources, infrastructure, or economic, social, or cultural assets in specific places that could be adversely affected.

Non-climatic trends and events, which are called *stressors*, can have an important effect on an exposed system. Examples are population growth or change of land-use; a larger percentage of sealed surface will in general increase the susceptibility to flooding events and thus the vulnerability of all exposed objects.

Different objects are more or less sensitive to a hazard. This is captured by the concept of *sensitivity*, defined as the degree to which an exposed object, species or system could be affected by the considered hazard. As such, sensitivity towards a hazard can be perceived as a property of an exposed object in regard to a specific hazard. Examples for sensitivity include the degree of surface sealing, age and density of a population, household-income, or elevation and density of buildings.

Coping capacity is defined as *"the ability of people, institutions, organisations, and systems, using available skills, values, beliefs, resources, and opportunities, to address, manage, and overcome adverse conditions in the short to medium term"* [17]. Examples include the draining capacity of sewer systems, the height of a dike, education and awareness of the population, and availability of early warning systems. In contrast, *adaptive capacity* takes a medium to long-term perspective and can be understood as the 'room to move' for adaptation: the capacity for increasing the coping capacity, reducing the sensitivity, and reducing the severity of impacts.

Vulnerability is derived from the interplay of stressors, sensitivity, and coping capacity. It contributes directly to the impact or consequences that a hazard causes to the exposed objects.[2]

A risk assessment takes into account the characteristics and intensity of the considered hazard, as well as the set of objects exposed to it and their vulnerabilities. To put it simply, there is less risk if, for example, a given system is not exposed to a hazard such as a flood or coastal storm surge. Similarly, risk can be reduced if a given system is able to increase its coping capacity and/or reduce its sensitivity to such hazards, i.e. it has adaptive capacity.

3 State of the Art

Activities between the climate change adaptation (CCA) and disaster risk reduction (DRR) communities often overlap and some gaps have traditionally existed when it comes to research, policy-making and practice. Both communities attempt to reduce the negative impacts of climate change and disasters, although different actors, time horizons, methodologies, and policy frameworks are involved. The H2020 research project *Platform for Climate Adaptation and Risk Reduction* [23] has addressed this lack of collaboration, coordination, and effective communication by creating a knowledge-exchange platform for these two communities to communicate more effectively and enhance dialogue between CCA and DRR stakeholders.

It is also remarkable how the risk concept has evolved in the framework of CCA approaches. In the last years, several research projects investigated the climate resilience of European cities and provided methods and tools to: (i) quantify climate change vulnerability and risks; (ii) assess costs of damage and losses, as well as costs and benefits of adaptation measures; (iii) design adaptation measures and pathways that increase a city's resilience; and (iv) support city decision-makers in developing and implementing climate adaptation and resilience plans and measures. Among these projects, perhaps the more remarkable ones would be the

[2]The RESIN project's working definitions are mostly standardised definitions and can be found in [8, 9].

following EU Research projects: The FP7 project *Reconciling Adaptation, Mitigation and Sustainable Development for Cities* [24] provided exploratory exercises for improving hazard and impact modelling, developed relevant information on impact-damage functions for selected hazards, offered an overview of adaptation options with a specific focus on cost-benefit analysis of health policies, and scouted pilot applications of the flexible pathway approach in local adaptation policies. The FP7 project *Bottom-Up Climate Adaptation Strategies Towards a Sustainable Europe* [1] provided a wide range of climate adaptation case studies, some of them addressing urban scale. The H2020 project *Smart Mature Resilience* [29] systematised a standardised framework and developed tools for resilience management with a strong focus on capacity building and governance, including assessment of key climate resilience factors like interdependencies.

There are also new H2020 research projects assessing climate effectiveness of nature-based solutions in cities, like *Green Cities for Climate and Water Resilience, Sustainable Economic Growth, Healthy Citizens and Environment* [13], which will contribute to a better management of climate change adaptation in cities.

In addition to the above-mentioned projects, additional sources need to be mentioned, which present structured methods for vulnerability and risk assessments. The German Association for International Collaboration (GIZ), together with Adelphi and EURAC developed the *Vulnerability Sourcebook* [3, 4], based on the Fourth Assessment Report of the IPCC, and the associated *Risk Supplement to the Vulnerability Sourcebook* [5], based on the changes promoted in the Fifth Assessment Report of the IPCC to provide guidance for indicator-based vulnerability and risk assessments.

In these projects and documents the vulnerability, risk, and urban adaptation concepts are presented and defined from the disaster risk management, critical infrastructure protection and climate change adaptation perspectives. Through the development of the IVAVIA method, the RESIN project makes a substantial contribution to the research field by:

- relating and aligning the climate risks approach from the Fifth Assessment Report of the IPCC and related analysis components (hazards, exposure and vulnerability) with traditional risk analysis components (probability and consequences);
- distinguishing between a qualitative and a quantitative stage in risk assessment, each with their own methods;
- standardising the steps to be taken in a full-fledged urban climate change risk assessment.

4 The RESIN Project – Overview

RESIN was an EU-funded research project running from 2015 to 2018 [25]. It developed standardised methods and decision support tools for developing local adaptation strategies. It was one of the first large-scale research projects based on the conceptual approaches of the Fifth Assessment Report of the Intergovernmental Panel on Climate Change. The change in risk and vulnerability concepts introduced in this report led the researchers to explore the combination of approaches from climate change adaptation and disaster risk management [10].

On this basis, the project has developed a suite of tools that support city climate adaptation officers and infrastructure managers in developing strategies and plans to be prepared for the impacts of climate change. The project follows a four-stage approach in planning for adaptation [6]:

1. Assessing climate risks
2. Developing adaptation objectives
3. Prioritizing adaptation options
4. Developing an implementation plan

Within each of these stages, the effort was oriented at standardising the approach and the tools needed. The method and tools described in this chapter support the first stage of adaptation planning: to assess the risk of climate change. Additionally, the RESIN project has developed a *European Risk Typology*. Developed around the spatial unit of NUTS3 regions, the *European Climate Risk Typology* allows cities and regions to strategically screen for the climate hazards that they face, and their levels of exposure and vulnerability to hazards. Such an overview can be used as a starting point for a more detailed risk assessment.

The RESIN project has further developed a library of adaptation options with, as far as possible, harmonized information on the effectiveness of the adaptation measures. The library supports stage 2 and 3 in the adaptation planning process and is linked to the impact chain modelling technique, described in the following section, on the level of climate threats and exposed sectors. All tools are linked together in the e-Guide, an internet-based guide and workspace for developing adaptation strategies, or parts thereof.

The RESIN methods and tools have been developed in a close co-creation process with the cities of Bilbao (Spain), Bratislava (Slovakia), Greater Manchester (United Kingdom), and Paris (France). Nevertheless, processes, methods, and tools are standardised and can be applied to all European urban population centres, while they at the same time can be tailored to the specific needs of a municipality, depending on the stage and maturity of the local adaptation processes.

5 Risk-Based Vulnerability Assessment: IVAVIA

IVAVIA is a standardised process for the assessment of climate change-related risks and vulnerabilities in cities and urban environments aimed at supporting practitioners and end-users through the risk-based vulnerability assessment process. The IVAVIA process consists of seven modules in three stages (Fig. 2): the qualitative stage, the quantitative stage, and the presentation of the outcome. Each module consists of three to five individual steps. The modular process has been adopted from the Vulnerability Sourcebook [3] and modified in order to realize a risk-oriented vulnerability assessment.

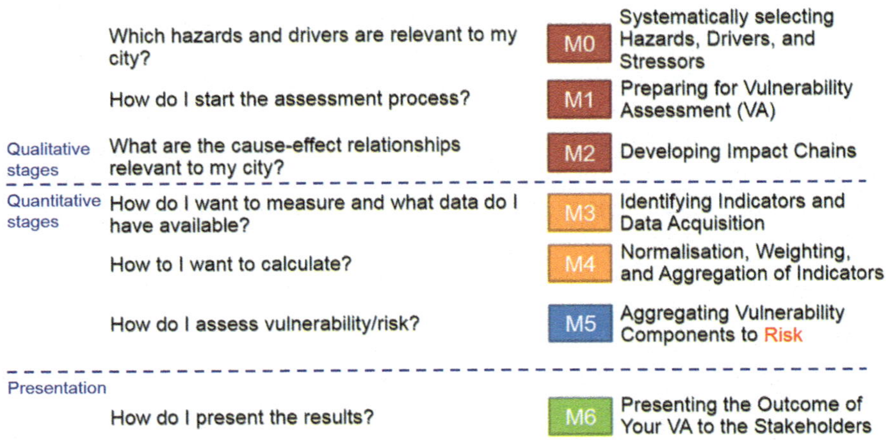

Fig. 2 Steps of the IVAVIA risk-based vulnerability assessment process

The decomposition of the process into modules and their steps is aimed at making it more manageable for end-users, many of whom may not be vulnerability/risk assessment experts but municipal employees tasked with such an analysis for the first time. Following the whole sequence is not mandatory – if an end-user is an expert in vulnerability/risk assessments, has material available from a previous assessment, lacks resources to conduct a complete assessment, wants to use different approaches to specific steps, or prefers to conduct steps within a module in a different order, they may opt for customizing IVAVIA and its modules to their needs.

Each step of the module descriptions contains information about input needed and output to be created. Following the IVAVIA process, a full qualitative and quantitative assessment would be covered by executing the modules in the given sequence, as each module generates input for the following ones. For a qualitative assessment, the process has only to be run up to Module M2 followed by immediately jumping to Module M6. In general, the amount of resources necessary for the assessment process varies widely, depending on the size of the studied area and the requested depth and scope of the evaluation.

The modules and steps are described in detail in the IVAVIA Guideline [26], addressed to local decision makers, with the more technical details of the process and reference information being covered by the IVAVIA Guideline Appendix [27] (Fig. 3).

5.1 Qualitativ Stage

Module M0, *'Systematically selecting hazards, drivers, and stressors'*, starts off the process with a systematic analysis and selection of hazards, drivers, and stressors

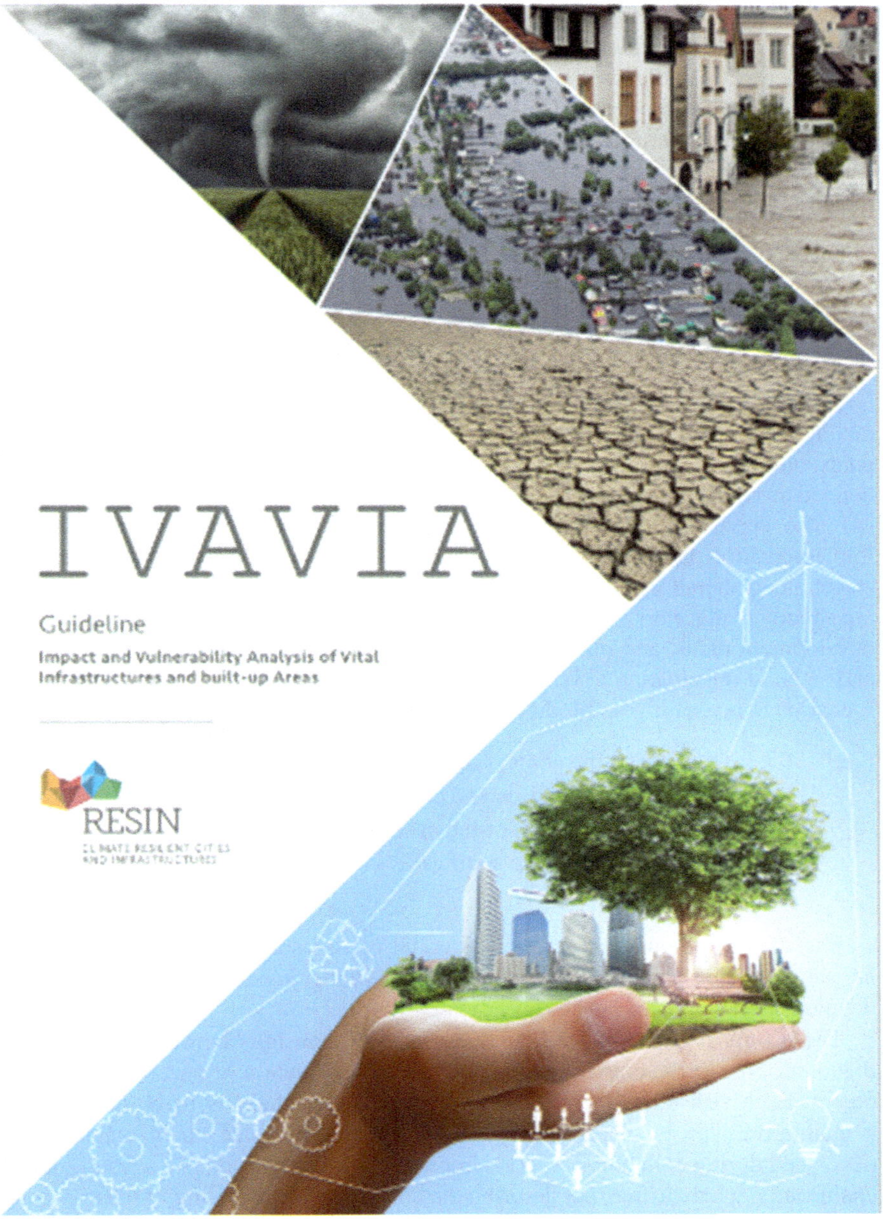

Fig. 3 Cover of the IVAVIA Guideline addressed to municipal decision makers and local stakeholders. The cover layout is © Fraunhofer, the cover images used are licensed from Fotolia (now Adobe Stock Photos) and modified by Fraunhofer

relevant to the region or urban area under examination. This serves as a base for the detailed planning of the assessment and ensures that the limited resources and budgets are spent on the most pressing current and future hazards, and that no threats or possible dependencies between different hazards are overlooked. In addition, a thorough documentation of the rationale for selecting hazards, drivers, and stressors ensures that future (re-) assessments can follow the same methodology, thus enabling result comparison. Module M0 consists of the following steps:

Step 0.1 Identify the hazards considered potentially relevant
0.2 Gather information on the identified hazards
0.3 Identify generally relevant drivers and stressors
0.4 Kick-off meeting and management decisions

As part of Module M1, *'Preparing for the vulnerability assessment'*, a common taxonomy is defined and communicated, and the overall objectives, scopes, participants and their roles and responsibilities, as well as the target audiences have to be defined in agreement and, ideally, in cooperation with the relevant stakeholders. M1 also serves to identify and gather relevant information to form a detailed implementation plan. The information needed for this step includes a list of relevant stakeholders including both institutions and individuals, measures and strategies that are already in place or to be considered (e.g. sector strategies, community or national development plans, and on-going adaptation measures), climatic, socio-economic, and sectoral information to be included, and a list of climate and city development scenarios to be examined. Module M1 consists of the following steps:

Step 1.1 Understand the context of the risk-based vulnerability assessment
1.2 Identify the objectives and expected outcomes
1.3 Determine the scope of the assessment
1.4 Develop the scenario settings
1.5 Prepare a work plan

Based on this foundation for the vulnerability assessment, impact chains (Fig. 4) are developed as part of Module M2, *'Developing impact chains'* (for a more detailed description see [20]).

These impact chains describe cause-effect-relationships between the elements that contribute to the consequences of a given combination of hazard and exposed object. Each risk component included in an impact chain is to be described in a qualitative way by specifying attributes. Usually, impact chain diagrams are developed during collaborative workshops with domain experts. As a result, impact chains are not exhaustive, but describe the common understanding of these experts. An important rule of thumb is: keep it simple! It is not possible (and needed) to cover each and every detail in an impact chain. Typically, the assessment starts with selecting a combination of hazard and exposed object, like the hazard 'pluvial flooding' and the exposed object 'road transport'. The more of such relevant combinations are assessed, the more comprehensive the assessment. Module M2 consists of five individual steps:

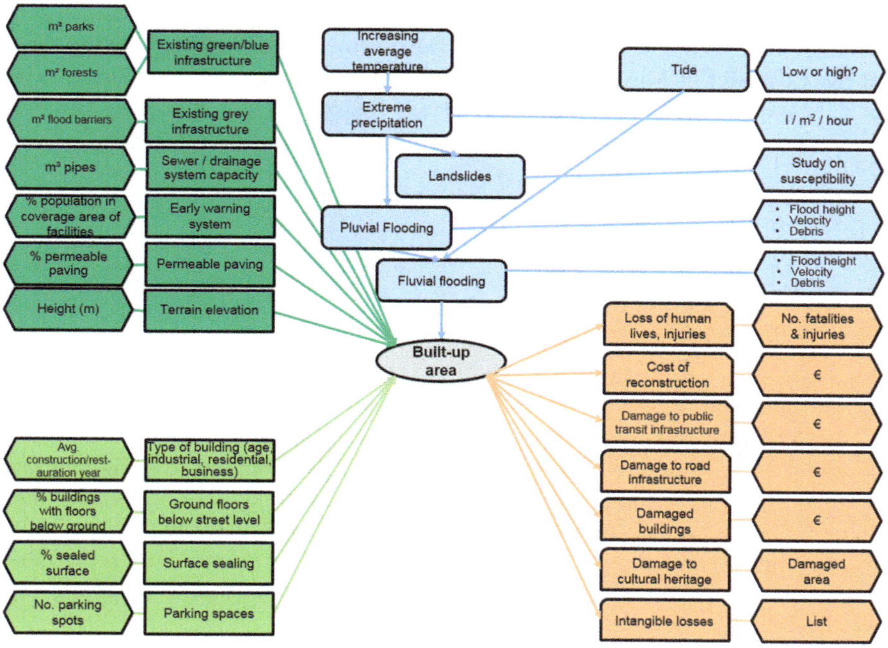

Fig. 4 Example impact chain for the hazard exposure combination 'flooding in built-up areas' for the city of Bilbao. Hazards and drivers in blue, exposed object in grey, coping capacity in green-blue, sensitivity in green, and impacts in orange [27]

Step	2.1	Determine exposure and hazard combinations
	2.2	Identify specific drivers and stressors
	2.3	Determine sensitivity
	2.4	Determine coping capacity
	2.5	Identify potential impact

5.2 *Quantitative Stage*

Module M3, *'Identifying indicators and data acquisition'*, describes the identification and definition of measurable indicators for the specified attributes of the generated impact chains. The indicator identification and data collection steps are highly dependent on each other. The availability of data is of critical importance for the quantitative stage: Without a feasible way for data acquisition, the best indicator would be inoperable. To this end, it is important to include domain experts with extensive knowledge about data availability. To ease the indicator selection process, established directories of standard indicators should be employed, for example, the annex of the Vulnerability Sourcebook (BMZ German Federal Ministry for Economic Cooperation and Development [4], pp. 14–17) or the annex of the Covenant of Mayors for Climate and Energy Reporting Guidelines (Neves et al. [21], pp. 61–67). Module M3 consists of five operational steps:

Step	3.1	Select indicators
	3.2	Check if the selected indicators are suitable
	3.3	Gather data
	3.4	Check data quality
	3.5	Manage data

Communicating a multitude of complex, multi-dimensional indicators in a comprehensive way is extremely complicated. Therefore, the calculated indicator values should be aggregated to composite scores (e.g. using weighted arithmetic mean [22]). The indicator values likely employ different measurement units and scales, and thus cannot be aggregated without being normalised (e.g. via min-max normalization [22]). In addition, the selected indicators may not necessarily have equal influence on their corresponding risk component, which should be reflected by assigning weights to them when combining them into composite scores. These issues are addressed in the course of Module M4, *'Normalisation, weighting, and aggregation of indicators'*, which consists of five steps:

Step	4.1	Determine the scale of measurement
	4.2	Normalise coping capacity and sensitivity indicator values
	4.3	Weight coping capacity and sensitivity indicators
	4.4	Aggregate coping capacity and sensitivity indicators
	4.5	Calculate vulnerability scores

Module M5, *'Aggregating vulnerability components to risk'*, covers the actual risk assessment, which is based on the well-established risk analysis process by the German Federal Office of Civil Protection and Disaster Assistance [2], assuring organisational, legal, and political interoperability. In this approach, impacts and probabilities are classified using discrete, ordinal classes (e.g. 'insignificant', 'minor', or 'disastrous' for impacts and 'very unlikely', 'likely', and 'very likely' for probabilities). The resulting impact and probability pairs, i.e. the risk scores, are then assigned to discrete, ordinal risk classes using a risk matrix. This matrix has one axis for the impact classes and one axis for the probability classes, and thus defines risk classes for every combination of the two. Module M5 consists of five steps:

Step	5.1	Define classification scheme
	5.2	Estimate hazard intensity and probability
	5.3	Estimate impacts/consequences
	5.4	Calculate risk scores
	5.5	Validate results

5.3 Presentation

The last Module M6, *'Presenting the outcomes of IVAVIA'*, concerns the systematic presentation of the IVAVIA process itself and its outcomes to all relevant stakeholders and funding bodies, including external vulnerability and risk assessment experts to assure external result validation. Best practices are shared, and supporting material, i.e. report and presentation templates are being provided, as well as graphs exported by the developed software tools. M6 consists of three steps:

Step 6.1 Plan your report
6.2 Describe the undertaken assessment process
6.3 Illustrate the findings

With the successful conclusion of Module M6 the risk-based vulnerability assessment process is complete. Building on this base the municipal stakeholders can now go on to systematically plan, and then finally implement adaptation measures.

5.4 Supporting IT Tools

Two integrated software tools were developed to support the IVAVIA work flow as part of the RESIN project. A web-based graphical *Impact Chain Editor* (ICE+) supports end-users by automatically arranging and colourising impact chain diagram components. ICE+ facilitates structured annotations and most importantly, it provides a rule-based engine for checking the validity of diagrams (e.g. hazard elements can only link to exposed objects or other hazard elements). As the elements and structure of the generated impact chains are stored in an online repository, the captured information is automatically available for further processing, and thus does not have to be manually converted to an appropriate format for transmission and storage. The tool also imports and exports XML, CSV, as well as MS Excel files, and provides standard graphics formats for presentation and print. ICE+ is accessible with all major web browsers; and includes comprehensive user management, providing each user or user group with their own account and workspace. ICE+ comes in seven different languages.

A PostgreSQL database was developed to serve as a repository for impact chain diagrams. In addition, it enables users to store data on the indicators used in impact chain diagrams in a simplified form sufficient for the (in-database) computation of indicator values for the calculation of composite sensitivity, coping capacity, and vulnerability scores.

The database connects to the web-based tool *Risk and Vulnerability Assessment System* (RIVAS), adapting the SIRVA tool [30], which was originally developed in the context of the RAMSES project. RIVAS reads impact chains and their indicators and calculates aggregated coping capacities, sensitivities, and vulnerabilities. It can be widely configured by end-users, who can select calculation methods like weighting and aggregation algorithms. RIVAS exports tables and maps to support the visualisation of results.

6 The RESIN City Case Studies

The RESIN city case studies were conducted as co-creation processes with the partnering tier-1 cities. Co-creation allows short-term feedback from stakeholders to developers already during development. Usually, projects use a linear process of eliciting requirements for a new tool or method, creating a specification, realising and implementing the tool/method, and evaluating it. Instead, co-creation comprises several cycles of lean versions of this process as well as iterative development and test of the tools/methods in close cooperation with the end-users. In total, co-creation takes longer, but partial results are available earlier and the final results can be more mature.

The co-creation process with Greater Manchester started in May 2016 and had the goal to develop and apply the qualitative part of the IVAVIA process. Therefore, two impact chain workshops have been conducted in Greater Manchester for developing impact chains for two different infrastructures. The city of Paris followed the Manchester activities and conducted its own stakeholder workshops with their urban development department in the district of Bercy-Charenton.

The Bratislava city case study started in November 2016 and fully commenced in early 2017. Initially it only comprised the selection and aggregation of indicators, but grew into a full risk-based vulnerability assessment after the city passed an action plan for the implementation of a climate change adaptation strategy in 2017 and took this opportunity to update and deepen an existing vulnerability assessment from 2014.

The co-creation process with Bilbao started in July 2016 and initially comprised a full risk-based vulnerability assessment on a neighbourhood scale. The results of this assessment were included in a climate change vulnerability report that Bilbao had to deliver in September 2017. Based on insights gained during this process, a higher resolution assessment was subsequently conducted.

7 IVAVIA Results for RESIN Cities

The IVAVIA process was adapted individually depending on the characteristics of each of the RESIN tier-1 cities, yielding different outcomes.

In order to select the impact chains to work on in Greater Manchester, stakeholders, who worked on climate resilience and critical infrastructure, completed a questionnaire to identify the priority themes and areas. Based on the results of this questionnaire, two hazard-exposure combinations were selected and workshops were held to develop impact chains.

The first impact chain workshop focused on the hazard-exposure combination *Pluvial flooding on transportation systems* and included local experts from the Low Carbon Hub, Transport for Greater Manchester, the Civil Contingences and Resilience Unit, as well as RESIN research partners. During the workshop it became clear that the theme of transportation systems was too broad, and several decisions were taken in order to further refine the scope, first to the road network and then to a 'main arterial road serving the city centre' in a hypothetical way.

The second impact chain workshop focused on the hazard-exposure combination *An extended period of hot dry weather on green infrastructure* and was attended by representatives from the Low Carbon Hub, the Civil Contingencies and Resilience Unit, ARUP, Natural England and City of Trees, as well as RESIN research partners. Rather than addressing a specific aspect of green infrastructure (e.g. an urban park, river corridor, a woodland), the decision was taken to look at this critical infrastructure theme from a more general perspective.

Greater Manchester stakeholders found that the impact chains were a good communication and awareness-raising tool. Additionally, the impact chains have been included in Greater Manchester's Preliminary Risk Assessment as part of its Rockefeller 100 Resilient Cities activities. The pluvial flooding impact chain was subsequently used as basis for semi-quantitative and quantitative risk assessments, which did not follow the IVAVIA approach.

The city case of Bratislava began with identifying potential stakeholders for a kick off workshop. In a joint decision with the tool developers it was decided to host the workshop in native language and prepare a ten-page summary of the IVAVIA guideline document in Slovak, to serve as a background document and reading material for the participants. The participants were mostly representatives of the relevant departments of the city administration, the national hydrometeorological institute, regional self-governmental authority, and providers of social care. During the workshop, participants focused on the three most relevant climate change hazards (pluvial floods, heatwaves, droughts) and the impacts they impose to citizens' health, quality of life, and urban infrastructure (including green infrastructure). The workshop was divided into two thematic sessions – the first one consisting of two breakout groups focusing on the effects of heatwaves and pluvial floods on health and wellbeing, while the last one was dedicated to the effects droughts have on green infrastructure. The participants listed altogether more than 90 attributes for all three hazard exposure combinations. In a follow-up step, city and research partners revised the developed impact chains and reduced the number of indicators to a manageable amount to ease data acquisition. Subsequently, the Office of the Chief City Architect with support of the Comenius University in Bratislava (local research partner of Bratislava in RESIN) collected relevant data, which was used by Fraunhofer IAIS to calculate vulnerabilities and risks. The first results of this process will be presented during a follow-up workshop in 2018 to get feedback from stakeholders on the validity of the outputs.

For the Bilbao city case an impact chain workshop was organised, which was attended by representatives from different departments of the municipality and RESIN research partners. During this workshop, attendees were given the opportunity to modify, delete or add elements to three draft impact chains, focusing on: *extreme precipitation on city traffic infrastructure, heatwave on public health, and flooding in built-up areas*. The draft impact chains were prepared by RESIN research partners and purposefully incomplete. The result was a qualitative vulnerability assessment of these areas of interest.

Based on these results, a quantitative vulnerability and risk mapping on a neighbourhood scale was produced, which was part of a climate change vulnerability report submitted by the municipality of Bilbao in 2017. To conduct the assessments, the Bilbao city council provided citywide spatial data for all sensitivity and coping capacity indicators defined in the created impact chains. The spatial data included, for example, the distribution of parks and forests across the city, building location, construction/restauration year, and number of floors, as well as position, length, and diameter of sewer pipes. Where necessary, the provided data was further processed (e.g. lengths and diameters of pipes where used to calculate volume). Subsequently, the indicator values where normalised to a scale from 0 ('optimal') to 1 ('critical') and aggregated to sensitivity and coping capacity scores. Weights for different indicators were chosen by the partners from the Bilbao city council based on their perceived importance.

The resulting composite scores were in turn normalised and aggregated to vulnerability scores using weighted arithmetic mean. Based on these results, risks were calculated. For example, for a 500 year flood expected worst-case impacts for all impact categories defined in the corresponding impact chain where derived using flood depth-damage functions (see [14, 19]). These functions combine flood depth and velocity to derive damage values (e.g. residential building damage per m^2), which were combined with the actual exposed objects (e.g. the surface area of residential buildings situated in flooded areas) and multiplied by the vulnerability scores to arrive at expected impacts. Finally, the expected impacts as well as the local probability were classified using discrete classes and combined into risk levels according to the risk analysis approach by the German Federal Office of Civil Protection and Disaster Assistance. In order to not equate losses of human lives and (monetary) material damages, the categorisation differentiated between material impacts (e.g. residential, commercial, and industrial building damages) and human impacts (e.g. fatalities and injuries).

While applying the IVAVIA method for Bilbao the partners realised that the initial scale was too coarse: Information about uneven spatial distributions of indicator data within neighbourhoods (e.g. significant elevation changes or an uneven distribution of green infrastructure across flood prone areas and non-flood prone areas) was lost by averaging across whole neighbourhoods. In addition, the initial resolution did not allow planning of specific adaptation measures. Therefore, a quantitative risk assessment, focused on the risk of flooding in built-up areas was conducted, using a regular grid with a cell size of 25 m $\times$ 25 m, raising the resolution from 39 neighbourhoods to approx. 66,000 cells. Figs. 5, 6 show the resulting vulnerability and material risks maps.

Having tailored methods and tools to help overcome the complexity of adaptation planning and gain new knowledge that can be transferred into feasible outcomes,

Fig. 5 Vulnerability map of Bilbao for flooding in built-up areas

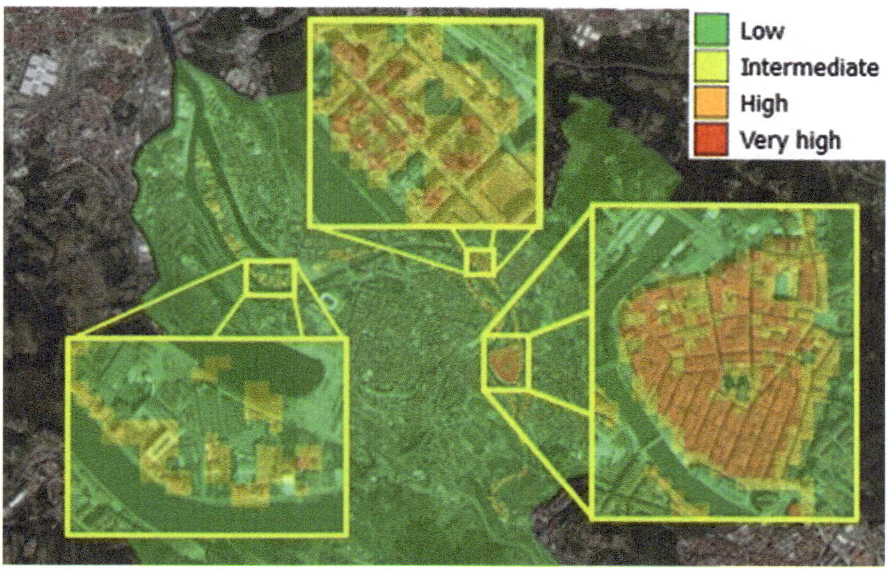

Fig. 6 Material risk map of Bilbao for flooding in built-up areas

such as vulnerability or risk maps, is crucial for selecting the right adaptation options. The IVAVIA tools and the concepts behind them were easily understood during the workshop. The joint design and visualisation of the impact chain diagrams helped the stakeholders get an understanding of the different components of the assessment and the different causalities. Overall, the IVAVIA process can be adjusted depending on the desired depth and with regard to available resources (time, certain skills, etc.). To make the whole process less time consuming, there are IVAVIA's supportive tools[3] to help with the calculations and producing other outcomes such as spatial visualisations (maps, impact chain diagrams, etc.).

8 Climate Change-Related Risk to Critical Infrastructures

Critical Infrastructure (CI) can be addressed with IVAVIA in the following ways: As practiced in several RESIN city case studies, CI can be the exposed object (exposure), that is, it is the immediate subject of the analysis. This can be done at different scales, e.g. a specific sector, like the road transport network, or a specific element, like a major road, or even a single element, like a pumping station. The depth of analysis of exposed objects, including CI, depends on the overall goals of the risk assessment and the available resources, data, and information. In many cases it is not necessary to acquire data and information from CI owners. IVAVIA yields an overall risk-based vulnerability assessment of the investigated CI (element) with regard to consequences of climate change. This result can also be used in more comprehensive risk analyses in the areas of critical infrastructure protection and resilience.

Impacts of consequences of climate change on CI may also be addressed in the investigation of impacts as elements of impact chains. Here, primary impacts and secondary impacts can be analysed by methods developed, for instance, in the CIPRNet project [28]. Secondary impacts may include cascading effects on other, dependent CI and socio-economic impacts as identified in the European Directive on CIP [11].

9 Conclusion

This chapter gave an introduction on risk-based vulnerability assessment of urban systems and infrastructure regarding the consequences of climate change, with a focus on the EU-funded research and innovation project RESIN and the embedded IVAVIA assessment approach.

After a brief introduction to essential concepts of risk and vulnerability assessments and the state of the art in related research projects, RESIN and its climate change adaptation planning approach were introduced. This approach consists of the four stages (1) assessing climate risks; (2) developing adaptation objectives; (3)

[3] See RESIN tool page at http://www.resin-cities.eu/resources/tools/

prioritizing adaptation options; and (4) developing an implementation plan, each supported by specific methods and tools. One of these methods, namely the IVAVIA method for assessing climate risks, was presented in more detail. The approach consists of seven modules across the three stages (1) qualitative assessment; (2) quantitative assessment; and (3) result presentation, which is supported by a suite of software tools and can be customised according to local conditions.

The applicability of the IVAVIA process was exemplarily demonstrated in four city case studies with the RESIN tier-1 cities Bilbao, Bratislava, Greater Manchester, and Paris. These case studies show that the IVAVIA process is a feasible means to analyse vulnerabilities regarding the impact of climate change in local urban contexts. Objects of investigation in the case studies included health and well-being of the population and vulnerable groups, as well as infrastructure like road transport, green infrastructure, and built infrastructure. While supporting its end-users with practical guidance, IVAVIA is flexible enough to be applicable to urban areas of different sizes and organisation, and suffering from different combinations of hazards.

Acknowledgments The authors thank their partners in the RESIN consortium for their valuable contributions during the development and test process. This paper is based in part upon work in the framework of the European project "RESIN – Climate Resilient Cities and Infrastructures".

This project has received funding from the European Union's Horizon 2020 research and innovation programme under grant agreement no. 653522. The sole responsibility for the content of this publication lies with the authors. It does not necessarily represent the opinion of the European Union. Neither the EASME nor the European Commission are responsible for any use that may be made of the information contained therein.

Without the great role model called "The Vulnerability Sourcebook", developed by GIZ and Eurac, IVAVIA would certainly have looked quite different. We would like to thank Till Below (GIZ) and Stefan Schneiderbauer (Eurac) as representatives of the numerous colleagues in their organizations who have created the Sourcebook in 2014. We are also grateful for the repeated collaboration and exchanges between GIZ and Eurac and RESIN partner Fraunhofer since 2016, which contributed to shaping both the Sourcebook supplement and the IVAVIA Guideline document.

References

1. BASE (2016) Bottom-up climate adaptation strategies towards a sustainable Europe. https://base-adaptation.eu/.Accessed 13 June 2018
2. BBK (2011) German federal office of civil protection and disaster assistance: method of risk analysis for civil protection. Wissenschaftsforum, Volume 8, Bonn. ISBN: 978-3-939347-41-5
3. BMZ German Federal Ministry for Economic Cooperation and Development (2014a) The vulnerability sourcebook. Concept and guidelines for standardised vulnerability assessments. Deutsche Gesellschaft für Internationale Zusammenarbeit (GIZ) GmbH, Bonn and Eschborn. http://www.adaptationcommunity.net/?wpfb_dl=203. Accessed 18 May 2018
4. BMZ German Federal Ministry for Economic Cooperation and Development (2014b) The vulnerability sourcebook annex. Deutsche Gesellschaft für Internationale Zusammenarbeit (GIZ) GmbH, Bonn and Eschborn. http://www.adaptationcommunity.net/?wpfb_dl=204. Accessed 18 May 2018

5. BMZ German Federal Ministry for Economic Cooperation and Development (2017) Risk supplement to the vulnerability sourcebook. Guidance on how to apply the Vulnerability Sourcebook's approach with the new IPCC AR5 concept of climate risk. Deutsche Gesellschaft für Internationale Zusammenarbeit (GIZ) GmbH, Bonn and Eschborn, and EURAC
6. Carter J, Connelly A (2016) RESIN conceptual framework, Deliverable D1.3. EU H2020 Project RESIN, The University of Manchester, Manchester, United Kingdom. http://www.resin-cities.eu/resources/framework-concept/
7. Carter J et al (2015) Weather and climate hazards facing European cities. RESIN state of the art report (3), University of Manchester. http://www.resin-cities.eu/resources/sota/hazards/. Accessed 18 May 2018
8. CIPedia (2018) Online glossary of terms in Critical Infrastructure Protection, including RESIN project glossary. http://cipedia.eu. Accessed 18 May 2018
9. Connelly A et al (2016) RESIN Glossary, Deliverable D1.2. EU H2020 Project RESIN, The University of Manchester, Manchester, United Kingdom
10. Connelly A, Carter J, Handley J, Hincks S (2018) Enhancing the practical utility of risk assessments in climate change adaptation. Sustainability 2018(10):1399. https://doi.org/10.3390/su10051399
11. European Council (2008) Council directive 2008/114/EC of 8 December 2008: identification and designation of European critical infrastructures and the assessment of the need to improve their protection. Techical report, Official Journal of the European Union, (L 345/75)
12. EUROSTAT (2016) Urban Europe – statistics on cities, towns and suburbs – 2016 edition. Eurostat statistical books, Luxembourg: Publications office of the European Union, 2016
13. GROWGREEN (2018) Green cities for climate and water resilience, sustainable economic growth, healthy citizens and environments. http://growgreenproject.eu/. Accessed 13 June 2018
14. Huizinga HJ, Moel H, Szewczyk W (2017) Global flood depth-damage functions. Methodology and the database with guidelines. EUR 28552 EN. https://doi.org/10.2760/16510
15. IPCC Intergovernmental Panel on Climate Change (2007) Climate change 2007: synthesis report. Contribution of working groups I, II and III to the fourth assessment report of the intergovernmental panel on climate change [Core Writing Team, Pachauri RK, Reisinger A (eds)]. IPCC, Geneva, Switzerland, 104 pp
16. IPCC Intergovernmental Panel on Climate Change (2012) Managing the risks of extreme events and disasters to advance climate change adaptation. Special Report of the Intergovernmental Panel on Climate Change. Cambridge University Press, Cambridge
17. IPCC Intergovernmental Panel on Climate Change (2014) Climate change 2014: synthesis report. Contribution of working groups I, II and III to the fifth assessment report of the intergovernmental panel on climate change [Core Writing Team, Pachauri RK, Meyer LA (eds)]. IPCC, Geneva, Switzerland, 151 pp
18. Klein RJT, Nicholls R (1999) Assessment of coastal vulnerability to climate change. Ambio 28:182–187
19. Kok M, Huizinga HJ, Vrouwenvelder ACWM, Barendregt A (2004) Standard method 2004. Damage and casualties caused by flooding. Highway and hydraulic engineering department
20. Lückerath D et al (2018) The RESIN Climate Change Adaptation Project and its Simple Modeling Approach for Risk-Oriented Vulnerability Assessment. In: Workshop 2018 ASIM/GI-Fachgruppen (ASM 2018), Proceedings of ASIM workshop, Hochschule Heilbronn, Germany, March 8–9, 2018, pp 21–26
21. Neves A, Blondel L, Brand K, Hendel-Blackford S, Rivas Calvete S, Iancu A, Melica G, Koffi Lefeivre B, Zancanella P, Kona A (2016) The Covenant of mayors for climate and energy reporting guidelines, EUR 28160 EN. https://doi.org/10.2790/586693
22. OECD (2008) Organisation for Economic Co-operation and Development – Handbook on constructing composite indicators: methodology and user guide, Technical Report. OECD Publishing, Paris
23. PLACARD (2018) Platform for climate adaptation and risk reduction. https://www.placard-network.eu/. Accessed 13 June 2018

24. RAMSES (2017) Reconciling adaptation, mitigation and sustainable development for cities. http://www.ramses-cities.eu/. Accessed 13 June 2018
25. RESIN (2018) EU H2020 Project RESIN – climate resilient cities and infrastructures website. http://resin-cities.eu/home. Accessed 18 May 2018
26. Rome E, Bogen M, Lückerath D, Ullrich O, Voss H, Voss N, Worst R (2017a) IVAVIA guideline, annex to deliverable D2.3 Realisation and implementation of IVAVIA. EU H2020 Project RESIN, Sankt Augustin, Germany
27. Rome E, Bogen M, Lückerath D, Ullrich O, Voss H, Voss N, Worst R (2017b) IVAVIA guideline appendix, annex to deliverable D2.3 Realisation and implementation of IVAVIA. EU H2020 Project RESIN, Sankt Augustin, Germany
28. Rome E, Doll T, Rilling S, Sojeva B, Voß N, Xie J (2017c) Chapter 10: The use of what-if analysis to improve the management of crisis situations. In: Setola R, Rosato V, Kyriakides E, Rome E (eds) Managing the complexity of critical infrastructures, a modelling and simulation approach. Springer. https://doi.org/10.1007/978-3-319-51043-9_10
29. SMR (2018) Smart mature resilience. http://smr-project.eu/home/. Accessed 13 June 2018
30. Tapia C, Abajo B, Feliu E, Mendizabal M, Martínez JA, Fernández JG, Laburu T, Lejarazu A (2017) Profiling urban vulnerabilities to climate change: An indicator-based vulnerability assessment for European cities. Ecol Indic 78:142–155. https://doi.org/10.1016/j.ecolind.2017.02.040
31. United Nations (2014) World urbanization prospects: the 2014 revision. https://esa.un.org/unpd/wup/publications/files/wup2014-report.pdf
32. Wolf S (2011) Vulnerability and risk: comparing assessment approaches. Natl Hazard 61:1099–1113. https://doi.org/10.1007/s11069-011-9968-4
33. Worldbank (2010) Cities and climate change – an urgent agenda. The international bank for reconstruction and development/World Bank, Washington DC, Urban Development Series. http://hdl.handle.net/10986/17381
34. Worldbank (2011) Hoornweg D, Freire M, Lee MJ, Bhada-Tata P, Yuen B (eds) Cities and Climate Change – Responding to an Urgent Agenda. The International Bank for Reconstruction and Development/World Bank, Washington DC, Urban Development Series. https://doi.org/10.1596/978-0-8213-8493-0

Part II
Dependencies & Network Analysis

Identification of Vulnerabilities in Networked Systems

Luca Faramondi and Roberto Setola

Abstract In last decades, thanks to the large diffusion of Information and communications technologies, the cooperation of distributed systems has been facilitated with the aim to provide new services. One of the common aspect of this kind of systems is the presence of a network able to ensure the connectivity among the elements of the network. The connectivity is a fundamental prerequisite also in the context of critical infrastructures (CIs), which are defined as a specific kind of infrastructures able to provide the essential services that underpin the society and serve as the backbone of our nation's economy, security, and health (i.e. transportation systems, gas and water distribution systems, financial services, etc). Due to their relevance, the identification of vulnerabilities in this kind of systems is a mandatory task in order to design adequate and effective defense strategies. To this end, in this chapter some of the most common methods for networks vulnerabilities identification are illustrated and compared in order to stress common aspects and differences.

Keywords Critical nodes · Network vulnerabilities · Optimization approach

1 Introduction

In the last years integration and cooperation of distributed systems has gained lager and larger relevance. Nowadays, distributed control systems are used instead of (or in conjunction with) centralized system to manage large scale systems geographically dispersed in order to improve efficacy and effectiveness. This because via direct integration of heterogeneous new services and new functionalities can be provided so as it is possible to move the "smartness" more close to the field allowing fast reactions and better capabilities to manage critical situations. Such

L. Faramondi (✉) · R. Setola
Campus Bio-Medico University, Rome, Italy
e-mail: l.faramondi@unicampus.it; r.setola@unicampus.it

D. Gritzalis et al. (eds.), *Critical Infrastructure Security and Resilience*,
Advanced Sciences and Technologies for Security Applications,
https://doi.org/10.1007/978-3-030-00024-0_5

systems, characterized by a strong dependence on communication networks, belong to the class of Cyber Physical Systems (CPS). If in one hand, this kind of systems overcome the limits of centralized systems thanks to the interaction of multiple subsystems, on the other hand, it suffers of the vulnerabilities induced by the communication infrastructure especially when it shared communication channels as internet.

The vulnerability of CPS is paramount, considering statistics about cyber incidents involving CIs in 2014 and 2015 in the United States [13]. In particular, 245 cyber incidents occurred in 2014, 32% of which involved CIs in the energy sector, while another 27% was relative to CIs in the manufacturing sector. In 2015 the number of incidents increases to 295 events 16% in the energy field and 33% in the manufacturing sector. It is worth mentioning that in the 38% of the above events, the source of the cyber incidents and the infection vector remains unknown.

Due to the increasing number of cyber incidents involving CIs, the research of novel approaches for networking systems vulnerability identification has long been a theme of discussion in order to provide new methodologies able to support more effective and efficient protection strategies to design more resilient infrastructures. One of the main goals is to preserve the network connectivity. Indeed, in several fields the connectivity has a fundamental role to guarantee the functionalities of the entire system [7].

Since the seminal works of Albert et al. [19] and Holme et al. [11] in the early 2000s, it has become evident that attacks that take into account the topological structure of the underlying network may have catastrophic consequences. In fact, knowing the topology of the network, an attacker can select more effectively the target sites in order to maximize the damage while keeping the cost of the attack at a minimum [2, 12, 16, 21, 25]. From a mathematical point of view, the research of the most suitable targets for a malicious attacker is equivalent to identify the critical nodes of the networks, i.e. those nodes whose removal largely compromises the connectivity of the network.

In the literature, several metrics have been adopted as a measure of nodes criticality. A first approach is based on the well known nodes centrality measures such as the node degree, the eigenvector centrality [3], and the betweenness [4]. A different approach is to consider the negative effects induced on the network the "removal" of a node or a link. In particular, a well-established approach is to focus on intentional attacks, considering a rational attacker that aims at maximizing the damage. This kind of approaches is based on the results of an optimization problem which simulate the behavior of an attacker.

As mentioned in [15], among other formalism, the *Critical Node Detection Problem* (CNP) [1, 22] proved to be particularly effective. Within the CNP, an attacker targets some of the nodes in the network (removing all their incident links) with the aim to minimize the *pairwise connectivity* (PWC) [23], that is, the number of pairs of nodes that are connected via a path after the attacked nodes have been removed. Specifically, the CNP assumes that up to a fixed number of k nodes can be attacked. Such an approach, however, requires a large number of Boolean decision variables [22] and a number of constraints that can be non-polynomial in the number of nodes of the network [22]; these factors limit the applicability of such a methodology. In [18] a dual problem is addressed, namely *Cardinality Constrained Critical Node Detection Problem* (CC-CNP), which constraints the largest connected component to be smaller than a user-defined value. In this case, the objective is to minimize the number of attacked nodes required to fulfill the constraint. In [24] the authors argue that the attacker decision process is intrinsically a multi-objective problem. They suggest that improvements can be obtained when not only the pairwise connectivity, but also the variance in cardinality among the connected components is minimized, i.e., the dimension of the "islands" obtained after removing some of the k nodes. Such an approach, however, suffers the same drawbacks of the standard CNP. Moreover, the two objectives are "scalarized"; such a scalarization is highly dependent on the specific priority between the objectives for the attacker, and thus it has limited validity, especially when the attacker behavior is not known a priori. A similar path is followed in [14], where the size of each connected component obtained as a result of the attack is constrained to be below a given bound. A slightly different formulation is described in [6], where the target consists in the minimization of the attack cost with a constrained connectivity of the solution.

More recent works based on optimization techniques are presented in [8, 9], and [10]. More precisely, in [8], a formulation named *Largest Partition Size Minimization* (LPM optimization) is provided. According to this approach, the attacker is not constrained to target a fixed number of nodes, and aims at dividing the network in a predetermined number of partitions while having two conflicting sub-objectives: minimize the number of attacked nodes and minimize the size of the largest component. Even such an approach, however, considers a scalarized objective function to balance the two clashing objectives.

In [9], an improvement is obtained by assuming that the number of partitions is not fixed a priori; the maximization of the number of disconnected components becomes an additional objective that has to be mediated with the minimization of the attacked nodes and the minimization of the size of the largest component. This approach is named *Partition Number and Size Optimization* (PNS optimization). More precisely, this formulation has $O(n^2)$ Boolean decision variables, and a polynomial number of constraints, hence the approach has lower complexity with respect to the approaches in [1, 22] while being more descriptive.

The approaches in [8, 9] are further improved in [10] by casting the problem in the framework of *Multi-Objective Optimization* (MOO). In this case the approach does not consider the optimizing a convex combination of the different objectives, noting that in such a way the result is tailored to a particular class of attackers (e.g., an attacker with large economic resources or an attacker with limited budget). Specifically with MOO the most critical nodes are identified on the base of all the possible combinations of the objectives. Thus, instead of inspecting the behavior of a particular class of attackers, MOO adopts a perspective that is independent from the specific class of attackers.

2 Preliminaries

In this section, for the seek of completeness, a brief collection of necessary preliminaries are collected. The preliminaries contain some definitions about graph theory and optimization problems.

2.1 *Notation*

In the following we denote by $|X|$ the cardinality of a set X; moreover, we represent vectors via boldface letters, and we use $\mathbf{k}_m$ to indicate a vector in $\mathbb{R}^m$ whose components are all equal to k.

2.2 *Graph Theory*

Let $G = \{V, E\}$ denote a *graph* with a finite number n of nodes $v_i \in V$ and e edges $(v_i, v_j) \in E \subseteq V \times V$, from node v_i to node v_j. A graph is said to be *undirected* if $(v_i, v_j) \in E$ whenever $(v_j, v_i) \in E$, and it is said to be *directed* otherwise; in the following we will consider undirected graphs. A graph $G = \{V, E\}$ is *connected* if each node can be reached by each other node by means of the links in E, regardless of their orientation. The *adjacency matrix* of a graph G is an $n \times n$ matrix A such that $A_{ij} = 1$ if $(v_j, v_i) \in E$ and $A_{ij} = 0$ otherwise. A *direct path* over a graph $G = \{V, E\}$, starting at a node $v_i \in V$ and ending at a node $v_j \in V$, is a subset of links in E that connects v_i and v_j, respecting the edge orientation and without creating loops. A directed graph that has a direct path from each vertex to every other vertex is said to be *strongly connected.* A connected component V_i of a graph is sub set of nodes of V such that each pair of nodes in V_i is connected. The *pairwise connectivity* (PWC) [10] of a graph G is an index that captures the overall degree of connectivity of a graph:

$$PWC(G) = \sum_{v_i, v_j \in V, v_i \neq v_j} p(v_i, v_j), \tag{1}$$

where $p(v_i, v_j)$ is 1 if the pair (v_i, v_j) is connected via a direct path in G, and is zero otherwise. In other words, the pairwise connectivity is the number of pairs of nodes that are connected via a path over G. Note that the PWC assumes its maximum value if G is a strongly connected. In this case the number of couples of nodes connected by a direct path is $n(n - 1)$, where n is the number of nodes in G.

2.3 Optimization Problems

An *Integer Linear Programming* (ILP)[20] problem is composed by a linear objective function and same constraints, moreover the solutions belong to the set of dimensional vectors n having integer components, $\mathbb{Z}^n$. An ILP problem is defined adopting the general form:

$$z = min(\mathbf{c}^T\mathbf{x})$$

$$A\mathbf{x} \leq \mathbf{b}$$

$$\mathbf{x} \in \mathbb{Z}^n$$

where z is the cost of the solution computed by the objective function $min(\mathbf{c}^T\mathbf{x})$, $c \in \mathbb{R}^n$ represents the vector of the costs associated to each decision variable x_i. The expression $A\mathbf{x} \leq \mathbf{b}$, represents the constraints. A is an $m \times n$ matrix whose entries $a_{i,j} \in \mathbb{R}$ and m is the number of the constraints. The constraints in an integer program form a polytope. However, the *feasible set* is given by the set of all integer-valued points within the polytope, and not the entire *polytope*. Therefore, the feasible region is not a convex set. Moreover, the optimal solution may not be achieved at an extreme point of the polytope; it is found at an extreme point of the convex hull of all feasible integral points. The naive way to solve an ILP problem is to simply remove the constraint that x is integer, solve the corresponding LP, that is the relaxation of ILP by neglecting the presence of the integer constraint, and then round the entries of the solution to the LP relaxation. But, not only may this solution not be optimal, it may not even be feasible, that is it may violate some constraints.

A *Multi-Objective Optimization problem* (MOO) consists in a optimization problems involving more than one objective function to be optimized simultaneously. Given a vector $\mathbf{x} \in \{0, 1\}^n$ representing n *decision variables*, a *MOO* problem can be expressed as follows

$$\min f(x) = \min \quad [f_1(\mathbf{x}), f_2(\mathbf{x}), \ldots, f_k(\mathbf{x})]^T, \quad \text{subject to } \mathbf{x} \in \mathscr{F}, \tag{2}$$

where $k \geq 2$ and the i-th objective is given by

$$f_i(\mathbf{x}) : \mathbb{R}^n \to \mathbb{R}, \qquad \text{for } i = 1, \ldots, k,$$

while $f(\mathbf{x}) \in \mathbb{R}^k$ is the *multi-objective function*. The set $\mathscr{F}$ represents the set of *feasible solutions* for the problem at hand. Moreover, the *multi-objective space* is defined as

$$\mathscr{Z} = \{\mathbf{x} \in \mathbb{R}^k : \exists\, \mathbf{x} \in \mathscr{F}, z = f(\mathbf{x})\}.$$

Within a MOO problem, therefore, the aim is to select a feasible solution $\mathbf{x}$ that minimizes at the same time all the different objectives f_i. Let us consider a solution $\mathbf{x}^*$ for which all the objectives $f_i(\mathbf{x}^*)$ are simultaneously minimized, and let us denote the associated multi-objective vector $f(\mathbf{x}^*)$ by z^{id}. Notice that, when there is no conflict among the objectives, we can solve Problem (2) by solving k scalar problems, thus obtaining z^{id} as the *ideal* multi-objective vector. Due to the conflicting nature of the objectives $f_i(\mathbf{x})$, however, it is realistic to assume that $z^{id} \notin \mathscr{Z}$. In most practical cases, therefore, there is a need to overcome the above naive definition of an optimal solution; a typical approach in the literature is to resort to the theory of *Pareto optimality* [5]. Let z^a and $z^b \in \mathscr{Z}$; we say that z^b is *Pareto-dominated* by z^a ($z^a \leq_P z^b$) if:

$z_i^a \leq z_i^b$ for each $i = 1, 2, \ldots, k$ and
$z_j^a < z_j^b$ at least for a value of $j \in \{1, \ldots, k\}$.

A solution vector $\mathbf{x}^* \in \mathscr{F}$ is a *Pareto optimal solution* if there is no other solution $\mathbf{x} \in \mathscr{F}$ such that:

$$f(\mathbf{x}) \leq_P f(\mathbf{x}^*). \tag{3}$$

The *Pareto front* $\mathscr{P}$ is the set of all possible Pareto optimal solutions $\mathbf{x}^*$ for the problem at hand, while we denote by $\mathscr{P}_f$ the set of values $f(\mathbf{x}^*)$, in the multi-objective space, which correspond to each $\mathbf{x}^* \in \mathscr{P}$.

3 Critical Nodes Identification

As introduced above, the investigation about structural vulnerabilities consists in the research of network items considered critical due to their relevance in the network. In most cases this kind of criticalities is discovered solving optimization problems. Often the optimization problems aim to the reduction of the network connectivity looking at a set of constraints useful to describe the specific problem in exam. This approach allows to simulate a malicious attacker that aims to reduce some connection-related index minimizing the attack cost. In the rest of this section, several methods based on this approach are shown in order to describe different attacker perspectives.

3.1 Critical Node Detection Problem

In the literature, a lot of approaches have been devoted to discover the presence of critical nodes in the networks. The authors of [1] present a mathematical model based on the ILP approach which provides optimal solutions for the classical *critical node detection problem* (CNP). Given an undirected graph $G = \{V, E\}$ and an integer k, the aim is to find a subset $V_C \subseteq V$ of the nework's nodes such that $|V_C| \leq k$, whose deletion minimizes the connectivity among the nodes in the induced subgraph $G(V \setminus V_C)$. The problem admits the following ILP formulation:

$$\min \sum_{i,j \in V} u_{ij}$$

that is the minimization of the network connectivity, where:

$$u_{ij} = \begin{cases} 1, \text{ if } i \text{ and } j \text{ are in the same component of } G(V \setminus A) \\ 0, \text{ otherwise} \end{cases}$$

The objective function of the proposed problem is related to the minimization of the nodes able to communicate via an undirected path among the induced subgraph $G(V \setminus V_C)$.

The constraints are the following:

(i) $u_{ij} + v_i + v_j \geq 1 \quad \forall (i, j) \in E,$
(ii) $u_{ij} + u_{jk} + u_{ki} \leq 1 \quad \forall (i, j, k) \in V,$
(iii) $u_{ij} - u_{jk} + u_{ki} \leq 1 \quad \forall (i, j, k) \in V,$
(iv) $-u_{ij} + u_{jk} + u_{ki} \leq 1 \quad \forall (i, j, k) \in V,$
(v) $\sum_{i \in V} v_i \leq k,$
(vi) $u_{ij} \in \{0, 1\} \quad \forall i, j \in V,$
(vii) $v_i \in \{0, 1\} \quad \forall i \in V.$

where the decision variables $v_i = 1$ if the i-th node is involved in the attack (i.e. it is in the partition V_C), and 0 otherwise. Moreover, such a formalism, has high memory and computational requirements: for instance, considering the formulation in [22], there is the need to consider a non-polynomial number of constraints, while in [22] the problem consists in choosing a value for $O(n^2)$ boolean decision variables.

3.2 *Cardinality Constrained Critical Node Detection Problem*

Another formulation based on the optimization approach is presented in [18] where a slightly modified problem is addressed, namely *Cardinality Constrained Critical Node Detection Problem* (CC-CNP). In this perspective, a maximum allowed connected graph component size L is specified and the objective is to minimize the number of attacked nodes required to fulfill this constraint. Given an integer L, the objective is to find a subset $V_C \subseteq V$ such that the largest connected component in the induced subgraph $G(V \setminus V_C)$ contains no more than L nodes. So the problem is restricted to the minimization of $|V_C| : |V_i| \leq L, 1 \leq i \leq T$, where T is the total number of connected components in the induced subgraph $G(V \setminus V_C)$. In order to describe this particular version of the problem a Boolean variable is introduced:

$$v_i = \begin{cases} 1, \text{if the node } i \text{ is deleted in the optimal solution} \\ 0, \text{otherwise.} \end{cases}$$

Finally the problem is expressed by the objective function:

$$\min \sum_{i \in V} v_i$$

that is the minimization of the attack cost. The constraints are the following:

(i) $u_{ij} + v_i + v_j \geq 1 \quad \forall (i, j) \in E$,
(ii) $u_{ij} + u_{jk} + u_{ki} \neq 2 \quad \forall (i, j, k) \in V$,
(iii) $\sum_{j \in V} u_{ij} \leq L \quad \forall i \in V, i \neq j$,
(iv) $u_{ij} \in \{0, 1\} \quad \forall i, j \in V$,
(v) $v_i \in \{0, 1\} \quad \forall i \in V$.

Constraints (i) and (ii) describe exactly the classical CNP formulation. The novelty is represented by the constraint (iii), here the maximum connectivity for each node is limited to L.

A variant of this formulation, known as *Component Cardinality Constrained Critical Node Problem* (3C-CNP) is presented in [14]. This variant of the CC-CNP considers a non-negative cost c_{ij} for each edge $(v_i, v_j) \in E$ and a weight w_i for each node $v_i \in V$. The 3C-CNP seeks to find V_C such that:

$$\min \sum_{v_i \in A} w_i,$$

moreover, in addition to the constraints (i), (ii), (iv), and (v) of the CC-CNP, another constraint is considered for each connected component V_i:

$$\sum_{v_i, v_j \in V_i} c_{ij} \leq L.$$

Thus, 3C-CNP consists in finding a set of nodes $V_C \subseteq V$ of minimal total weight, such that the total connection cost of each connected component V_i in the induced subgraph is no more than L.

3.3 β-Vertex Disruptor

A different ILP approach known as *β-vertex disruptor* is introduced in [6]. The problem consists in the research of subset $V_C \subseteq V$ with the minimum cardinality, such that the connectivity in $G(V \setminus V_C)$, obtained by removing the nodes in V_C from V, is not more than:

$$\beta\binom{n}{2} = \beta\frac{n!}{2(n-2)!}$$

As introduced in the definitions of CNP and CC-CNP approaches, the decisional variables are $u_{i,j}$ and v_i. The objective function consists in the minimization of removed nodes in V_C

$$\min \sum_{i \in V_C} v_i$$

while the constraints are:

(i) $u_{ij} \leq v_i + v_j \quad \forall (i,j) \in E,$
(ii) $u_{ij} + u_{jk} \geq u_{ik} \quad \forall (i,j,k) \in V, i \neq j \neq k,$
(iii) $\sum_{i<j} u_{ij} \geq (1-\beta)\binom{n}{2}$
(iv) $v_i \leq u_{ij}, \quad i \neq j,$
(v) $u_{ij} \in \{0,1\} \quad \forall i,j \in V,$
(vi) $v_i \in \{0,1\} \quad \forall i \in V.$

The constraint (ii) is the *triangle inequality* which implies that if the node v_i and v_j are connected and, and v_j and v_k are connected, then also v_i and v_k are connected (i.e. v_i, v_j, and v_k are in the same connected partition V_i). Constraint (iii) limits the network connectivity respect to the bound defined by β. Constraint (i) implies that if two nodes, v_i and v_j, are neighbors and none of them is removed, then they remain connected. Finally, constraint (iv) implies that if a node is removed then it is disconnected from any other nodes.

3.4 Large Partition Minimization

The approach for the discovery of critical nodes presented in [8] is slightly different from the previous approaches. In this formulation the objective function is a linear combination of two sub-objective. In more details, this model represents the perspective of an attacker that aims to remove nodes from the network in order divide the network in a fixed number of partitions m by minimizing the size of the largest connected partition. In this way the proposed model tends to provide solution characterized by balanced partitions in terms of number of nodes.

Concerning the decision variables, the proposed approach requires $O(mn)$ Boolean variables. In more details, the variables $\mathbf{x}^{(1)}, \ldots, \mathbf{x}^{(m)}, \mathbf{c} \in \{0,1\}^n$, such that $x_j^{(i)} = 1$ when i-th node is assigned to i-th partition and zero otherwise. The entries $c_i = 1$ if the node is involved in the attack, zero otherwise.

More precisely, the objective function is composed by two sub-objectives as described in Eq. (4): the minimization of the largest partition and the minimization of the weighted attack cost (i.e. the number of attacked nodes). Note that there will be a removal cost p_i for each attacked node c_i node. The parameter α is introduced in order shifting the focus between the two sub-objectives, in this way the model is able to reproduce several attack strategies. In more details if α is close to 0, then the solution will be focused to the minimization of the largest partition size, otherwise, for values of α close to 1, the focus is on the minimization of the attack cost.

$$min\Big(\alpha \mathbf{p}^T \mathbf{c} + (1-\alpha) \max_{i=1,\dots,m} \left(\mathbf{1}_n^T \mathbf{x}^{(i)}\right)\Big) \tag{4}$$

Note that, according to the proposed formulation, the nodes that belong to the same partition are not necessarily connected each other. In more details the problem provides that, as described in the classical CNP, each node has to be assigned to just one set (see Eq. (5)), and the nodes assigned to a partition V_i are not directly connected to the nodes in the others partitions (see Eq. (6)).

$$\sum_{i=1}^{m} \mathbf{x}^{(i)} \leq \mathbf{1}_n. \tag{5}$$

$$(v_a, v_b) \notin E, \text{ for } v_a \in V_i, v_b \in V_j, \quad \forall i, j = 1, \dots, m. \tag{6}$$

Additional constraints are necessary to ensure that empty partition are not hallowed and at least a node must be attacked (Eq. (7)).

$$\begin{array}{c} \mathbf{1}_n^T \mathbf{x}^{(i)} \geq 1, \quad \forall i = 1, \dots, m; \\ \mathbf{1}_n^T \sum_{i=1}^{m} \mathbf{x}^{(i)} \leq n-1, \end{array} \tag{7}$$

3.5 Partition Number and Size Minimization

Another optimization problem for the discovery of network vulnerabilities is presented in [9]. Here, the attacker aims at divide the network by maximizing the number of partitions, by keeping the attack cost to the minimum, and by reducing the size of the largest partition. Similarly to the approach described in Sect. 3.4, the proposed problem requires $O(n^2)$ Boolean variables.

The decision variables are defined according to the following scheme: $x_j^{(i)} = 1$ when node v_j is assigned to the partition V_i, moreover c_j is defined as a vector of Boolean variables such that $c_j = 1$ if v_j is involved in the attack and $c_j = 0$ otherwise. In more details, the objective function is a linear combination of three sub-objectives as described in Eq. (8).

$$\min \Big\{ -\alpha_1 \sum_{i=1}^{n-1} \mathbf{1}_n^T \mathbf{x}^{(i)} + \alpha_2 \max_{i=1,\dots,n-1} \left(\mathbf{1}_n^T \mathbf{x}^{(i)}\right) - \alpha_3 \sum_{i=1}^{n-1} t_i \Big\} \tag{8}$$

where $t_i = 0$ if partition V_i is empty, otherwise $t_i = 1$.

More precisely, the three sub-objectives represent the minimization of the attack cost, the minimization of the size of the largest partition, and the maximization of the partitions number. Notice that, according to the formulation described in Sect. 3.4, also in this scheme, the preferences of the attacker can be represented by changing the values of the weights α_1, α_2, and α_3 (such that $\alpha_1 + \alpha_2 + \alpha_3 = 1$).

Concerning with the constraints, this formulation requires $O(n^2 e)$ constraints, where e is the number of edges in the graph. In more details they are the same of the formulation presented in Sect. 3.4. Notice that only a set additional constraints is required in order to describe the relations between the decisional variables $x_j^{(i)}$ and the variables t_i. As described in Eq. (9), if a node v_j is assigned to the partition i, then the variable t_i must be set to 1.

$$x_1^{(i)} + \ldots + x_n^{(i)} \geq t_i \qquad i = 1 \ldots n-1. \tag{9}$$

Notice that, according to the objective function in Eq. (8), the number of non-empty partition is maximized. In this way if $t_i = 1$ at least one node must be assigned to the i-th partition.

3.6 Multi Objective Optimization Approach

An innovative approach for the research of critical nodes in a graph has been presented in [10]. Differently from the previous approaches, in this case the problem is approached as Multi Objective Optimization problem. Indeed, it considers two conflicting objectives: the minimization of the network connectivity (f_1) and the reduction of the total cost of the attack (f_2) as defined in Eq. (10).

$$\min f(\mathbf{x}) = \min[f_1(\mathbf{x}), f_2(\mathbf{x})]^T, \tag{10}$$

Note that the proposed approach requires n boolean decision variables x_i, such that, $x_i = 1$ if the i-th node is involved in the attack, 0 otherwise. In more details, the two objectives are defined in Eq. (11):

$$f_1 = PWC(G(V \setminus V_C)) \qquad f_2 = \frac{\mathbf{c}^T \mathbf{x}}{\mathbf{1}^T \mathbf{c}}. \tag{11}$$

Where V_C is the set of nodes involved in the attack. In the attack cost (f_2), the vector $\mathbf{c} \in \mathbb{R}^n$ is introduced to describe the removal cost for each node of the network. Due to the conflicting nature of the two objective functions, the proposed schema is unable to find an unique optimal solution. Instead, each solution that belongs to the Pareto front is characterized by an attack cost and a connectivity value as defined in Eq. (11). In this way, tit is possible to consider a set of multiple attackers' classes each one characterized by different preferences in terms of budget and attack strategy. Thanks to the analysis of each solution in the Pareto front, it is possible to recognize those nodes of the network which more often appear as targets in the different attack strategies, i.e. in attack plans with different objectives and budgets. As described in [10], the frequency with which a given node is targeted in the solutions belonging to the Pareto front is used as a measure of its criticality. More details about this metrics are discussed in Sect. 4.2.

4 Node Importance Metrics

In this section, vulnerability and importance metrics are discussed in order to presents two classes of indices able to capture the relevance of the different nodes in a network, i.e. whose removal from the network largely compromises the network connectivity. In more details, in Sect. 4.1, node centrality metrics based on network structure and topology are defined, while in Sect. 4.2 vulnerability metrics for critical nodes based on optimization problems are discussed.

4.1 Node Centrality Measures

The *eigenvector centrality* [3] is a measure of the influence of a node in a network. It assigns relative scores to all nodes in the network based on the assumption that being connected to highly influential nodes makes a node influential. Let us consider a graph $G = \{V, E\}$ with n nodes and let A the adjacency matrix. The eigenvector centrality for a vertex v_i can be defined as:

$$\eta_{v_i} = \frac{1}{\lambda} \sum_{t \in \mathcal{N}(v)} \eta_t = \frac{1}{\lambda} \sum_{t \in V} a_{t,v} \eta_t$$

where $\mathcal{N}(v)$ is a set of the neighbors of v, λ is a constant, and $a_{t,v}$ are the entries of the matrix A.

Another node centrality measure is the *betweenness centrality* [4], it is based on the number of shortest paths. The betweenness centrality for a node v is the sum over all distinct node pairs s, t of the fraction of minimum paths from s to t that pass trough v. The betweenness centrality of a node v is given by the expression:

$$g(v) = \sum_{s \neq v \neq t} \frac{\sigma_{st}(v)}{\sigma_{st}}$$

where σ_{st} is the total number of shortest paths from node s to node t and $\sigma_{st}(v)$ is the number of those paths that pass through v.

The adoption of these measures in the context of network vulnerability is well addressed in [17], where the network connectivity is studied with respect to multiple attacks. In more details the attacker aims at disconnecting the network by removing the nodes in descending order of eigenvector centrality or betweenness centrality.

4.2 Criticality Measures

As introduced in Sect. 3, in the literature, several approaches for the evaluation of network vulnerabilities are based on optimization problems. The biggest difference between this class of metrics and the evaluations based on structural properties (see Sect. 4.1) is that the solution of the optimization problems depends on the fixed parameters of the formulation, (e.g. the attack cost k for the CNP formulation, the size of the largest connected components L in CC-CNP formulation, the number of desired partitions m in LPM, etc). Moreover, while the evaluation based on centrality measures provides a ranking of important nodes, the solution of optimization problems consists only in a subset of nodes without identify a specific value for each element of the network. In [9] and [10], with the aim to overcome this limit, an in depth analysis has been performed. In more details, the approach consists of the research of vulnerabilities by solving optimization problems for different values of fixed parameters, in this way, the behaviors of multiple attackers with different preferences are evaluated. This approach is based on the research of recurrent schemes in the attack plans of multiple attackers which are different in terms of attack preferences.

Take now into account the classical CNP formulation described in Sect. 3.1, it requires a fixed attack cost. The research of recurrent attack schemes is performed by computing multiple solutions by varying the attack cost k in the range $[0, \ldots, n]$. Let n_S the number of different attack costs considered in the range $[0, \ldots, n]$, it consists also in the number of optimal collected solutions characterized by different values of the parameter k. The optimal solution of the CNP optimization problem are collected as column vector[1] in the matrix $X \in \{0; 1\}^{n \times n_S}$ (see Eq. 12).

$$X = [\mathbf{v}^1, \ldots, \mathbf{v}^{n_S}] \tag{12}$$

The vector whose entries represents the criticality indices is defined as:

$$\chi = \frac{1}{n_S} X \mathbf{1}_n \tag{13}$$

it consists in the normalized frequency with which each node is involved in the attack plans with respect to the number of considered attack plans n_S.

The same definition of criticality index is also adopted for the CC-CNP. According to the optimization problem presented in Sect. 3.2, this formulation requires a fixed upper-bound about the size of the largest connected component of the graph after the removal of critical nodes. With the aim to identify common targets for different attackers (i.e. different values of the parameter L) the problem is solved for multiple values of $L \in \{0, \ldots, n-1\}$. Let n_S be the considered number

[1]Notice that the decision variables collected in the vectors of the matrix X consists of the subset of decision variables that represents the nodes deletion.

of different values of L, the matrix $X \in \{0, 1\}^{n \times n_S}$ and the critical indices, for the CC-CNP formulation, are respectively defined according to the Eqs. (12) and (13).

Concerning the β-Vertex Disruptor approach, for a given admissible connectivity upper-bound, it returns the minimum number of nodes to attack with the aim to reduce the connectivity to the desired value. Also in this case the research of recurrent attack schemes is performed by solving the optimization problem by considering multiple values of β in the range $[0, \ldots, 1]$.

A similar approach is also adopted for LPM and PNS problems. In this case the attack cost and the size of the largest connected components are free parameters. The multiple evaluations are computed by analyzing the solutions for multiple values of m (i.e. the number of final partitions) for LPM formulation, and for multiple combination of the parameters α_1, α_2, and α_3 for PNS formulation. Note that according to the formulation described in Sect. 3.5, the parameters α_1, α_2, and α_3 are used to shift the focus of the attack among the sub-objectives of the attacker.

Differently from the previous approaches, the MOO problem, due to its multi-objective form, it provides a set of optimal solutions (i.e. the Pareto Front) without considering fixed parameters. The frequency with which a given node is targeted in any one of the solution belonging to the Pareto front, is used to estimate the node criticality. In this case the frequency is normalized with respect to the number of optimal solutions that belong to the Pareto front.

5 Metrics Comparison

In this section the methodologies for the discovery of network vulnerabilities, based on node centrality measures (see Sect. 4.1) and criticality metrics (see Sect. 4.2) are compared in order to highlight their differences and common aspects. To this end, the IEEE24 Power System, which is represented by a direct graph having $n = 24$ vertices and $e = 35$ edges has been considered. Methods based on ILP approaches (i.e. CNP, CC-CNP, β-Vertex Disruptor, LPM, and PNS) handle the intrinsic multi-objective nature of the critical node detection problem by constraining one of the possible degrees of freedom (e.g. attack cost, partitions size, number of partitions, and network connectivity). In order to compare such methods with MOO approach and with the measures based on structural properties (e.g. betweenness and eigenvector centrality), the whole set of possible solutions by varying the constrained value and evaluating the different obtained results has been explored.

In more details, concerning the CNP approach, the results about the criticality measures are based on 25 solution of this approach by varying the attack cost ion the range $[0, \ldots, 24]$. The same approach has been adopted also for the CC-CNP and the β-Vertex Disruptor, respectively analyzed by considering 24 and 21 optimal solutions by varying the size of the largest connected component, in the range $[0, \ldots, 23]$, and the desired network connectivity in the range $[0, \ldots, 1]$. 10 optimal solutions have been considered for the LPM optimization problem, in this case the solution are associated to different values of the partition number m in

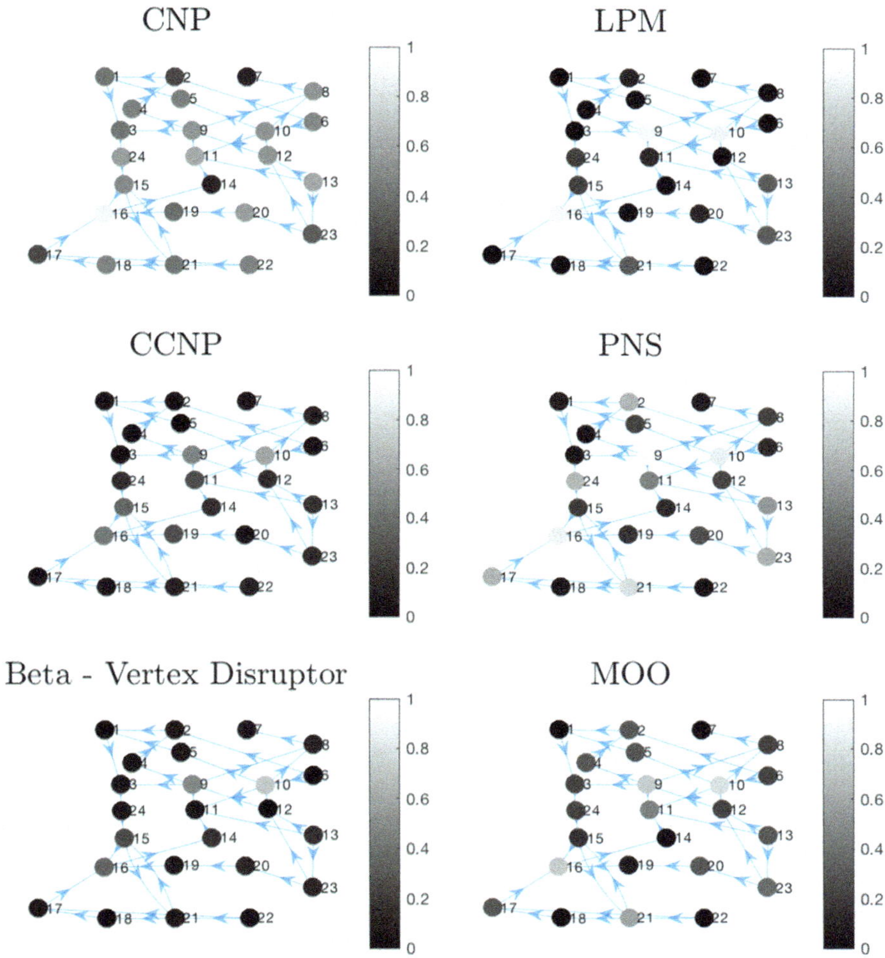

Fig. 1 IEEE24 colormap. Criticality indices based on CNP, CC-CNP, β-Vertex Disruptor, LPS, PNS, and MOO solutions. The nodes color are related to the criticality indices computed considering equal attack costs. High values are associated to light colors

the range $[1, \ldots, 10]$. Finally, the results about the PNS approach are evaluated by considering 66 different combinations of the parameter α_1, α_2, and α_3.

The results about the criticality indices, as defined by Eq. (13), considering optimization approaches, are represented in the colormaps in Fig. 1, while the values of betweenness and eigenvector centrality are provided in Fig. 2.

The results shown in Fig. 1 highlight the presence of a subset of critical nodes identified by multiple approaches. With reference to Fig. 1, concerning the results of methods based on optimization problems, the nodes 9, 10, and 16 are frequently

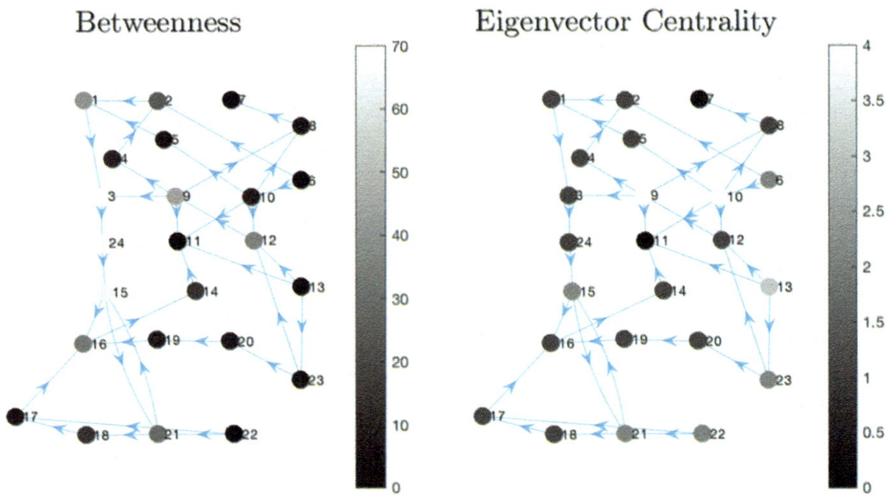

Fig. 2 IEEE24 colormap. Values of betweenness and eigenvector centrality. The nodes color are related to the criticality indices computed considering equal attack costs. High values are associated to light colors

involved in the attacks plans. This result suggests that each approach based on optimization problem can be considered as a different point of view of the same phenomenon. The unique difference among these results is in the distribution of the criticality indices over the network. Despite the approaches converge to the identification of the same critical nodes, each approach proposes a different evaluation of criticality with respect to the other nodes. The results associated to LPM approach highlight the presence of a subset of nodes that are considered strongly critical with respect to the other nodes of the network while the other approaches propose a more homogeneous distribution of criticality indices. In Fig. 2 the values of node centrality are shown. It is evident that the two evaluations catch different characteristics of the networks. In more details, while the betweenness highlights a set of most critical nodes different from the set identified in Fig. 1, the eigenvector centrality, according to the results in Fig. 1 identifies the node 9 and 10 as the most critical element of the network. More details about the criticality value of each node are collected in Table 1.

6 Conclusions

In this chapter the problem of networks vulnerabilities identification has been analyzed by assuming the perspective of a malicious attacker and by comparing two classes of approaches. The results suggest that, despite some minor difference on the evaluation, the approaches converge on the identification of a common subset

Table 1 IEEE24 power network. Ranking in descending order of criticality with respect to centrality measures and optimization methods

CNP		CCNP		β-Vert.		LPM		PNS		MOO		Centrality measures			
Id	χ_i	Id	χ_i	Id	χ_i	Id	χ_i	Id	χ_i	Id	χ_i	Id	Bwn	Id	Eig. Cent.
16	0.88	10	0.58	10	0.71	16	0.90	9	0.93	10	0.78	3	70	10	4
13	0.60	9	0.45	9	0.47	10	0.90	16	0.86	16	0.71	24	67	9	4
11	0.60	16	0.41	16	0.33	9	0.90	10	0.84	9	0.71	15	65	13	3
24	0.56	15	0.33	15	0.23	23	0.30	21	0.77	21	0.57	9	44	23	2
20	0.56	19	0.25	14	0.14	21	0.30	24	0.65	11	0.42	1	39	22	2
12	0.52	11	0.25	13	0.14	13	0.30	17	0.63	23	0.35	12	34	21	2
10	0.52	13	0.16	23	0.09	24	0.20	2	0.63	20	0.28	16	32	15	2
9	0.52	24	0.12	20	0.09	15	0.20	23	0.62	17	0.28	21	28	6	2
8	0.52	23	0.12	8	0.09	20	0.10	13	0.53	13	0.28	2	23	24	1
15	0.48	14	0.12	24	0.04	11	0.10	11	0.42	5	0.28	18	14	20	1
6	0.48	12	0.12	21	0.04	2	0.10	20	0.25	4	0.28	14	14	19	1
22	0.44	22	0.08	19	0.04	22	0	12	0.21	2	0.28	4	9	18	1
18	0.44	21	0.08	17	0.04	19	0	8	0.21	24	0.21	19	7	17	1
5	0.44	8	0.08	11	0.04	18	0	5	0.21	12	0.21	10	7	16	1
4	0.44	20	0.04	7	0.04	17	0	15	0.18	8	0.21	20	6	14	1
21	0.40	18	0.04	5	0.04	14	0	19	0.15	6	0.21	17	6	12	1
3	0.40	17	0.04	4	0.04	12	0	14	0.12	3	0.21	8	6	8	1
1	0.40	7	0.04	3	0.04	8	0	6	0.12	15	0.14	5	5	5	1
19	0.36	6	0.04	2	0.04	7	0	7	0.07	19	0.07	23	4	4	1
23	0.32	4	0.04	22	0	6	0	1	0.07	22	0	22	0	3	1
17	0.24	3	0.04	18	0	5	0	3	0.06	18	0	13	0	2	1
2	0.24	2	0.04	12	0	4	0	22	0.04	14	0	11	0	1	1
14	0.08	1	0.04	6	0	3	0	18	0.04	7	0	7	0	11	0
7	0.08	5	0	1	0	1	0	4	0.04	1	0	6	0	7	0

of suitable targets. Future works will be devoted to the investigation of networks protection plans based on the results of the aforementioned approaches. Another interesting study, in the field of networks vulnerability identification, is the analysis of hybrid networks composed by heterogeneous nodes. In this scenario each node belong to a specific class of nodes whose relevance in the network is specified.

References

1. Arulselvan A, Commander CW, Elefteriadou L, Pardalos PM (2009) Detecting critical nodes in sparse graphs. Comput Oper Res 36(7):2193–2200
2. Berezin Y, Bashan A, Danziger MM, Li D, Havlin S (2015) Localized attacks on spatially embedded networks with dependencies. Sci Rep 5:8934
3. Bonacich P (2007) Some unique properties of eigenvector centrality. Soc Netw 29(4):555–564

4. Brandes U (2001) A faster algorithm for betweenness centrality. J Math Soc 25(2):163–177 ISO 690
5. Censor Y (1977) Pareto optimality in multiobjective problems. Appl Math Optim 4:41–59
6. Dinh TN, Xuan Y, Thai MT, Park EK, Znati T (2010) On approximation of new optimization methods for assessing network vulnerability. In: INFOCOM, 2010 Proceedings IEEE, San Diego, pp 1–9. IEEE
7. Esposito JM, Dunbar TW (2006) Maintaining wireless connectivity constraints for swarms in the presence of obstacles. In: Proceedings 2006 IEEE International Conference on Robotics and Automation, ICRA'06, Orlando, pp 946–951. IEEE
8. Faramondi L, Oliva G, Pascucci F, Panzieri S, Setola R (2016) Critical node detection based on attacker preferences. In: 2016 24th Mediterranean Conference on Control and Automation (MED), Athens, pp 773–778. IEEE
9. Faramondi L, Setola R, Panzieri S, Pascucci F, Oliva G (2017) Finding critical nodes in infrastructure networks. Int J Crit Infrastruct Prot 20:3–15
10. Faramondi L, Oliva G, Panzieri S, Pascucci F, Schlueter M, Munetomo M, Setola R (2018) Network structural vulnerability: a multiobjective attacker perspective. IEEE Trans Syst Man Cybern Syst (99):1–14
11. Holme P, Kim BJ, Yoon CN, Han SK (2002) Attack vulnerability of complex networks. Phys Rev E 65(5):056109
12. Huang X, Gao J, Buldyrev SV, Havlin S, Stanley HE (2011) Robustness of interdependent networks under targeted attack. Phys Rev E 83:065101
13. ICS-CERT USDHS (2014) ICS-monitor incident response activity. National Cybersecurity and Communications Integration Center
14. Lalou M, Tahraoui MA, Kheddouci H (2016) Component-cardinality-constrained critical node problem in graphs. Discret Appl Math 210:150–163
15. Lalou M, Tahraoui MA, Kheddouci H (2018) The critical node detection problem in networks: a survey. Comput Sci Rev 28:92–117
16. Louzada VHP, Daolio F, Herrmann HJ, Tomassini M (2015) Generating robust and efficient networks under targeted attacks. In: Dariusz K, Damien F, Bogdan G (eds) Propagation phenomena in real world networks. Springer, Cham/Heidelberg/New York/Dordrecht/London, pp 215–224
17. Lu ZM, Li XF (2016) Attack vulnerability of network controllability. PloS One 11(9):e0162289
18. Pullan W (2015) Heuristic identification of critical nodes in sparse real-world graphs. J Heuristics 21(5):577–598
19. Réka A, Hawoong J, Barabási A (2000) Error and attack tolerance of complex networks. Nature 406(6794):378–382
20. Schrijver A (1998) Theory of linear and integer programming. Wiley, Chichester/New York
21. Shao S, Huang X, Stanley HE, Havlin S (2015) Percolation of localized attack on complex networks. New J Phys 17(2):023049
22. Shen Y, Nguyen NP, Xuan Y, Thai MT (2013) On the discovery of critical links and nodes for assessing network vulnerability. IEEE/ACM Trans Netw (TON) 21(3):963–973
23. Sun F, Shayman MA (2007) On pairwise connectivity of wireless multihop networks. Int J Secur Netw 2(1–2):37–49
24. Ventresca M, Harrison KR, Ombuki-Berman BM (2015) An experimental evaluation of multi-objective evolutionary algorithms for detecting critical nodes in complex networks. In: European Conference on the Applications of Evolutionary Computation, Copenhagen, pp 164–176. Springer
25. Wu J, Deng HZ, Tan YJ, Zhu DZ (2007) Vulnerability of complex networks under intentional attack with incomplete information. J Phys A Math Theor 40(11):2665

Game-Theoretic Decision Making for the Resilience of Interdependent Infrastructures Exposed to Disruptions

Yiping Fang and Enrico Zio

Abstract This chapter addresses the challenges associated with assessing and improving the resilience of interdependent critical infrastructure systems under potential disruptive events. A specific set of analytical tools are introduced based on quantitative models of infrastructure systems operation and their functional interdependencies. Specifically, the game-theoretic attacker-defender and defender-attacker-defender modeling techniques are applied to assessing the resilience of interdependent CI systems under worst-case disruptions, and advising policymakers on making pre-disruption decisions for improving the resilience of interdependent infrastructures. A case of interdependent power and gas systems is presented to show the proposed model and highlight the significance of protecting interdependent CIs.

Keywords Critical infrastructure · Interdependencies · Resilience · Game theoretic models

1 Introduction

The phrase, "critical infrastructure protection (CIP)," did not appear in print until in 1997, when the "Marsh report" [1] provided the first definition of infrastructure as "a

Y. Fang
Chaire on Systems Science and the Energy Challenge, Fondation Electricité de France (EDF), Laboratoire Génie Industriel, CentraleSupélec, Université Paris-Saclay, Gif-Sur-Yvette Cedex, France
e-mail: yiping.fang@centralesupelec.fr

E. Zio (✉)
Chaire on Systems Science and the Energy Challenge, Fondation Electricité de France (EDF), Laboratoire Génie Industriel, CentraleSupélec, Université Paris-Saclay, Gif-Sur-Yvette Cedex, France

Energy Department, Politecnico di Milano, Milan, Italy
e-mail: enrico.zio@centralesupelec.fr; enrico.zio@polimi.it

D. Gritzalis et al. (eds.), *Critical Infrastructure Security and Resilience*, Advanced Sciences and Technologies for Security Applications,
https://doi.org/10.1007/978-3-030-00024-0_6

network of independent, mostly privately-owned, man-made systems that function collaboratively and synergistically to produce and distribute a continuous flow of essential goods and services". Then, critical infrastructures (CIs) are defined as network systems that provide life-essential services [2] and whose incapacity or destruction can have a debilitating impact on the health, safety, security, economics, and social well-being, including the effective functioning of governments [3, 4].

CI systems, usually distributed on large geographical extensions, are complex collections of many interacting elements (or subsystems) having an internal dynamic structure and comprising a unified whole. More importantly, different CIs do not operate in isolation of one another – the Internet requires electricity, transportation networks often use sophisticated control and information systems, the generation of electricity requires fuels, and so forth. CIs are physically, geographically, cyber and logically dependent and interdependent, thus called interdependent CIs [5, 6].

On one side, the interdependencies can improve the operational efficiencies of CI systems, but on the other side they can also create new vulnerabilities by providing new hazards and introducing additional channels for failure propagation within and across different CIs, i.e., the disruption of one part of a CI may trigger a domino effect causing the loss of functionality of other key services, as seen in various recent disasters ranging from hurricanes to large-scale power outages and terrorist attacks [7, 8].

By recognizing the significance of these issues, many governments and organizations have initiated interdependent CIs protection plans aiming at strengthening the security and resilience of national/regional interdependent CIs, such as issuing "Critical infrastructure resilience: final report and recommendations" in 2009 in USA [9]; publishing "Australian Government's Critical Infrastructure Resilience Strategy" in 2010 in Australia [10]; issuing "Climate Resilient Infrastructure: Preparing for a Changing Climate" in 2011 in UK [11]; initiating the European Programme for Critical Infrastructure Protection and lanuching a Thematic Area to address it systematically since 2006 in European Union [3]. In these plans, the concept of infrastructure "resilience" has been highlighted.

"Resilience" has many definitions, without a broadly accepted one, even only focusing on CIs [12–15]. A complicating aspect in previous attempts to define resilience is the recognition that "resilience is a family of related ideas, not a single thing" [16]. Zolli and Healy [17] provide perhaps the most comprehensive discussion of the concepts of resilience. Recently, other authors have also provided fairly comprehensive surveys and summaries of the growing literature on resilience and its relationship to the study of risk, specifically for engineered infrastructure systems [12, 18, 19]. Although there are no unique resilience definition and no common resilience metric, there still exist some consensuses. Basically, resilience is recognized as the capability of a system to withstand internal/external stresses and to recover from them. The main difference for various resilience definitions and metrics is that such capability to face adverse events can be considered (and computed) with reference to the time needed to recover, to the time slot in which urban services do not work, to the number of citizens reallocated, to the urban efficiency loss, and so forth [12, 13, 20, 21]. Nevertheless, these are all factors directly related to the system functionality and to its ability to guarantee continuity, even when the global equilibrium is compromised. Thus, a distinguishing feature of resilience is the adaptation in the way that components work together to achieve persistence in the ability of a system to function over time, and in the presence of disruptions.

In this chapter, we consider the challenges associated with assessing and improving the resilience of interdependent CI systems under potential disruptive events. We describe a specific set of analytical tools based on quantitative models of infrastructure systems operation and their functional interdependencies. Specifically, we are interested in (1) assessing the resilience of multiple interdependent CIs, (2) identifying critical vulnerabilities that threaten their continued function, and (3) advising policymakers on making pre-disruption decisions for CI resilience improvement. We apply the game-theoretic attacker-defender (AD) and defender-attacker-defender (DAD) modeling techniques [22] to assess the worst-case disruptions to system function and to identify the most effective defensive measures against them.

The remainder of this chapter is organized as follows. Section 2 begins with the quantitative CI operation and interdependency models. Section 3 discusses the detailed formulation of the optimization framework for assessing and improving the resilience of interdependent CIs. Section 4 illustrates how to apply this framework to a specific example. Concluding remarks are provided in Sect. 5.

2 Operational Models of Interdependent Infrastructures

A CI system can be viewed as a collection of interconnected components that work together to accomplish a particular, domain-specific function. It achieves this through either human or automated decision making that responds to the demands placed on the system to deliver the best possible service in any given situation. This decision-making process is usually termed the operation of the system, and an operational model of a system is used to quantitatively evaluate the service performance of a system by explicitly embracing this decision making in its formulation.

2.1 Optimization-Based System Operation Model

The operation of modern infrastructure systems is fundamentally driven by the demands that are placed on their functionality. The system as a whole needs to "work", i.e., providing service to its users, which are often seen as objectives (e.g., minimize unmet demand of service) and, then, measured in terms of system functionality. In addition, the operation of the CI system is restricted by what is possible, due to physical, economic, or regulatory constraints, e.g., the amount of electric power that a transmission line carries cannot exceed its capacity. In this respect, constrained optimization [23] is ideally suited to model this type of decision problem: system operators make decisions, in an optimum way, about the behavior of the system in pursuing these objectives (what we want the system to do) while subject to its constraints (what the system can do).

In constrained optimization models of CI system operations, potential courses of actions are modeled by decision variables, and the solution to a particular problem indicates decisions that should be taken to reconcile objectives and constraints in an optimum manner with regard to the specified objective. Importantly, this model technique is naturally suited to represent disruptions to CI systems as changes to input data [24]. For example, the operation of an electric power transmission network can be modeled by linear programming (LP) based on the direct circuit (DC) representation, taking available generation units, transmission lines and buses, and identifies the set of power flows that minimizes unmet demand [25]. If the system loses a transmission line in a disruption, we simply need to leave the damaged transmission line "out" of the model (e.g., using an indicator variable to represent its unusable state [13, 25]) and resolve the same operation model (or slightly modified model, e.g., give more weight to the quality of system service rather than the cost of system operation in the objective function when facing disruption); then, the solution to this modified problem will indicate the best possible response of the system.

For illustration purpose, a commonly used network flow-based approach [26] is used here to model the operation of interdependent CIs, where each CI is modeled as a network and their interdependencies are represented via inter-links. Formally, the set of CIs of concern is denoted by $\boldsymbol{\kappa}$. Each infrastructure system k in $\boldsymbol{\kappa}$ is modeled by a network $G^k(N^k, L^k)$ described by a collection of nodes N^k and edges L^k. Each link $l \in L^k$ in CI network k has an associated capacity $\overline{f}_l^k$ representing the maximal amount of flow that can pass through it, while each node $n \in N^k$ has a supply capacity $\overline{s}_n^k$ and a required demand $\widehat{d}_n^k$ of flow for its nominal operation. Flow distributes through the CI networks according to the flow capacities of the links and supply capacities of the nodes, following flow conservation.

For CI network $k \in \boldsymbol{\kappa}$, its resilience to a disruptive event is regarded as the system functionality level immediately after the event, normalized by the total satisfied demand level

$$R^k = \frac{\sum_{n \in N^k} d_n^k}{\sum_{n \in N^k} \widehat{d}_n^k} \tag{1}$$

where d_n^k denotes the satisfied flow at node $n \in N^k$. Then, the overall resilience of interdependent CIs under this event is represented by the weighted sum of the resilience of each CI network, expressed by

$$R = \sum_{k \in \kappa} w^k R^k \tag{2}$$

where w^k is the weighting factor for the resilience of CI network k.

Then, the mathematical formulation of the operation model (OM) of CI network $k \in \boldsymbol{\kappa}$ is represented by

$$OM(k) : \max_{\boldsymbol{o}^k \in \mathbb{O}^k} R^k \tag{3}$$

where the system operators seek to maximize the total satisfied demand level. Set $\mathbb{O}^k$ represents the feasible space for decision variable $\boldsymbol{o}^k$. Different feasible operation spaces $\mathbb{O}^k$ may be formulated for different CI systems with various physical, economic, and/or regulatory constraints. An example of formulation of $\mathbb{O}^k$ by applying the network flow approach is given as follows:

$$\mathbb{O}^k = \left\{ \boldsymbol{o}^k : \left[s_n^k, f_l^k, d_n^k \right] \middle| 0 \le s_n^k \le \overline{s}_n^k, \forall n \in N^k \right. \tag{4}$$

$$0 \le d_n^k \le \widehat{d}_n^k, \forall n \in N^k \tag{5}$$

$$-\overline{f}_l^k z_l^k \le f_l^k \le \overline{f}_l^k z_l^k, \forall l \in L^k \tag{6}$$

$$\left. s_n^k - \sum_{l \in L^k | o(l) = n} f_l^k + \sum_{l \in L^k | d(l) = n} f_l^k - d_n^k = 0, \forall n \in N^k \right\} \tag{7}$$

where constraint (4) bounds the output of flow generation at node n to its capacity. Constraint (5) ensures that the real satisfied demand cannot exceed the required demand for each node. Constraint (6) limits the flow across link l in network k to its capacity. The term z_l^k in (6) models the operation status of link l in network k, i.e., $z_l^k = 1$ if link l is operating; $z_l^k = 0$, otherwise. Finally, constraint (7) guarantees flow conservation at each node, where $o(l)$ indicates the origin or sending node of line l and $d(l)$ indicates the destination or receiving node of line l. The direction of a transmission line is predefined and given as input to the model.

If there is a centralized agent who is in charge of making decisions about the behavior of interdependent CIs, the operation model (OM) can be represented by

$$OM : \max_{\boldsymbol{o} \in \mathbb{O}} R \tag{8}$$

where $\boldsymbol{o} = \bigcup_{k \in \boldsymbol{\kappa}} \boldsymbol{o}^k$. The objective function is now modified to the overall resilience of all CI networks in $\boldsymbol{\kappa}$. It is noted that $\mathbb{O} = \bigcup_{k \in \boldsymbol{\kappa}} \mathbb{O}^k$ does not necessarily hold when we consider the interdependencies among different CIs, i.e., additional constraints may be posed to the operations of individual CI systems. For example, load shedding for a substation bus in an electrical power system is allowed when considering only the power system itself; this may not be permitted (e.g., due to regulatory constraints) when this bus provides power to some critical compressor stations of a national gas transmission system.

2.2 Infrastructure Interdependency Model

Different types of interdependencies exist among CI networks. Rinaldi et al. [5] defined four principal classes of interdependencies: physical, cyber, geographic, and logical. Physical interdependency means the state of one CI depends on the material output(s) of the other; cyber interdependency means the state of one CI depends on information transmitted by the information infrastructure; geographical interdependency means a local environmental event can create state changes in multiple CIs; logical interdependency means the state of each CI depends on the state of the other via a mechanism that is not a physical, cyber, or geographic connection. For a detailed and comprehensive discussion about CI interdependency, interested readers can refer to recent surveys [6, 27].

For illustration purpose, we discuss here how to model CI interdependency quantitatively by referring to interdependent power and gas networks (IPGNs). For IPGNs, typical connections include: (i) *sink-source* connections where a gas city gate can fuel a gas turbine engine, which is an electric generator, (ii) *sink-sink* connections where a city gate requires some energy from an electrical load to regulate its valves, and (iii) *sink-transmit* connections where compressors consume electricity from an electrical load to increase the pressure on a gas pipeline, as sufficient line pressure is a feasibility requirement for the gas network.

All these interdependencies can be modeled by defining a set of ordered components pairs (i,j) associated with node i in one CI network and component (node or line) j in another network, where the interdependency relation for (i,j) works if the flow demand of node i is fully satisfied [28–30]. We use the following notations to facilitate explanation:

$L_n^{k,nbr}$ Set of neighboring lines of node $n \in N^k$, i.e., $L_n^{k,nbr} = \{l | l \in L^k : o(l) = n \text{ or } d(l) = n\}$

$F_{i,j}^{k \to m}$ Set of ordered pairs (i,j) associated with node i in CI network k and node j in CI network m, and node j is operational only when the demand of flow of node i in network k can be fully satisfied

$M_{i,j}^{k \to m}$ Set of ordered pairs (i,j) associated with node i in CI network k and line j in CI network m, and line j operates with its full capacity when the demand of flow of node i in network k is fully satisfied; otherwise line j operates with a reduced capacity $\overset{\sim m}{f}_j$

For the former two types of interdependencies in IPGNs, component j will be completely failed if the interdependency relation for (i,j) does not work. The *sink-transmit* connections in IPGNs are modeled as capacity reduction, i.e., the capacity of line j is reduced if the interdependency relation for (i,j) does not work [31].

For this, we define a binary variable $\delta_{ij}^{k\to m}$ to represent the interdependency from node i in network k to component (node or line) j in network m: $\delta_{ij}^{k\to m} = 1$ if the interdependency works normally and $\delta_{ij}^{k\to m} = 0$ otherwise. For each ordered pair $(i, j) \in F_{i,j}^{k\to m} \cup M_{i,j}^{k\to m}$, the interdependency works normally, i.e., $\delta_{ij}^{k\to m} = 1$, only if the demand level at node i in network k is fully satisfied, i.e., $d_i^k = \widehat{d}_i^k$, as described by the following constraint:

$$d_i^k - \delta_{ij}^{k\to m}\widehat{d}_i^k \geq 0, \forall (i, j) \in F_{i,j}^{k\to m} \cup M_{i,j}^{k\to m} \tag{9}$$

For each node j in the ordered pair$(i, j) \in F_{i,j}^{k\to m}$, the flow generation is bounded by zero or its generation capacity, as stated by constraint (10), and its demand level is bounded by zero or the required demand, as stated by constraint (11):

$$g_j^m - \delta_{ij}^{k\to m}\overline{g}_j^m \leq 0, \forall (i, j) \in F_{i,j}^{k\to m} \tag{10}$$

$$d_j^m - \delta_{ij}^{k\to m}\widehat{d}_j^m \leq 0, \forall (i, j) \in F_{i,j}^{k\to m} \tag{11}$$

Furthermore, if node j is not functioning, all its attached lines will not work and the flow on these lines should be zero, as described by constraint (12):

$$-\delta_{ij}^{k\to m}\overline{f}_l^m \leq f_l^m \leq \delta_{ij}^{k\to m}\overline{f}_l^m, \forall (i, j) \in F_{i,j}^{k\to m}, l \in L_j^{m,nbr} \tag{12}$$

Finally, constraint (13) models the *sink-transmit* interdependencies in IPGNs; the capacity of line j in network m decreases from its normal level $\overline{f}_j^m$ to a reduced level $\overset{\sim m}{f}_j$ ($\overset{\sim m}{f}_j < \overline{f}_j^m$) if the demand of its dependent node i in network k is not fully satisfied ($\delta_{ij}^{k\to m} = 0$):

$$\begin{aligned} &-\delta_{ij}^{k\to m}\overline{f}_j^m - \left(1 - \delta_{ij}^{k\to m}\right)\overset{\sim m}{f}_j \\ &\leq f_{jt}^m \leq \delta_{ij}^{k\to m}\overline{f}_j^m + \left(1 - \delta_{ij}^{k\to m}\right)\overset{\sim m}{f}_j, \forall (i, j) \in M_{i,j}^{k\to m} \end{aligned} \tag{13}$$

Until now, we have shown that the interdependency relations in IPGNs can be formally represented by constraints (9, 10, 11, 12, and 13). These constraints can, then, be added into the operation models of the interdependent networks, i.e., model (8) if we are considering the context of the centralized decision making. The feasible operation space $\mathbb{O}$ is, therefore, given by:

$$\mathbb{O} = \left\{ o : \left[s_n^k, f_l^k, d_n^k \right] | (4) - (7), (9) - (13), \forall k \right\} \tag{14}$$

3 System Resilience Under Disruptions

3.1 Impact Models of Disruptions

In practice, CI systems face various types of internal/external shocks, e.g., technical failures, accidents, natural hazards, and deliberate attacks. The study of failures in engineering systems has yielded an extensive literature on system reliability and probabilistic risk analysis [32–34]. However, the concept of resilience is usually discussed in the context of high-impact low-probability (HILP) events [35, 36], i.e., the risks that are difficult or even impossible to foresee (e.g., due to a lack of statistically evident historical data of the event); therefore, probabilistic assessment may not be applicable in this case. Furthermore, for deliberate threats induced by an intelligent, goal-oriented terrorist, probabilities may not be suitable for modeling the behavior of the adversary [37]. Brown and Cox [38] show that probabilistic assessment of terrorism risk can even lead to misleading results.

Instead of focusing on the source of a disruption, we look at the problem form the point of view of the system functionality. Specifically, we consider disruptions as the simultaneous losses of one or more system components and assess the performance of CIs under the worst-case disruptions. To identify the worst-case disruptions, a hypothetical intelligent adversary (an attacker) is considered to have perfect knowledge and capable of using limited resources to intentionally damage the CIs. From the point of view of system operators, the attacker is not necessarily a real human being. Instead, it could be mother nature, a terrorist, simple bad luck, or anything else that causes the simultaneous loss of components; the operators are concerned with doing the best they can to maintain the functionalities of CIs following the loss of these components. We emphasize that the purpose of assuming a personalized attacker here is simply to identify worst-case disruptions, not to model the actual behavior of any particular adversary.

Formally, the damage of CI systems in a disruption is represented by the state variables of the systems components, e.g., z_l^k for network line $l \in L^k$ where $z_l^k = 0$ if link l is attacked; $z_l^k = 0$ otherwise, as explained in constraint (6). It is noted that here we consider only the failure of network links since the failure of a network node is equivalent to the simultaneous failures of all the links connecting to it. Then, the impact of disruptions to interdependent CI systems is represented by the following attacker-defender (AD) model [13, 22, 24, 25]

$$\min_{z \in \mathbb{Z}} \max_{o \in \mathbb{O}(z)} R \tag{15}$$

where the state variable z is now determined by the attacker, and $\mathbb{Z}$ represents the set of all possible links attacks. The system operators still face the same functionality maximization problem, i.e., the operation model (8), whose feasible operation space $\mathbb{O}(z)$ is now a function of the system state z obtaining from the attacker's behavior. In other words, after the realization of the attacks, the systems will adapt their behaviors to maintain continuity of functionality in presence of the disruptions caused by the attacks.

3.2 Resilience Assessment

The above-introduced AD model can be used to assess the resilience of interdependent CI systems to the worst-case disruptions. Before that, we should carefully define the constraints on z, to avoid that the obvious "absolute worst-case" turns out to be that with the simultaneous loss of all system components that leads to complete failure of the systems. A straightforward idea would be to limit the maximum number of lost components by a cardinality constraint, as follows:

$$\sum_{k \in \kappa} \sum_{l \in L^k} \left(1 - z_l^k\right) \leq B_A \tag{16}$$

where B_A characterizes the disruption "magnitude" of the attack in terms of the maximum number of links that can simultaneously fail in the attack. This parameterization is useful because it allows considering different levels of disruptions and assessing the best achievable worst-case functionality of CI systems as a function of the disruption "magnitude" B_A, obtaining the so-called "resilience curve" [24].

Furthermore, the cardinality constraint (16) can be generalized to any notion of "budget" by specifying a cost associated with attacking each component in the systems. Furthermore, any available information of the attacker's intent of attacking, or on the disruptive event's threat profile to the systems, can be carefully formulated in terms of additional constraints on z to narrow down the space $\mathbb{Z}$. For instance, the impact of a natural hazard like a hurricane on CI system components is usually quantified, in a probabilistic manner, based on the physical model of the hurricane threat (e.g., gust wind speed) [39] and the fragility models of system components [40]. The resulting failure probabilities of system components can be related to their binary damage state variables z through Shannon's information theory. Interested readers can refer to Ref. [41] for a detailed formulation of this model.

3.3 Resilience Improvement

The usefulness of resilience assessment is limited unless it is used to guide the planning for the resilience improvement of interdependent CIs: to build and enhance resilience of the CI systems is the ultimate goal. In the context of the AD model, this means improving the functionality of CI systems under the worst-case simultaneous losses of system components. Nevertheless, doing so will require investment on certain actions, e.g., hardening and upgrading weak system components to increase their chances of survival under disruptions. To quantify this pre-disruption decision, the AD model is extended to the so-called defender-attacker-defender (DAD) model, as follows [22, 24, 25, 41, 42]:

$$\max_{y\in\mathbb{Y}}\min_{z\in\mathbb{Z}}\max_{o\in\mathbb{O}(y,z)} R \tag{17}$$

where $\boldsymbol{y}$ is a decision variable representing defensive investments and $\mathbb{Y}$ represents the set of all feasible investments. These investment decisions potentially change the set of feasible system operations $\boldsymbol{o} \in \mathbb{O}(\boldsymbol{y}, \boldsymbol{z})$. The first level problem in (17) is to identify the optimal set of network lines to protect so that the overall resilience of the interdependent CIs is maximized. The worst case system disruptions and the successive adaptive actions are considered in the middle-low level problem $\mathcal{H}(\boldsymbol{y}) = \min_{z\in\mathbb{Z}}\max_{o\in\mathbb{O}(y,z)} R$, which is almost identical to the prior model (15), except that the feasible system operation space $\mathbb{O}(\boldsymbol{y}, \boldsymbol{z})$ now depends also on the investment decisions $\boldsymbol{y}$.

For illustrative purpose, this chapter considers a typical ex-ante resilience strategy, i.e., protecting CI network lines. Protected lines are assumed to be invulnerable and cannot be damaged in a disruption. Other possible resilience improvement actions like constructing new components [25] can be easily incorporated into this analysis framework. Formally, we let binary variable y_l^k represent the investment decision that $y_l^k = 1$ if link l in network k is protected, 0 otherwise. The ability to invest in improvements is constrained by limited resources. Therefore, the set of feasible investments $\mathbb{Y}$ can be represented by

$$\mathbb{Y} = \left\{ \boldsymbol{y} |\, y_l^k \in \{0,1\}, \forall l \in L^k, k \in \boldsymbol{\kappa} \;\; \sum_{k\in\kappa}\sum_{l\in L^k} c_l^{k,P} y_l^k \le B_P \right\} \tag{18}$$

where $c_l^{k,P}$ denotes the cost of protecting link l in network k, and B_P parametrizes the total protection budget.

The feasible system opeartions space $\mathbb{O}(\boldsymbol{y}, \boldsymbol{z})$ can now be specified by considering the real function state of a network link l in network k: if the link is protected $y_l^k = 1$, it will be always functional no matter if it is attacked ($z_l^k = 0$) or not ($z_l^k = 1$); otherwise, its function state will depend on whether it is attacked. Therefore, the real function state of the link can be represented by $\left[y_l^k + \left(1 - y_l^k\right) z_l^k\right]$, and $\mathbb{O}(\boldsymbol{y}, \boldsymbol{z})$ is given by

$$\begin{aligned}\mathbb{O}(\boldsymbol{y}, \boldsymbol{z}) = \Big\{ \boldsymbol{o} : \left[s_n^k, f_l^k, d_n^k\right] |(4)(5)(7)\,, (9)-(13), \forall k \; -\overline{f}_l^k \left[y_l^k + \left(1 - y_l^k\right) z_l^k\right] \\ \le f_l^k \le \overline{f}_l^k \left[y_l^k + \left(1 - y_l^k\right) z_l^k\right], \forall l \in L^k, \forall k \Big\}\end{aligned} \tag{19}$$

The max-min-max formulation (17) configures a mixed-integer nonlinear tri-level programming problem, whose solution is challenging. Due to the presence of binary variables $\delta_{ij}^{k\rightarrow m}$ in the third level, the second and third level min-max problems cannot be merged into a single min problem using the KKT conditions (or the strong duality) of the third level max problem [43]. In this regard, sophisticated decomposition or approximation methods are required for the model solutions, e.g., the recently developed "Column-and-Constraint Generation" (C&CG) method [44], is proven to be effective in dealing with mixed integer programming recourse problems [13, 25, 42].

4 Numerical Example

This section presents a simple numerical study involving IPGNs, adapted from [42]; the network layouts of the two systems are shown in Fig. 1. The interdependency relations are described as follows: the gas node g8 depends on the power demand node p11; the gas node g7 depends on the power demand node p10; the gas node g1 depends on the power demand node p4; the gas node g3 depends on the power demand node p9; the power generation node p1 depends on the gas demand node g9 [42].

For simplicity, we assume that protecting one link in the interdependent CIs needs one unit of protection resources and set the cardinality constraint (16) to all possible link attacks. The weighting factor w^k is set as 0.5 for the resilience of both the power and gas systems.

We first investigate the resilience assessment of the IPGNs. Figure 2 illustrates the worst-case system disruptions by attacking from one to five links and Fig. 3

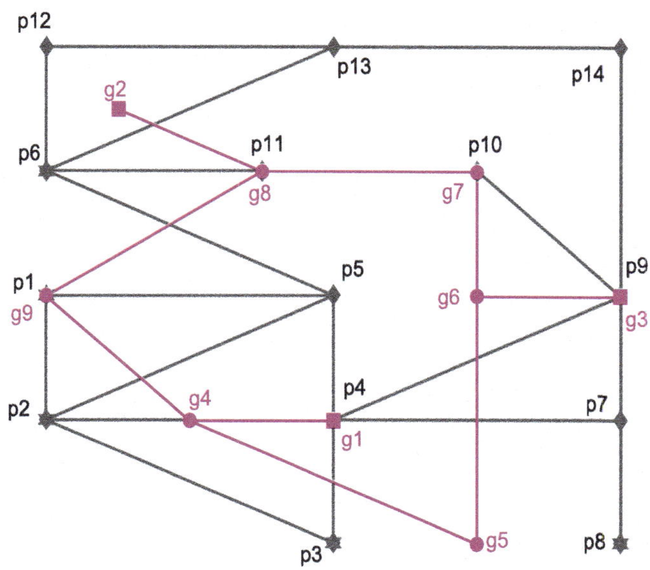

Fig. 1 The layout of the interdependent power and gas systems [42]

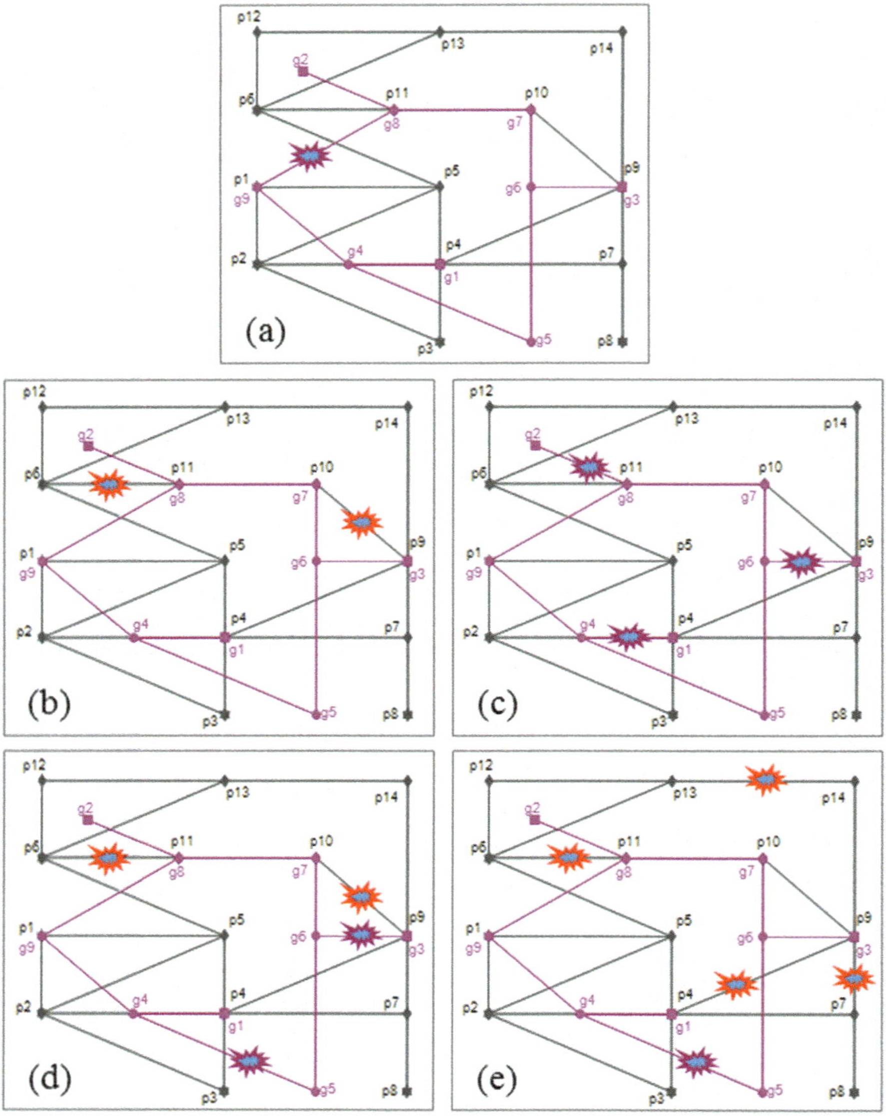

Fig. 2 Worst-case link attacks. (**a**) The worst-case single link attack is of link g8-g9, resulting in a combined power and gas system resilience R = 0.782, i.e., 21.8% power and gas demands cannot be satisfied. (**b**) The worst-case two-link attack is of links p6-p11 and p9-p10, resulting in R = 0.586. (**c**) The worst-case three-link attack is of links g1-g4, g2-g8, and g3-g6, resulting in R = 0.451. (**d**) The worst-case four-link attack is of links g3-g6, g4-g5, p6-p11, and p9-p10, resulting in R = 0.429. (**e**) The worst-case five-link attack is of links p4-p9, p6-p11, p7-p9, p13-p14, and g4-g5, resulting in R = 0.352

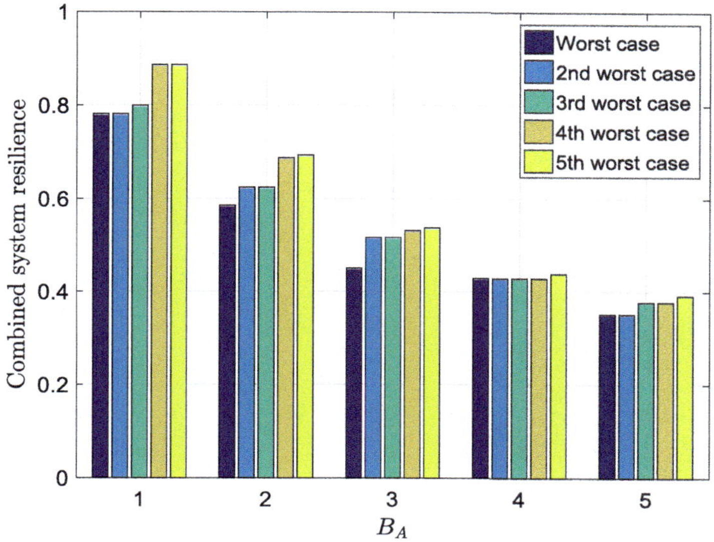

Fig. 3 The combined power and gas system resilience associated with the worst-case, the second-worst through the fifth-worst attacks for each attack budget

shows the combined power and gas systems resilience associated with the worst attack disruptions, and the second worst (i.e., rank order 2) through fifth-worst (i.e., rank order 5) combination of system resiliences for each attack budget. These second-worst through fifth-worst results were obtained by adding a new constraint that eliminates the previous solution. From the Figure, it is possible to see that the combined system resilience generally decreases as the attack budget increases for the worst case attack, which is expected. Furthermore, the second-worst attacks do not necessarily have strictly larger resilience than the worst cases, e.g., for the cases $B_A = 1$, 4 and 5. In other words, the identified worst-case scenarios are not unique but are accompanied by some equally bad ones, implying that defending against only one of the worst cases is not likely to improve the overall system resilience to attacks.

Second, when the protection investment is considered, we solve the DAD model for different combinations of protection budget B_P and attack budget B_A. Figure 4 shows the combined power and gas resilience as a function of the attack budget B_A under different B_P. From the Figure, it can be seen that in the case of no defense, the resilience decreases almost linearly with the increase of B_A, which can be mitigated by increasing the protection budget B_P, i.e., $B_P = 2$, 4, 6 and 8. However, due to the non-uniqueness of the worst case attack for some attack budgets, the improvement of system resilience is not always promising. For example, the combined system resilience is increased by only 2.3% when B_P is increased from 0 to 2 for $B_A = 1$, compared to the average improvement of 28.4% for other attack scenarios under the same increase of B_P.

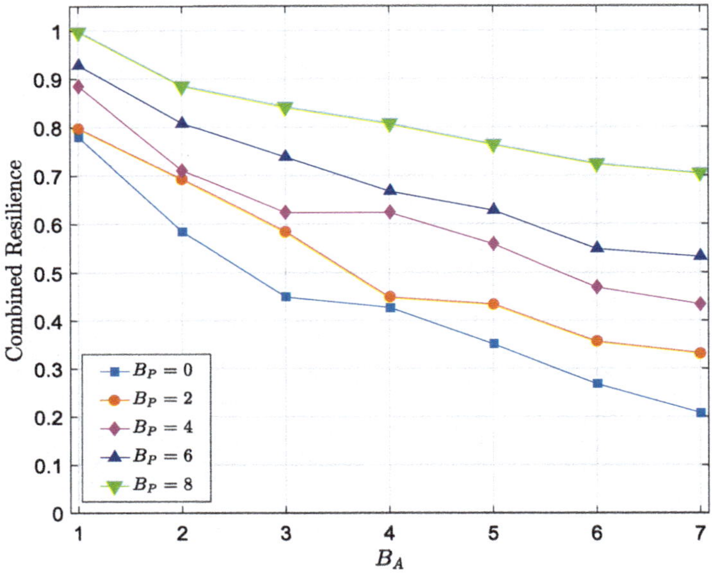

Fig. 4 The combined resilience of the interdependent power and gas systems

Then, we investigate the importance of considering interdependency for system protection decisions. In practice, a coordinated protection agency for different CIs may not exist. Thus, each system makes its own protection decisions without considering the interdependencies. To investigate this case, we assume there is a governor who distributes the budget evenly to the power and gas systems, and each of them protects itself separately without considering the interdependencies among them, while the attacker disrupts the two systems by recognizing the interdependencies. We call this strategy "separate protection" to differentiate it from the "coordinated protection" where the interdependent systems are protected as a whole. Figure 5a shows the combined power and gas system resilience as a function of the attack budget B_A for the separate protection and the coordinated protection when the protection budget $B_P = 4$. It is clearly shown that the combined resilience values in the case of separate protection are always smaller than those in the case of coordinated protection. The difference of the combined system resilience between the two cases can reflect the importance of considering interdependencies in interdependent CIs protection. Figure 5b presents the difference of the combined system resilience between the two cases for different protection budget B_P. From this Figure, it can be seen that when B_P is relatively small, the difference of the combined system resilience is relatively insignificant, e.g., under or around 0.1 when $B_P = 2$; when B_P increases, the difference becomes increasingly significant. These results highlight the significance of protecting interdependent CIs as a whole against potential disruptions, especially when the protection budget is relatively high.

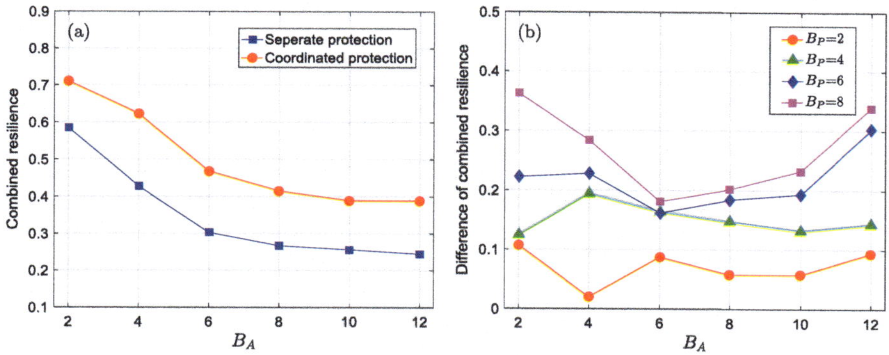

Fig. 5 (**a**) The combined system resilience curves as a function of the attack budget B_A for the separate protection and the coordinated protection when $B_P = 4$; (**b**) The combined resilience difference between the separate protection and the coordinated protection as a function of the attack budget B_A when $B_P = 2, 4, 6$ and 8

5 Concluding Remarks

This chapter has introduced a set of quantitative models of operation of interdependent CI systems and their functional interdependencies. The game-theoretic AD and DAD models are introduced and advocated to be used for assessing and improving the resilience of interdependent CIs under worst-case disruptions. By assuming an intelligent attacker and exploiting its optimization, these multi-level defender-attacker models aim to estimate a worst case damage scenario for any feasible protection strategy. It is noted that the tri-level DAD game takes the identical form of two-stage adaptive robust optimization (ARO) [45, 46], albeit the DAD game model and the two-stage ARO have different origins. This modeling framework has been successfully applied to identify the optimum resilience strategies for electric power grids [25, 47, 48], rail systems [49], commodity distribution networks [24], and facility networks [50].

Although in the present models we restrict the adaptive behavior of the systems to the normative decisions (i.e., only network flow can be re-dispatched), the framework is also flexible enough to incorporate other adaptive behaviors/decisions in the presence of disruption, to the extent that one can describe the way in which this might happen. For example, we have shown in [13] that the decisions about the repair sequence of damaged components under limited repair resources can be carefully defined and incorporated into the third level system operation model after disruptions, resulting in a more comprehensive consideration of system resilience.

By considering the simultaneous losses of system components, the present model is agnostic about the source of a disruption, providing a rapid and objective way of calculating the consequence of damage to any set of components, and can, therefore, be used to identify vulnerabilities and to evaluate the improvement in resilience provided by any protection plan. Furthermore, as we have mentioned at the end of Sect. 3.2, when we are able to calculate the failure probabilities of system components, this information can be carefully formulated as additional constraints on z, e.g., through Shannon's information theory, to narrow down the space $\mathbb{Z}$ and obtain the "most-likely" (informed by the failure probabilities) worst case disruptions.

Finally, our results of the numerical example demonstrate the significance of having a centralized decision maker to protect interdependent CIs as a whole against potential disruptions. However, in practice, many CI systems are owned or operated by the private sector and a centralized decision-making agent does not exist. Therefore, in terms of future research, it would be interesting to investigate whether and how different kinds of interaction/collaboration mechanisms among these independent decision-makers will improve the resilience of individual CI systems and all the interdependent CIs as a whole.

References

1. Ellis J et al (1997) Report to the President's Commission on critical infrastructure protection
2. Moteff J, Copeland C, Fischer J (2003) Critical infrastructures: what makes an infrastructure critical? Library of Congress Washington DC, Congressional Research Service
3. Zio E (2016) Challenges in the vulnerability and risk analysis of critical infrastructures. Reliab Eng Syst Saf 152:137–150
4. Kröger W, Zio E (2011) Vulnerable systems. Springer Science & Business Media
5. Rinaldi SM, Peerenboom JP, Kelly TK (2001) Identifying, understanding, and analyzing critical infrastructure interdependencies. IEEE Control Syst 21(6):11–25
6. Ouyang M (2014) Review on modeling and simulation of interdependent critical infrastructure systems. Reliab Eng Syst Saf 121:43–60
7. Vespignani A (2010) Complex networks: the fragility of interdependency. Nature 464(7291):984
8. Buldyrev SV et al (2010) Catastrophic cascade of failures in interdependent networks. Nature 464(7291):1025
9. Council, N.I.A (2009) Critical infrastructure resilience: final report and recommendations: National Infrastructure Advisory Council
10. Government A (2010) Australian government's critical infrastructure resilience strategy
11. Environment, S.o.S.f (2011) Food and rural affairs by command of her majesty, climate resilient infrastructure: preparing for a changing climate
12. Hosseini S, Barker K, Ramirez-Marquez JE (2016) A review of definitions and measures of system resilience. Reliab Eng Syst Saf 145:47–61
13. Ouyang M, Fang Y (2017) A mathematical framework to optimize critical infrastructure resilience against intentional attacks. Comput Aided Civ Inf Eng 32(11):909–929
14. Fang Y (2015) Critical infrastructure protection by advanced modelling, simulation and optimization for cascading failure mitigation and resilience. Ecole Centrale Paris
15. Fang Y-P, Pedroni N, Zio E (2016) Resilience-based component importance measures for critical infrastructure network systems. IEEE Trans Reliab 65(2):502–512
16. Westrum R (2017) A typology of resilience situations. In Resilience engineering. CRC Press, pp 67–78
17. Zolli A, Healy AM (2013) Resilience: why things bounce back. Simon and Schuster
18. Park J et al (2013) Integrating risk and resilience approaches to catastrophe management in engineering systems. Risk Anal 33(3):356–367

19. Ayyub BM (2014) Systems resilience for multihazard environments: definition, metrics, and valuation for decision making. Risk Anal 34(2):340–355
20. Francis R, Bekera B (2014) A metric and frameworks for resilience analysis of engineered and infrastructure systems. Reliab Eng Syst Saf 121:90–103
21. Franchin P, Cavalieri F (2015) Probabilistic assessment of civil infrastructure resilience to earthquakes. Comput Aided Civ Inf Eng 30(7):583–600
22. Brown G et al (2006) Defending critical infrastructure. Interfaces 36(6):530–544
23. Bertsekas DP (2014) Constrained optimization and Lagrange multiplier methods. Academic
24. Alderson DL, Brown GG, Carlyle WM (2015) Operational models of infrastructure resilience. Risk Anal 35(4):562–586
25. Fang Y, Sansavini G (2017) Optimizing power system investments and resilience against attacks. Reliab Eng Syst Saf 159:161–173
26. Lee EE II, Mitchell JE, Wallace WA (2007) Restoration of services in interdependent infrastructure systems: a network flows approach. IEEE Trans Syst Man Cybern Part C Appl Rev 37(6):1303–1317
27. Pederson P et al (2006) Critical infrastructure interdependency modeling: a survey of US and international research. Ida Nat Lab 25:27
28. Ouyang M (2017) A mathematical framework to optimize resilience of interdependent critical infrastructure systems under spatially localized attacks. Eur J Oper Res 262(3):1072–1084
29. Gong J et al (2014) An interdependent layered network model for a resilient supply chain. Omega 46:104–116
30. González AD et al (2016) The interdependent network design problem for optimal infrastructure system restoration. Comput Aided Civ Inf Eng 31(5):334–350
31. Coffrin C, Van Hentenryck P, Bent R (2012) Last-Mile Restoration for Multiple Interdependent Infrastructures. In AAAI
32. Zio E (2009) Reliability engineering: old problems and new challenges. Reliab Eng Syst Saf 94(2):125–141
33. Enrico Z (2007) An introduction to the basics of reliability and risk analysis, vol 13. World Scientific
34. Aven T, Zio E (2011) Some considerations on the treatment of uncertainties in risk assessment for practical decision making. Reliab Eng Syst Saf 96(1):64–74
35. Deng PG, Fei F (2012) Infrastructure resilience for high-impact low-chance risks. Proc Inst Civ Eng 165(6):13
36. Panteli M, Mancarella P (2015) The grid: Stronger, bigger, smarter?: presenting a conceptual framework of power system resilience. IEEE Power Energ Mag 13(3):58–66
37. Council, N.R (2010) Committee to review the department of Homeland Security's approach to risk analysis. Review of the department of Homeland Security's approach to risk analysis. National Academies Press, Washington, DC
38. Brown GG, Cox LAT Jr (2011) How probabilistic risk assessment can mislead terrorism risk analysts. Risk Anal 31(2):196–204
39. Davis C et al (2008) Prediction of landfalling hurricanes with the advanced hurricane WRF model. Mon Weather Rev 136(6):1990–2005
40. Booker G et al (2010) Estimating cellular network performance during hurricanes. Reliab Eng Syst Saf 95(4):337–344
41. Fang Y, Sansavini G, Zio E (2018) An Optimization-based framework for the identification of vulnerabilities in electric power grids exposed to natural hazards. Under Review
42. Fang Y, Zio E (2017) Optimizing the resilience of interdependent infrastructure systems against intentional attacks. In: ICSRS 2017
43. Thiele A, Terry T, Epelman M (2009) Robust linear optimization with recourse. Rapport technique, pp 4–37
44. Zhao L, Zeng B (2012) An exact algorithm for two-stage robust optimization with mixed integer recourse problems. Submitted, available on Optimization-Online. org
45. Bertsimas D, Brown DB, Caramanis C (2011) Theory and applications of robust optimization. SIAM Rev 53(3):464–501

46. Ruiz C, Conejo AJ (2015) Robust transmission expansion planning. Eur J Oper Res 242(2):390–401
47. Alguacil N, Delgadillo A, Arroyo JM (2014) A trilevel programming approach for electric grid defense planning. Comput Oper Res 41:282–290
48. Yuan W et al (2016) Robust optimization-based resilient distribution network planning against natural disasters. IEEE Trans Smart Grid 7(6):2817–2826
49. Alderson DL et al (2011) Solving defender-attacker-defender models for infrastructure defense. Naval Postgraduate School Monterey CA Dept of Operations Research
50. Losada C, Scaparra MP, O'Hanley JR (2012) Optimizing system resilience: a facility protection model with recovery time. Eur J Oper Res 217(3):519–530

Smallest Pseudo Target Set Identification and Related Problems Using the Implicative Interdependency Model

Arun Das, Chenyang Zhou, Joydeep Banerjee, Anisha Mazumder, and Arunabha Sen

Abstract Critical infrastructures such as the power grid and the communication network form a complex interdependent system where the failure of a small set of entities can trigger a *cascading event* resulting in the failure of a much larger set of entities. Recognizing the need for a deeper understanding of the interdependence between such critical infrastructures, in the last few years several interdependency models have been proposed and analyzed. However, most of these models are over-simplified and fail to capture the complex interdependencies that may exist in such networks. The more recently proposed *Implicative Interdependency Model* (IIM) overcomes the limitations of existing models and is able to capture complex relationships that may exist between entities of heterogeneous interdependent networks. In this chapter we outline some of the problems studied using this model and present a detailed study of the *Smallest Pseudo Target Set Identification Problem* in the IIM setting. We divide the problem into four classes, and show that it is solvable in polynomial time for one class, and is NP-complete for others. We provide an *approximation algorithm* for the second class, and for the most general class, we provide an optimal solution using an Integer Linear Program, and a heuristic solution. We evaluate the efficacy of our heuristic using power and communication network data of Maricopa County, Arizona. The experiments show that our heuristic almost always produces near optimal results.

Keywords Interdependent Networks · Critical Infrastructure Networks · Implicative Interdependency Model · Network Robustness and Resiliency

---

A. Das (✉) · C. Zhou · J. Banerjee · A. Mazumder · A. Sen
School of Computing, Informatics and Decision System Engineering, Arizona State University, Tempe, AZ, USA
e-mail: arun.das@asu.edu; czhou24@asu.edu; joydeep.banerjee@asu.edu; amazumde@asu.edu; asen@asu.edu

D. Gritzalis et al. (eds.), *Critical Infrastructure Security and Resilience*, Advanced Sciences and Technologies for Security Applications,
https://doi.org/10.1007/978-3-030-00024-0_7

1 Introduction

The last few years have seen a heightened awareness in the research community that the critical infrastructures of the nation do not operate in isolation. Instead, these infrastructures are closely coupled together and form a complex ecosystem of interdependent networks where the well being of one infrastructure depends heavily on the well being of another. A case in point is the interdependent relationship between the electric power grid and the communication network. Power grid entities, such as the SCADA systems that control power stations and sub-stations, are reliant on the communication network to send and receive control signals. On the other spectrum, communication network entities, such as routers and base stations are reliant on electric power. Understanding the impact of cascading failures in the power grid, a not so uncommon phenomena, becomes even more complex when the coupling between the power grid and communication network entities are considered. This coupling, or interdependence, allows not only entities in the power network, such as generators and transmission lines, to trigger power failure, but also communication network entities, such as routers and optical fiber lines, can potentially trigger failures in the power grid.

As failures in such interdependent systems can be introduced either by Nature (such as from the impact of hurricanes, wild fires, or earthquakes), or by Man (such as from terrorist attacks, EMP attacks, or human error), a thorough analysis of the robustness and resiliency of the system is needed to be able to assess and mitigate the impact of faults on such interdependent networks. For instance, techniques to identify the *critical entities* of interdependent infrastructures, the *root cause of failure*, *progressive recovery* from failure, and *entity hardening* techniques are all essential tools that must be part of the arsenal of network designers and operators of interdependent infrastructures. As a first step in developing such techniques, there is a need to develop an interdependency model that is able to capture the complex relationships that may exist between such interdependent systems. Although several interdependency models have been proposed and analyzed in the last few years, most of such models have several limitations especially when modeling real-world interdependencies that may exist between critical infrastructures. To overcome the limitations of existing models, in this chapter we elaborate the recently proposed *Implicative Interdependency Model* (IIM) [14] that is able to capture such complex interdependencies. We then proceed to briefly outline some of the problems studied using this model and present a detailed study of the *Smallest Pseudo Target Set Identification Problem* (SPTSIP) in the IIM setting.

The study of the SPTSIP is motivated by the inherent characteristic of interdependent systems where a failure involving a small set of entities of the interdependent network can trigger a *cascading event* that can result in the failure of a much larger set of entities. This creates a potential scenario where an adversary with an intent to jeopardize a specific set of entities E', or *real targets*, now no longer needs to destroy E' directly. Instead, the adversary can take advantage of the cascading failure process by identifying a smaller set of entities E'', or *pseudo targets*, whose failure eventually leads to the failure of E' due to the cascade. Thus, the objective of the adversary is to identify the smallest set of pseudo targets E'' whose failure eventually causes E' to fail. In this chapter we refer to this problem as the *Smallest Pseudo Target Set Identification Problem*, and in the IIM setting, categorize the problem in four classes. We show that one class of the problem is solvable in polynomial time, whereas for others it is NP-complete. For the second class of the problem we provide an *approximation algorithm*, and for the most general form of the problem we provide an optimal solution using Integer Linear Programming, and also provide a polynomial time heuristic solution. Finally, we evaluate the efficacy of our heuristic using real-world power and communication network data of Maricopa County, Arizona. The experiments show that our heuristic almost always produces near optimal results.

The rest of this chapter is organized as follows: In Sect. 2 a brief overview of the existing interdependent network models is outlined, and in Sect. 3 we present the Implicative Interdependency Model (IIM) and briefly discuss some of the problems studied in the IIM setting. In Sect. 4 we present a detailed study of the Smallest Pseudo Target Set Identification Problem, and finally conclude the chapter is Sect. 5.

2 Related Work

Recognizing the need for a deeper understanding of the interdependent relationships between multilayered networks, in the last few years, significant efforts have been made by the research community to achieve this goal. Accordingly, a number of models have been proposed and analyzed [2–4, 7, 11–13, 15, 17]. However, many of the proposed models are overly simplistic and fail to capture the complex interdependencies that may exist between entities of heterogeneous networks such as the power grid and communication network. We provide an overview of some of these models and highlight their limitations below.

Motivated by the 2003 electricity blackout in Italy, in [3], Buldyrev et al. proposed a graph based interdependency model with a power network graph A, and a communication network graph B. The authors assume that (i) the number of nodes in the power network is equal to the number of nodes in the communication network (i.e. $|A| = |B|$), and (ii) there exists a one-to-one dependency between a node in the power network and a node in the communication network. The model also makes an implicit assumption that the power (communication) nodes are homogeneous with respect to functionality, i.e., there is no distinction between power-plant, sub-station or load nodes (or cell towers and routers). The authors describe a cascading failure process (the rules by which the nodes and edges of the graph are removed), and using this model, they compute the percolation threshold for existence of a giant connected component. The assumptions about this one-to-one dependency between the power and communication network nodes, and the homogeneous nature of the nodes are unrealistic as the model fails to capture the complex interdependencies that may exist between the entities of an interdependent network. The authors also opine in a subsequent paper (in [7]), that a single node in one network may be dependent on multiple nodes in the other network.

In [13], Rostato et al. model the power flow in the power grid, and the data flow in the communication network separately. They then analyze the effect of failures in the communication network, caused by failures in the power grid using a coupling model between the two infrastructures. The authors construct graphs for the power grid and communication network from the Italian high voltage electric transmission network (HVIET), and the high-bandwidth backbone of the Italian Internet network (GARR). For the power network, the model considers the DC power flow model [16], and for the communication network, a probabilistic packet routing model is considered for sending data packets from randomly generated source and destination nodes. A dependency between the two networks is setup by associating a node from the communication network to the closest load node from the power network (in terms of Euclidean distance). It may be noted that the dependency considered in this model is one directional, i.e., for a communication node to be operational it is dependent on a power network node, however, the power network node is not dependent on the communication node for its survival. In the event of a failure, a load re-dispatching process is initiated on the power network, and a communication network node remains operational as long as the load node it is connected to is dispatched with power greater than a computed threshold. Although the model proposed in [13] is realistic to a point, the dependency model is a one way dependency model and fails to represent the interdependency that may exist between the power and communication networks.

Although a number of interdependency models have been proposed and analyzed in the recent past [3, 4, 7, 11–13, 15, 17], most of these models are over simplified and fail to capture complex interdependencies that may involve a combination of conjunctive and disjunctive relationships between network entities. For instance, suppose the power network entities such as *power generators, transmission lines* and *substations* are denoted by the set $A = \{a_1, a_2, \ldots, a_n\}$ and the entities of the communication network, such as *routers, fiber optic lines* and *base stations* are denoted by the set $B = \{b_1, b_2, \ldots, b_m\}$. Due to the topological design of the power-communication networks, it may so happen that an entity a_i is *operational* if: (i) the entities b_j *and* b_k *and* b_l are operational, *or* (ii) b_m *and* b_n are operational, *or* (iii) b_p is operational. Graph based interdependency modeling, such as in [3, 4, 7, 11–13, 15, 17] cannot capture such interdependencies involving both conjunctive and disjunctive terms. In the following section we outline the *Implicative Interdependency Model* [14] that is able to capture such complex interdependencies using Boolean Logic and overcomes the limitations of existing graph based approaches. We then proceed to briefly outline some of the problems studied using this model.

3 Implicative Interdependency Model (IIM)

The *Implicative Interdependency Model* (IIM) proposed in [14], is an entity based model that allows representation of complex dependency relations between entities of interdependent multilayer network systems. The dependent relationships between the network entities are represented using Boolean Logic and are termed as *Implicative Interdependency Relations* (IDRs). For instance, in a sample *Interdependent Power-Communication Network* (IPCN), if the power network entities are the set of A type entities, $A = \{a_1, \ldots, a_n\}$ and the communication network entities are the set of B type entities, $B = \{b_1, \ldots, b_m\}$. If power network entity a_i is *operational* when (i) the entities b_j *and* b_k *and* b_l are operational, *or* (ii) b_m *and* b_n are operational, *or* (iii) b_p is operational, the corresponding IDR would be of the form $a_i \leftarrow b_j b_k b_l + b_m b_n + b_p$. It may be noted that the IDRs only provide a *necessary condition* for entities (such as a_i) to be *operational*. In other words, a_i may *fail* independently and may be *inoperable* even when the conditions given by the corresponding IDR are *satisfied*.

Table 1 outlines a set of IDRs representing a sample IPCN where the power network and communication network entities are represented by the sets $A = \{a_1, a_2, a_3, a_4\}$ and $B = \{b_1, b_2, b_3\}$ respectively. The IDRs represent a set of necessary Boolean conditions that need to be satisfied for an entity to be operational. In Table 1, entity b_1 is operational if either a_2 is operational, or both a_1 and a_3 is operational. It may be noted that although in the IDRs of this example, A (B) type entities appear on either the left hand side or the right hand side of an IDR, the IIM does not require that A (B) type entities appear only on one side of an IDR. In other words, an IDR can also be of the form $a_i \leftarrow a_q b_j b_k b_l + a_r b_m b_n + b_p + a_s$ implying that A (B) type entities may depend on A (B) type entities. The conjunction of entities, such as $a_r b_m b_n$, is also referred to as a *minterm*.

The interdependencies expressed through IDRs govern the failure cascade process in IIM, where the failure of a set of entities can trigger further failures due to the interdependencies shared between the entities. This cascading failure process is illustrated with the help of an example: For the IPCN system of Table 1, Table 2 shows a time-stepped cascading failure of entities triggered by the failure of $\{a_2, b_3\}$ at time step 0.

In Table 2, the cascading failure process initiated by the failure of a subset of A type entities at time step 0 (denoted by A_d^0), and a subset of B type entities (denoted by B_d^0), till the system reaches its final steady state is shown diagrammatically in Fig. 1. Accordingly, an interdependent multilayer network can be viewed as a

Table 1 A sample *Interdependent Power-Communication Network* (IPCN)

Power network	Comm. network
$a_1 \leftarrow b_1 b_2$	$b_1 \leftarrow a_1 a_3 + a_2$
$a_2 \leftarrow b_1 + b_2$	$b_2 \leftarrow a_1 a_2 a_3$
$a_3 \leftarrow b_1 + b_2 + b_3$	$b_3 \leftarrow a_1 + a_2 + a_3$
$a_4 \leftarrow b_1 + b_3$	–

Table 2 Time stepped failure propagation for the sample IPCN of Table 1 when entities $\{a_2, b_3\}$ fail at time step 0, or $A_d^0 = \{a_2\}$, $B_d^0 = \{b_3\}$. A value of 1 denotes entity failure

	Time steps (t)						
Entities	0	1	2	3	4	5	6
a_1	0	0	1	1	1	1	1
a_2	1	1	1	1	1	1	1
a_3	0	0	0	0	1	1	1
a_4	0	0	0	0	1	1	1
b_1	0	0	0	1	1	1	1
b_2	0	1	1	1	1	1	1
b_3	1	1	1	1	1	1	1

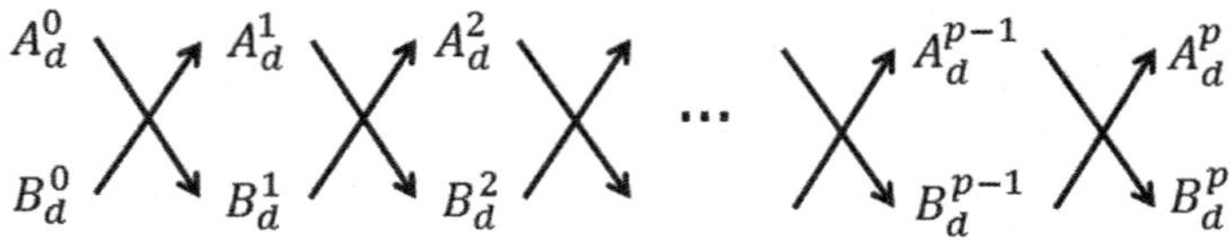

Fig. 1 Cascading failures reach steady state after p time steps

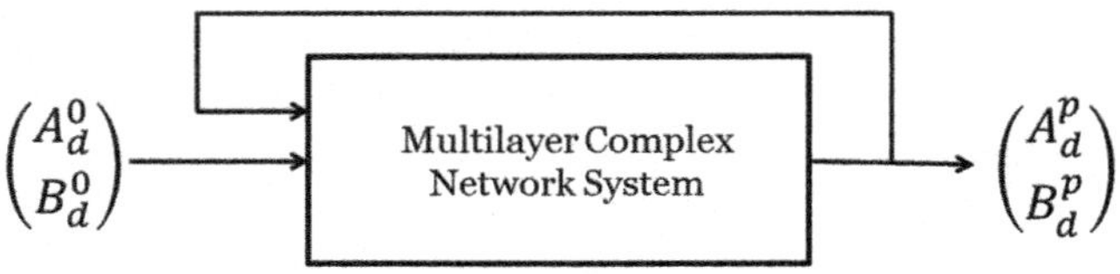

Fig. 2 Interdependent multilayer network as a closed loop control system

"closed loop" control system as shown in Fig. 2. Finding the steady state after an initial failure in this case is equivalent to computing the *fixed point* [6] of a function $\mathcal{F}(.)$ such that $\mathcal{F}(A_d^p \cup B_d^p) = A_d^p \cup B_d^p$, where p represents the number of steps when the system reaches the steady state. In the sample cascade of Table 2, at time step $p = 4$ the system reaches a steady state after the initial failure of $\{a_2, b_3\}$ at time step 0, i.e., $A_d^0 = \{a_2\}$, $B_d^0 = \{b_3\}$ and $A_d^4 = \{a_1, a_2, a_3, a_4\}$, $B_d^4 = \{b_1, b_2, b_3\}$.

The dependency relationships, or IDRs, used to represent the interdependent system can be formed either by careful analysis of the underlying system as was done in [2], or by consultation with the subject matter experts of the complex systems.

3.1 *Problems Studied Using the Implicative Interdependency Model*

In this section we provide an overview of some of the problems studied using the Implicative Interdependency Model.

3.1.1 Identification of the $\mathcal{K}$-Most Vulnerable Entities

In [14], the authors study the problem of identifying the $\mathcal{K}$ *Most Vulnerable Nodes* in an interdependent network. The set of $\mathcal{K}$ entities in a multilayer interdependent network is defined to be most vulnerable, if failure of these $\mathcal{K}$ entities triggers the failure of the largest number of other entities. The goal of the $\mathcal{K}$ *Most Vulnerable Nodes* ($\mathcal{K}$MVN) problem is to identify this set of $\mathcal{K}$ nodes, given an interdependent network with A and B type entities, where $n = |A|, m = |B|$, and a set of IDRs that represent the dependencies between the network entities. This is equivalent to identifying $A_d^0 \subseteq A$, $B_d^0 \subseteq B$, that *maximizes* $|A_d^p \cup B_d^p|$, $p = n + m - 1$, subject to the constraint $|A_d^0 \cup B_d^0| \leq \mathcal{K}$. The authors show that the $\mathcal{K}$MVN problem can be solved in polynomial time for some special cases, whereas for others, the problem is NP-complete. A technique is provided utilizing Integer Linear Programming to compute the solution to the $\mathcal{K}$MVN problem. The authors conduct experiments using the proposed technique on real world power grid and communication network data after synthetically generating dependencies between the network entities.

3.1.2 Root Cause Analysis of Failures in Interdependent Networks

In an interdependent network system, an *initial failure* of a set of entities may introduce *triggered failures* into the system through failure propagation due to the nature of interdependencies shared among the interdependent entities. In such a setting where an initial failure set can cascade and result in a much larger *combined failure* set, it may be required to identify the initial failure set from the combined failure set. In [5], the authors study the problem of identifying this initial failure set, or the *Root Cause of Failure*, from the combined failure set in an interdependent network. Formally, given (i) an interdependent network with a set of A and B type entities, where $n = |A|, m = |B|$, (ii) a set of IDRs representing the dependencies between the network entities, and (iii) a set of failed entities at time step p, $A_d^p \cup B_d^p$, where $A_d^p \subseteq A$, $B_d^p \subseteq B$, $p = n+m-1$. The *Root Cause of Failure* (RCF) problem is to identify the set of entities $A_d^0 \subseteq A$, $B_d^0 \subseteq B$ whose failure at time step 0 results in the failure of $A_d^p \cup B_d^p$ at time step p, such that $|A_d^0 \cup B_d^0|$ is *minimized*. The authors show that the RCF problem can be solved in polynomial time for some special cases, whereas for others, the problem is NP-complete. The authors provide an optimal solution using Integer Linear Programming and an *approximation algorithm* for the most general case of the problem. Using real world power and communication network data from Maricopa County, Arizona, the authors compare the results of their approximation algorithm to the optimal solution and present their results.

3.1.3 Progressive Recovery from Failure in Interdependent Networks

In an interdependent network setting with A and B type entities, where the failure of $A_d^0 \cup B_d^0$ entities at time step 0 results in the failure of $A_d^p \cup B_d^p$ at time step

p, the goal of the *Progressive Recovery* (PRREC) problem studied in [10] is to determine the *repair sequence* of failed elements $A_d^0 \cup B_d^0$, such that the *system utility* is *maximized* over the entire recovery process. Formally, given (i) an interdependent network with a set of A and B type entities, where $n = |A|, m = |B|$, (ii) a set of IDRs representing the dependencies between the network entities, (iii) a set of utility values $u(e_i)$, for each entity $e_i \in A \cup B$, and (iv) the set of failed entities at time step 0, $A_d^0 \cup B_d^0$, $A_d^0 \subseteq A$, $B_d^0 \subseteq B$. The authors define *System Utility at Instance of Time* t or $SUIT(t)$ as the total utility derived from the entities that are operable at time step t, i.e., if $A_d^t \subseteq A$ ($B_d^t \subseteq B$) denotes the inoperable entities that belong to set A (B) at time step t, $SUIT(t)$ is computed as follows:

$$SUIT(t) = \sum_{a_i \in A \setminus A_d^t} u(a_i) + \sum_{b_i \in B \setminus B_d^t} u(b_i)$$

The authors also define *System Utility Over Time interval* 0 *to* t, or $SUOT(t)$ as:

$$SUOT(t) = \sum_{i=0}^{t} SUIT(i)$$

The objective of the PRREC problem is to identify a repair sequence of $A_d^0 \cup B_d^0$ entities such that $SUOT(t), t = n + m - 1$ is *maximized*. The authors show that the problem can be solved in polynomial time for some special cases, whereas for others, the PRREC problem is NP-complete. Two *approximation algorithms* with performance bounds of 2 and 4 respectively are presented in the paper. The authors also provide an optimal solution to the problem utilizing Integer Linear Programming and a polynomial time heuristic solution. They then present their evaluation of the efficacy of the heuristic solution compared to the optimal solution with both synthetic data and real world data.

3.1.4 Entity Hardening Problem in Interdependent Networks

As critical infrastructure networks can be subject to adversarial attacks by hostile actors who may want to target the network entities that cause the maximum widespread impact, there is a need for network operators to be able to identify network entities that should be protected, or "*hardened*", to withstand the adversary's attack and prevent the impact of failure. In [1], the authors study the *Entity Hardening Problem* (ENP) in interdependent networks which addresses this problem. The authors define "*entity hardening*" as the ability of the network operator to ensure that an adversary cannot render a network entity from an *operative* (operational) to an *inoperative* (failed) state. They assume that the adversary is resourceful and is capable of identifying the most vulnerable network entities of the network (Sect. 3.1.1), whose failure causes the maximum number of network entities to fail due to the shared interdependencies. They further assume that the adversary possesses resources to destroy at most $\mathcal{K}$ entities, and the network operator is aware of the adversary's targets. In such a setting, given the network operator's entity hardening budget of k, $k < \mathcal{K}$, i.e., the operator can harden (prevent from failure) at most k entities, the objective of the ENP is to identify the k entities in the network such that the impact of the adversary's attack on the $\mathcal{K}$ entities is minimized.

Formally, the ENP is stated as follows: given (i) an interdependent network with a set of A and B type entities, where $n = |A|$, $m = |B|$, (ii) a set of IDRs representing the dependencies between the network entities, (iii) the set of the most vulnerable entities of the system $A_d^{'0} \cup B_d^{'0}$, where $|A_d^{'0} \cup B_d^{'0}| \leq \mathcal{K}$, $A_d^{'0} \subseteq A$, $B_d^{'0} \subseteq B$, whose failure at time step 0 causes the maximum number of entities to fail at time step $p = n + m - 1$, and (iv) the network operator's entity hardening budget k, $k < \mathcal{K}$. The objective of the ENP is to identify the set of entities $A_d^0 \cup B_d^0$, $|A_d^0 \cup B_d^0| \leq k$, $A_d^0 \subseteq A$, $B_d^0 \subseteq B$ which when hardened, *minimizes* the entities that fail at time step $p = n + m - 1$, when $A_d^{'0} \cup B_d^{'0}$ fail at time step 0. It may be noted that if an entity is hardened and is part of both A_d^0 (B_d^0) and $A_d^{'0}$ ($B_d^{'0}$), that entity does not fail. The authors show that the ENP can be solved in polynomial time in some special cases, whereas for some others, the problem is NP-complete. They provide an *approximation algorithm* for one case, and show that the problem is *inapproximable* in the general case and provide a heuristic soltion. From data collected from real world power and communication networks, the authors compare the results of their heuristic solution to the optimal solution (computed through an Integer Linear Program) and present their results.

4 Smallest Pseudo Target Set Identification Problem (SPTSIP)

In this section we present a detailed study of the *Smallest Pseudo Target Set Identification Problem* in the IIM setting. As noted earlier, the interdependent relationships between the entities of an interdependent network implies that a failure involving a small set of entities can trigger a *cascading event* that can result in the failure of a much larger set of entities. This creates a potential scenario where an adversary with an intent to jeopardize a specific set of entities E', or *real targets*, now no longer needs to destroy E' directly. Instead, the adversary can take advantage of the cascading failure process by identifying a smaller set of entities E'', or *pseudo targets*, whose failure eventually leads to the failure of E' due to the failure cascade. Thus, the objective of the adversary is to identify the smallest set of pseudo targets E'' whose failure eventually causes E' to fail. We refer to this problem as the *Smallest Pseudo Target Set Identification Problem*, or SPTSIP. The problem is formally analyzed in subsequent sections.

4.1 Problem Formulation and Computational Complexity Analysis

In this section we formally state the SPTSIP in the IIM setting, and analyze its complexity for different types of dependency relations. We formulate the SPTSIP as follows:

The Smallest Pseudo Target Set Identification Problem

INSTANCE: Given:

(i) the set A and B representing the entities of the power and communication networks respectively with $n = |A|$, $m = |B|$
(ii) a set of dependency relations or IDRs, between A and B type entities
(iii) the set of *real targets* $E' \subseteq A \cup B$
(iv) positive integer K

QUESTION Is there a subset $E'' \subseteq A \cup B$ of *pseudo targets*, with $|E''| \leq K$, whose failure at time step 0, triggers a cascade of failures resulting in failure of the real target set E' by time step $p = n + m - 1$?

We outline some assumptions for the SPTSIP: First, we assume that an entity $e_i \in A \cup B$ can fail by itself and not due to its dependencies, only at time step 0. Any failures after time step 0 occur due to the cascade effect of entities that failed at time step 0. Second, we assume that dependent entities immediately fail in the next time step, i.e. if $e_i \leftarrow e_j e_k$, and e_k fails at time step $p - 1$, then e_i fails at p. Third, time step $p = n + m - 1$ is a sure end of any failure cascade that begins at time step 0 as there are at most $n + m$ entities and we assume that entities cannot become operational once they fail. Finally, the pseudo target set E'' does not have to be unique.

It may be noted that the SPTSIP and the Root Cause of Failure (RCF) problem of [5] are considerably different. In the RCF problem, given a failure set F, $F \subseteq A \cup B$ the objective is to find the minimum number of entities F', $F' \subseteq F$ such that when F' entities fail at time step 0, F fails at time step p. It may be noted that for solving the RCF problem it is sufficient to look at the entities in F and their corresponding IDRs to find a solution F'. The set of entities in $(A \cup B) \setminus F$ can completely be ignored for computing F'. However, in the case of the SPTSIP, for the real target set E' that must fail by time step p there is no requirement that $E'' \subseteq E'$ and E'' can be any subset of $A \cup B$. This modification to the problem considerably changes the techniques required for tackling the SPTSIP.

To analyze the complexity of the SPTSIP, we categorize the type of IDRs encountered in interdependent networks in terms of the number of minterms they contain, and the size of each minterm. We analyze the complexity of each of these categories as follows:

Algorithm 1: Case I optimal algorithm for SPTSIP

Data:
1. Set of network entities $A \cup B$, with $n = |A|$ and $m = |B|$
2. A set S of IDRs of the form $y \leftarrow x$, where $x, y \in A \cup B$
3. A set of *real targets* E'

Result: The smallest set of *pseudo targets* E'' such that if E'' fails at time step 0, the real target set E' fails by time step $p = n + m - 1$

1 **begin**
2   Construct a directed graph $G = (V, E)$, where $V = A \cup B$. For each IDR $y \leftarrow x$ in S, where $x, y \in A \cup B$, introduce a directed edge $(x, y) \in E$;
3   For each node $x_i \in V$, construct a transitive closure set C_{x_i} as follows: If there is a path from x_i to some node $y_i \in V$ in G, then include y_i in C_{x_i}. As $|A| + |B| = n + m$, we have $n + m$ transitive closure sets C_{x_i}, $1 \leq i \leq (n + m)$. Each x_i is termed as the *seed entity* for the transitive closure set C_{x_i};
4   Remove all the transitive closure sets which are proper subsets of some other transitive closure set;
5     $E'' \leftarrow \emptyset$;
6   **while** $E' \neq \emptyset$ **do**
7   For entity $e_j \in E'$, find set C_{x_i} such that $e_j \in C_{x_i}$;
8   Include *seed entity* x_i in E'';
9         $E' \leftarrow E' \setminus C_{x_i}$;
10   **return** E''

4.1.1 Case I: Problem Instance with One Minterm of Size One

For Case I the IDR's are represented as: $x_i \leftarrow y_j$, where x_i and y_j are elements of the set A (B) and B (A) respectively. In the example $a_i \leftarrow b_j$, $x_i = a_i$, $y_1 = b_j$. As noted in [14], a conjunctive implication of the form $a_i \leftarrow b_j b_k$ can be written as two separate IDRs $a_i \leftarrow b_j$ and $a_i \leftarrow b_k$. However, this case is considered in Case II and not in Case I. This exclusion implies that the entities that appear on the left hand side of an IDR in Case I are unique. For Case I, Algorithm 1 presents a polynomial time algorithm for the solution of the SPTSIP.

Algorithm 1 Time Complexity: Since $|A|+|B| = n+m$. Step 2 takes $O(n+m+|S|)$ time, where S is the set of Case I type IDRs. Step 3 can be executed in $O((n+m)^3)$ time. Step 4 takes at most $O((n+m)^3)$ time. The *while* loop in Step 6 takes at most $O(n(n+m))$. Therefore the overall complexity of Algorithm 1 is $O((n+m)^3)$.

Theorem 1 *For each pair of transitive closure sets C_{x_i} and C_{x_j} produced in Step* 3 *of Algorithm* 1*, either $C_{x_i} \cap C_{x_j} = \emptyset$ or $C_{x_i} \cap C_{x_j} = C_{x_i}$ or $C_{x_i} \cap C_{x_j} = C_{x_j}$, where $x_i \neq x_j$.*

Proof The theorem is proved by contradiction, assume there exists a pair of transitive closure sets C_{x_i} and C_{x_j} such that $C_{x_i} \cap C_{x_j} \neq \emptyset$, $C_{x_i} \cap C_{x_j} \neq C_{x_i}$ and $C_{x_i} \cap C_{x_j} \neq C_{x_j}$. Let $x_k \in C_{x_i} \cap C_{x_j}$, this implies that there exists a path $P1$ from x_i to x_k, as well as a path $P2$ from x_j to x_k. Thus there exists some x_l such that $x_l \in P1$ and $x_l \in P2$. Without loss of generality, assume that x_l is the first node in $P1$, also, as $x_l \in P2$, x_l has an in-degree greater than 1. This implies that there is more than one IDR in set S such that x_l appears on the left hand side of these IDRs. This is a contradiction as it violates the definition of Case I type IDRs and hence the theorem is proved.

Theorem 2 *Algorithm 1 gives an optimal solution for the SPTSIP in an interdependent network for Case I type IDRs.*

Proof Theorem 1 proves that every pair of the transitive closure sets created in Step 3 of Algorithm 1 are either disjoint or is a proper subset of the other, in Step 4 of the algorithm all transitive closure sets that are proper subsets of some other transitive closure set are removed. This implies that the remaining sets are all necessarily disjoint, and for every $e_i \in E'$, e_i belongs to exactly one transitive closure set. This necessitates that the *seed entity* x_k of the transitive closure set C_{x_k} that e_i belongs to, must be included in the solution. This is done in the *while* loop of Step 6. To prove the optimality claim we need to show that the number of seed entities chosen by the algorithm is minimum. If we assume that the number of seeds chosen is not minimum, then some C_{x_i} chosen by the algorithm must necessarily be a proper subset of another closure. This contradicts Theorem 1, and hence Algorithm 1 always returns the optimal solution.

4.1.2 Case II: Problem Instance with One Minterm of Arbitrary Size

For Case II, the IDR's are represented as:
$x_i \leftarrow \prod_{k_1=1}^{l} y_{k_1} \prod_{k_2=1}^{q} x_{k_2}$ (with $x_i \neq x_{k_2} \forall x_{k_2}, 1 \leq k_2 \leq q$), where x_i, x_{k_2} are elements of set A (B) and y_{k_2} is an element of set B (A). The size of the minterm is given as $l + q$. In the example $a_r \leftarrow b_u b_v a_s$. $l + q = 3$, $x_i = a_r$, $y_1 = b_u$, $y_2 = b_v$ and $x_1 = a_s$.

Theorem 3 *The SPTSIP for Case II is NP Complete*

Proof We prove that the SPTSIP for Case II is NP-complete by giving a transformation for the *Set Cover* (SC) problem [8]. An instance of the set cover problem is specified by a universal set $S = \{s_1, \ldots, s_{n+m}\}$ and a set of subsets S', $S' = \{S_1, \ldots, S_q\}$, where $S_i \subseteq S, \forall i, 1 \leq i \leq q$. In the set cover problem one wants to know whether there exists a subset of $S'' \subseteq S'$ such that $\bigcup_{S_i \subseteq S''} S_i = S$ and $|S''| \leq Q$, for some specified integer Q. From an instance of the SC problem we create an instance of the SPTSIP in the following way: For every $s_i \in S$, we create an IDR of the form $s_i \leftarrow \prod_{s_i \in S_j} S_j$, $\forall S_j \in S'$. We set the real target set $E' = S$ and $K = Q$. It can now easily be verified that the instance of the SC problem has a set cover of size Q, iff in the created instance of SPTSIP the failure of K entities at time step 0 triggers a cascade of failures resulting in the failure of all entities in the set E' by time step $p = n + m - 1$.

We now define the following:

Definition 1 *Kill Set of a set of Entities* $\mathcal{P}$: The *Kill Set* of a set of entities $\mathcal{P}$, denoted by $KillSet(\mathcal{P})$, is the set of all entities in the multilayer network (including

Algorithm 2: Case II Approx. Algorithm for SPTSIP

Data:
1. Set of network entities $A \cup B$, with $n = |A|$ and $m = |B|$
2. A set of IDRs of the form $y \leftarrow \prod_{i=1}^{q} x_i$, where $x_i, y \in A \cup B, \forall 1 \leq i \leq q$
3. A set of *real targets* E', with $M = |E'|$

Result: Set of entities $E'' \subseteq A \cup B$ such that failure of E'' entities in time step 0 results in failure of E' entities by time step $p = n + m - 1$.

1 **begin**
2     $U \leftarrow \emptyset$;
3     $DEP_i \leftarrow \emptyset,\ S_i \leftarrow \emptyset, \quad \forall i = 1, \ldots, M$;
4     $KillSet_j \leftarrow \emptyset, \quad \forall j = 1, \ldots, n + m$;
5   **foreach** $e_i \in E'$ **do**
6   **foreach** *entity* $e_j \in$ *IDR* $e_i \leftarrow \prod_{j=1}^{q} e_j$ **do**
7             $DEP_i \leftarrow DEP_i \cup \{e_j\}$;
8         $U \leftarrow U \cup \{i\}$;
9         $S_i \leftarrow U \cup \{i\}$;
10   **foreach** $e_i \in A \cup B$ **do**
11         $KillSet_i \leftarrow KillSet(e_i)$;
12   **for** $d = 1$ *to* M **do**
13   **if** $KillSet_i \cap DEP_d \neq \emptyset$ **then**
14                 $S_i \leftarrow S_i \cup \{d\}$;
15     $E'' \leftarrow \emptyset$;
16   **while** $U \neq \emptyset$ **do**
17   Select $S_i, i = 1, \ldots, M$ that maximizes $|S_i \cap U|$;
18         $E'' \leftarrow E'' \cup \{e_i\}$;
19         $U \leftarrow U \setminus S_i$;
20   **return** E''

$\mathcal{P}$) that fail by $p = n + m - 1$ time steps as a consequence of: (i) the failure of $\mathcal{P}$ entities at time step 0, and (ii) the interdependency relationships (IDRs) shared between the entities of the network.

In Algorithm 2 we present an approximation algorithm for the SPTSIP with Case II type IDRs.

Theorem 4 *The approximation solution produced by Algorithm* 2 *for Case II type IDRs is at most* $O(ln(M))$ *times the optimal, where* $M = |E'|$

Proof Algorithm 2 implements a greedy approach for solving a set cover problem. We set up the set cover problem the following way: First, in Steps 5–9, for each entity $e_i \in E'$ we construct dependency DEP_i as the set of entities out of which at least one entity must fail for e_i to fail, thus "unsatisfying" the dependency. In Step 9 we also account for dependency DEP_i getting unsatisfied due to failure of e_i itself. The universe U contains the indexes of each of these M dependencies. Next, in Steps 10–14, for each entity $e_i \in A \cup B$ we compute $KillSet(e_i)$ and construct set S_i that contains the index of the dependencies in $DEP_j, j = 1, \ldots, M$ that has a non-empty intersection with $KillSet(e_i)$. This implies that with the failure of e_i and the ensuing failure propagation, DEP_j gets unsatisfied. With the universe set of U and the subsets $S_i, i = 1, \ldots, M$, the greedy technique for set cover is used in Steps 16–19 that yields a known approximation factor of $O(ln(M))$ times the optimal solution [9].

Algorithm 2 Time Complexity: To construct the dependencies in Steps 5–9 at most M IDRs will be traversed each with at most $n + m$ entities hence these steps take $O(M(n + m))$ time. In Steps 10–14 computing the kill set of each of the $n + m$ entities and comparing it to each of the M dependencies of maximum size $n + m$ requires $O(M(n+m)^3)$ time. And finally, the greedy set cover in Steps 16–19 takes $O(M\ log(n + m))$. Overall the complexity of Algorithm 2 is $O(M(n + m)^3)$.

4.1.3 Case III: Problem Instance with an Arbitrary Number of Minterms of Size One

For Case III an IDR has the following form:
$x_i \leftarrow \sum_{k_1=1}^{l} y_{k_1} + \sum_{k_2=1}^{q} x_{k_2}$ (with $x_i \neq x_{k_2} \forall x_{k_2}, 1 \leq k_2 \leq q$), where x_i, x_{k_2} are elements of set A (B) and y_{k_2} is an element of set B (A). The size of the minterm is given as $l + q$. In the example $a_r \leftarrow b_u + b_v + a_s$. $l + q = 3$, $x_i = a_r$, $y_1 = b_u$, $y_2 = b_v$ and $x_1 = a_s$.

Theorem 5 *The SPTSIP for Case III is NP Complete*

Proof We prove that the SPTSIP for Case III is NP-complete by giving a transformation for the *Vertex Cover* (VC) problem [8]. An instance of the vertex cover problem is specified by an undirected graph $G = (V, E)$ and an integer R. In the vertex cover problem, one wants to know whether there is a subset $V' \subseteq V$ such that $|V'| \leq R$, and for every edge $e \in E$, at least one end vertex of e is in V'. From an instance of the VC problem we create an instance of the SPTSIP in the following way: From the graph $G = (V, E)$ for each vertex $v_i \in V$ that has adjacent nodes (say) v_j, v_k and v_l, we create an IDR $v_i \leftarrow v_j + v_k + v_l$. We set the real target set $E' = V$ and $K = R$. It can now be verified that the instance of the VC problem has a vertex cover of size R, iff in the created instance of SPTSIP the failure of K entities at time step 0 triggers a cascade of failures resulting in the failure of all entities in the set E' by time step $p = |V| - 1$.

4.1.4 Case IV: Problem Instance with an Arbitrary Number of Minterms of Arbitrary Size

This is the most general case where IDRs have arbitrary number of minterms of arbitrary size.

Theorem 6 *The SPTSIP for Case IV is NP Complete*

Proof As both Case II and Case III are special cases of Case IV, the SPTSIP for Case IV is NP-Complete as well.

4.2 Algorithms for the SPTSIP

In this section we propose an optimal solution for the SPTSIP using Integer Linear Programming (ILP), and a polynomial time heuristic solution.

4.2.1 Optimal Solutions for the SPTSIP Problem

We formulate an optimal solution for the SPTSIP with an ILP that uses two variables x_{it} and y_{jt}. Where $x_{it} = 1$, when entity $a_i \in A$ is in a failed state at time step t, and 0 otherwise. And, $y_{jt} = 1$, when entity $b_j \in B$ is in a failed state at time step t, and 0 otherwise.

The objective function can now be formulated as follows:

$$min \sum_{i=1}^{n} x_{i0} + \sum_{j=1}^{m} y_{j0} \tag{1}$$

Where $n = |A|$ and $m = |B|$. The constraints are as follows:
Failure Consistency Constraints: $x_{it} \geq x_{i(t-1)}, \forall t, 1 \leq t \leq p$, these constraints ensure that if an entity a_i fails at time step t, it continues to remain in a failed state for all subsequent time steps. A similar constraint applies for y_{it} variables [14].
Failure Propagation Constraints: These constraints govern the failure cascade process caused by the dependencies shared between the network entities. The correctness of these constraints is established in [14], we outline an overview of these constraints here for consistency. For any Case IV type IDRs of the form $a_i \leftarrow b_j b_k b_l + b_v b_u + b_q$ the subsequent steps are followed to model the failure propagation:

Step 1: Transform the IDR to a disjunctive form of size one minterms, i.e. $a_i \leftarrow c_1 + c_2 + b_q$.

Step 2: For each of the c type minterms create constraints to model the failure cascade for individual c type minterms, i.e., for $c_1 \leftarrow b_j b_k b_l$ introduce $c_{1t} \leq y_{j(t-1)} + y_{k(t-1)} + y_{l(t-1)}, \forall t, 1 \leq t \leq p$.

Step 3: For each transformed IDR from Step 1, for example $a_i \leftarrow c_1 + c_2 + b_q$, introduce a constraint of the form $N \times x_{it} \leq c_{1(t-1)} + c_{2(t-1)} + b_q, \forall t, 1 \leq t \leq p$, where N is the number of minterms in the transformed IDR, in this example $N = 3$.

Prior to the transformation of Step 1 if an IDR does not contain any disjunctions (Case II), then Step 3 is skipped, or if it does not contain any conjunctions (Case III), then Step 2 is skipped.

Real Target Set Failure Constraints: $x_{ip} = 1, \forall a_i \in E'$, and $y_{ip} = 1, \forall b_i \in E'$, these constraints ensure that all entities of the real target set E' are in a failed state at time step p.

Adhering to the above constraints, the objective in (1) minimizes the total number of entities that need to fail at time step 0 so that E' entities fail by time step p.

4.2.2 Heuristic Solution

We first outline the following definition:

Definition 2 *Kill Impact of a set of Entities* $\mathcal{P}$: The Kill Impact of a set of entities $\mathcal{P}$, denoted by $KillImpact(\mathcal{P})$, is defined as the contribution of $\mathcal{P}$ entities in causing the failure of entities in E'. It may be noted that any entity $e_i \in E'$ can fail due to two reasons: (i) when e_i itself fails at time step 0, or (ii) when at least one entity in all the minterms of e_i's IDR fail in some time step. $KillImpact(\mathcal{P})$ captures these two aspects by computing the impact of failure of $\mathcal{P}$ entities on E' based on: (i) the number of entities that fail in E' at time step p when $\mathcal{P}$ entities fail at time step 0, and (ii) the number of minterms in the IDR of each entity $e_i \in E'$ that get affected at time step p when $\mathcal{P}$ entities fail at time step 0. For a given set of $\mathcal{P}$ entities, and the set of minterms $MT_i = \{mt_1, mt_2, \ldots, mt_{|MT_i|}\}, mt_j \subseteq A \cup B$, for each entity $e_i \in E'$, to compute $KillImpact(\mathcal{P})$ we first compute $impact_i$ as the impact of failure of $\mathcal{P}$ on e_i as follows:
If $e_i \in KillSet(\mathcal{P}), impact_i = 1$, else if $e_i \notin KillSet(\mathcal{P})$:

$$impact_i = \frac{\left|\bigcup_{mt_j \cap KillSet(\mathcal{P}) \neq \emptyset} mt_j\right|}{|MT_i|}, \quad \forall mt_j \in MT_i$$

We then compute $KillImpact(\mathcal{P})$ as follows:

$$KillImpact(\mathcal{P}) = \frac{\sum_{i=1}^{|E'|} impact_i}{|\mathcal{P}|}$$

In Algorithm 3, we present a heuristic technique to solve the SPTSIP for the general case of the problem. The general approach for Algorithm 3 is to greedily select a set of entities that provide the maximum benefit towards reaching the objective of failing E'. In Steps 5–7, for each entity $e_j \in A \cup B$ we compute how frequently e_j appears in all minterms in the set of IDRs, we also compute $KillImpact(e_j)$. Next, in Steps 8–14, for each entity $e_i \in E'$, we examine each of the minterms of e_i's IDR and select the highest frequency entity of each minterm to construct set k_i and compute $KillImpact(k_i)$. In Step 15 we choose the most impactful set of entities from the total $KImpact$ sets constructed. Intuitively, this selection of a higher kill impact set for inclusion implies more failures in the target set. Also, since our objective is to minimize the size of the entities selected, a set

Algorithm 3: Case IV Heuristic for SPTSIP

Data:
1. Set of network entities $A \cup B$, with $n = |A|$ and $m = |B|$
2. A set S of IDRs of type Case IV (general case)
3. A set of *real targets* E'

Result: A set of *pseudo targets* E'' such that when E'' fails at time step 0, the real target set E' fails by time step $p = n + m - 1$

1 **begin**
2   $E'' \leftarrow \emptyset$, $failed \leftarrow \emptyset$;
3   **while** $E' \nsubseteq failed$ **do**
4     $KImpact \leftarrow \emptyset$;
5   **foreach** *entity* $e_j \in A \cup B$ *and* $e_j \notin failed$ **do**
6   Compute $frequency_j$ as the number of times e_j appears in a minterm for all IDRs in S;
7       $KImpact \leftarrow KImpact \cup (e_j, KillImpact(e_j))$;
8   **foreach** *entity* $e_i \in E'$ *and* $e_i \notin failed$ **do**
9   Let idr in S be the IDR of entity e_i;
10       $k_i \leftarrow \emptyset$;
11   **foreach** *minterm* MT *in* idr **do**
12   Select entity $e_j \in MT$ with largest $frequency_j$ from all entities in MT;
13         $k_i \leftarrow k_i \cup e_j$;
14       $KImpact \leftarrow KImpact \cup (k_i, KillImpact(k_i))$;
15   Select tuple $(failSet, failVal) \in KImpact$ where $failVal \geq val, \forall (set, val) \in KImpact$;
16     $E'' \leftarrow E'' \cup failSet$;
17     $failed \leftarrow KillSet(E'')$;
18   Remove IDR of entity e_k from S, $\forall e_k \in failed$;
19   For each IDR in S remove all minterms that contain entity e_k, $\forall e_k \in failed$;
20   **return** E''

with the largest impact to size ratio is preferred. Finally, the algorithm proceeds to update E'' and $failed$ set of entities, and prunes the IDR set and minterm set in Steps 16–19. This greedy selection process repeats until $E' \subseteq failed$. The heuristic ensures that for every iteration of the *while* loop in Step 3 the E'' set increases in such a way that at least one additional entity in E' fails than the previous iteration, thus moving closer to the objective. Algorithm 3 runs in polynomial time, specifically it runs in $O(M(n + m)^4)$ time, where $M = |E'|$. In Sect. 4.3, our experiments show that Algorithm 3 almost always produces the optimal result.

4.3 Experimental Results

We now present experimental results for the SPTSIP and compare the optimal solution computed using an ILP, with the proposed heuristic algorithm. The experiments were conducted on power and communication network data of Maricopa County,

Arizona. The power network data was obtained from Platts (www.platts.com), and the communication network data obtained from GeoTel (www.geo-tel.com). This data consisted of 70 power plants, 470 transmission lines, 2, 690 cell towers, 7, 100 fiber-lit buildings and 42, 723 fiber links. We identified five non-intersecting geographical regions, and from the consolidated power and communication network data of each region, we set up interdependencies between the network entities using the rules outlined in [14]. For continuity, we briefly outline an overview of these rules here: For each generator to be operational, either (i) the nearest cell tower must be operational, or (ii) the nearest fiber-lit building and the fiber link connecting the generator to the fiber-lit building must be operational. For each fiber-lit building and cell tower to be operational, at least one of the two nearest generators and the connecting transmission lines must be operational. The transmission lines and the fiber links have no dependencies.

The optimal solutions were obtained by solving Integer Linear Programs using the IBM CPLEX Optimizer 12.5. For each of the five regions $R1$ through $R5$, real target sets of entities of sizes 5, 10, 15 and 20 were chosen from the set of all power and communication entities of that region. For each real target set the optimal and heuristic solutions were computed, and these results are presented in Fig. 3. Our experiments showed that for the five regions considered, in the worst case the heuristic solution differed from the optimal by a factor of 0.16, in the best case was equal to the optimal, and on an average was within a factor of 0.02 of the optimal solution.

5 Conclusion

In this chapter we elaborated the recently proposed *Implicative Interdependency Model* (IIM) that is able to capture complex interdependencies that may exist in real world interdependent infrastructures, such as the power grid and the communication network. We then briefly outlined some of the problems studied using this model including the *$\mathcal{K}$ Most Vulnerable Nodes* problem, the *Root Cause of Failure* problem, the *Progressive Recovery* problem, and the *Entity Hardening* problem. We also presented a detailed study of the *Smallest Pseudo Target Set Identification Problem* in the IIM setting. We classified the problem into four classes and showed that the problem is solvable in polynomial time for the first class, whereas for others it is NP-complete. We provided an *approximation algorithm* for the second class, and for the general class we provided an optimal solution using an Integer Linear Program, and also provided a polynomial time heuristic technique. Finally, we evaluated the efficacy of our heuristic using power and communication network data of Maricopa County, Arizona. Our experiments showed that our heuristic almost always produced near optimal results.

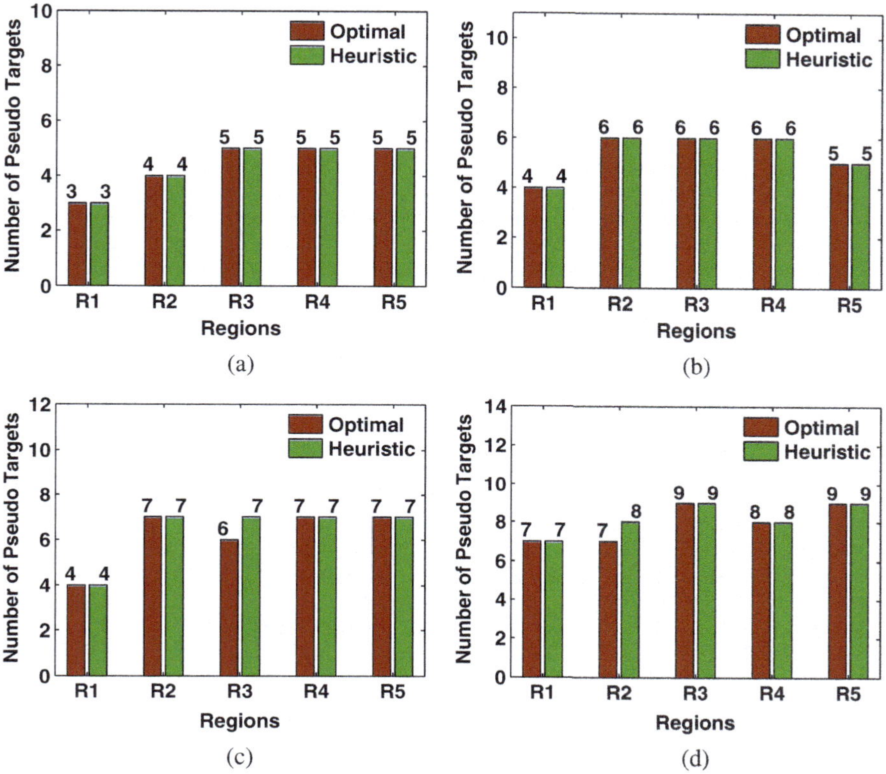

Fig. 3 Comparison of optimal and heuristic approaches for computing pseudo targets (E''), for given real targets (E') of sizes 5, 10, 15 and 20, on five geographical regions of Maricopa County, Arizona. (**a**) Real target set size: 5. (**b**) Real target set size: 10. (**c**) Real target set size: 15. (**d**) Real target set size: 20

References

1. Banerjee J, Das A, Zhou C, Mazumder A, Sen A (2015) On the entity hardening problem in multi-layered interdependent networks. In: 2015 IEEE Conference on Computer Communications WIDN Workshop (INFOCOM WKSHPS), Hong Kong, China, pp 648–653. IEEE
2. Bernstein A, Bienstock D, Hay D, Uzunoglu M, Zussman G (2014) Power grid vulnerability to geographically correlated failuresanalysis and control implications. In: 2014 Proceedings IEEE INFOCOM, Toronto, pp 2634–2642. IEEE
3. Buldyrev SV, Parshani R, Paul G, Stanley HE, Havlin S (2010) Catastrophic cascade of failures in interdependent networks. Nature 464(7291):1025–1028
4. Castet JF, Saleh JH (2013) Interdependent multi-layer networks: modeling and survivability analysis with applications to space-based networks. PloS One 8(4):e60402
5. Das A, Banerjee J, Sen A (2014) Root cause analysis of failures in interdependent power-communication networks. In: 2014 IEEE Military Communications Conference (MILCOM), Baltimore, MD, USA, pp 910–915. IEEE

6. Fudenberg D, Tirole J (1991) Game theory. Translated into Chinesse by Renin University Press, Bejing: China. MIT Press, Cambridge, MA
7. Gao J, Buldyrev SV, Stanley HE, Havlin S (2011) Networks formed from interdependent networks. Nat Phys 8(1):40–48
8. Garey MR, Johnson DS (1979) Computer and intractability. A guide to the NP-completeness. WH Freeman and Company, New York
9. Kleinberg J, Tardos É (2006) Algorithm design. Pearson Education, Boston
10. Mazumder A, Zhou C, Das A, Sen A (2014) Progressive recovery from failure in multi-layered interdependent network using a new model of interdependency. In: Conference on Critical Information Infrastructures Security (CRITIS). Limassol, Cyprus, Springer
11. Nguyen DT, Shen Y, Thai MT (2013) Detecting critical nodes in interdependent power networks for vulnerability assessment. IEEE Trans Smart Grid 4(1):151–159
12. Parandehgheibi M, Modiano E (2013) Robustness of interdependent networks: the case of communication networks and the power grid. arXiv preprint arXiv:1304.0356
13. Rosato V, Issacharoff L, Tiriticco F, Meloni S, Porcellinis S, Setola R (2008) Modelling interdependent infrastructures using interacting dynamical models. Int J Crit Infrastruct 4(1):63–79
14. Sen A, Mazumder A, Banerjee J, Das A, Compton R (2014) Identification of k most vulnerable nodes in multi-layered network using a new model of interdependency. In: NetSciCom Workshop (INFOCOM WKSHPS), Conference on Computer Communications, Toronto, ON, Canada, pp 831–836. IEEE
15. Shao J, Buldyrev SV, Havlin S, Stanley HE (2011) Cascade of failures in coupled network systems with multiple support-dependence relations. Phys Rev E 83(3):036116
16. Wood AJ, Wollenberg BF (2012) Power generation, operation, and control. Wiley, New York
17. Zhang P, Peeta S, Friesz T (2005) Dynamic game theoretic model of multi-layer infrastructure networks. Netw Spat Econ 5(2):147–178

Leveraging Network Theory and Stress Tests to Assess Interdependencies in Critical Infrastructures

Luca Galbusera and Georgios Giannopoulos

Abstract Many modern critical infrastructures manifest reciprocal dependencies at various levels and on a time-evolving scale. Network theory has been exploited in the last decades to achieve a better understanding of topologies, correlations and propagation paths in case of perturbations. The discipline is providing interesting insights into aspects such as fragility and robustness of different network layouts against various types of threats, despite the difficulties arising in the modeling of the associated processes and entity relationships. Indeed, the evolution of infrastructures is not, in general, the straightforward outcome of a comprehensive a priori design. Rather, factors such as societal priorities, technical and budgetary constraints, critical events and the quest for better and cost-effective services induce a continuous change, while new kinds of interdependencies emerge. As a consequence, mapping emerging behavior can constitute a challenge and promote the development of innovative approaches to analysis and management. Among them, stress tests are entering the stage in order to assess networked infrastructures and reveal the associated operational boundaries and risk exposures. In this chapter, we first overview key developments of network science and its applications to primary infrastructure sectors. Secondly, we address the implementation of network-theoretical concepts in actions related to resilience enhancement, referring in particular to the case of stress tests in the banking sector. Finally, a discussion on the relevance of those concepts to critical infrastructure governance is provided.

Keywords Critical infrastructures · Network theory · Stress tests · Interdependencies · Resilience

---

L. Galbusera · G. Giannopoulos (✉)
European Commission, DG Joint Research Centre (JRC), Directorate E – Space, Security and Migration, Technology Innovation in Security Unit, Ispra, VA, Italy
e-mail: luca.galbusera@ec.europa.eu; georgios.giannopoulos@ec.europa.eu

D. Gritzalis et al. (eds.), *Critical Infrastructure Security and Resilience*, Advanced Sciences and Technologies for Security Applications,
https://doi.org/10.1007/978-3-030-00024-0_8

1 Introduction

One of the peculiar traits of many modern critical infrastructures (CIs) is their networked nature. In the first place, many of the sectors considered as critical for our societies individually display this hallmark feature, for instance due to their spatially distributed layout or as they consist of interacting facilities, systems and functions. Secondly, a web of dependencies emerges when observing that sectors do not operate in isolation and that critical services are typically the result of orchestrated work by multiple actors, often within competitive markets and under tight constraints. See also [102] for a discussion on historical developments of large-scale critical infrastructures and emerging complexity.

In consideration of both the above-mentioned aspects, networked critical infrastructures (NCIs) hold a central interest in today's legislation and forward-looking policies. For instance, the *Communication from the Commission to the Council and the European Parliament – Critical Infrastructure Protection in the fight against terrorism*[1] (2004 EC-CIP Communication), issued in 2004, pointed out both a number of key sectors and the high degree of connectedness and interdependence of European CIs. The subsequent initiation of a European Programme for Critical Infrastructure Protection (EPCIP) and a Critical Infrastructure Warning Information Network (CIWIN) was accompanied by several policy documents[2] which raised public awareness. Worldwide, a number of similar initiatives related to critical infrastructure protection stress the importance of a comprehensive and relational analysis and management. In this sense, we can observe that novel sectors are being counted as CIs and that the relationships between the so-called hard and soft infrastructure is subject to rethinking, also taking into account the diverse pace of sectoral transformations.

The research community is accompanying the ongoing policy efforts, for instance by developing models for NCIs as well as methods to analyze and optimize their performance against different events such as natural disasters and man-made hazards (e.g. terrorism, malicious attacks, cyber events). For the purpose of scientific investigation, NCIs display various factors of complexity. Seminal in this domain is the contribution proposed in [99], wherein infrastructures were qualified

[1] https://eur-lex.europa.eu/legal-content/EN/ALL/?uri=CELEX:52004DC0702

[2] Including:

- *COM/2005/0576 final*: https://eur-lex.europa.eu/legal-content/EN/ALL/?uri=celex:52005DC0576.
- *Council Directive 2008/114/EC*: https://eur-lex.europa.eu/legal-content/EN/ALL/?uri=CELEX:32008L0114.
- *Council Decision 2007/124/EC,Euratom*: https://eur-lex.europa.eu/legal-content/EN/ALL/?uri=CELEX:32007D0124.
- *Commission Staff Working Document SWD(2013)318/F1*: http://ec.europa.eu/transparency/regdoc/rep/10102/2013/EN/10102-2013-318-EN-F1-1.PDF.
- *Council Decision (EU, Euratom) 2015/457*: https://eur-lex.europa.eu/legal-content/EN/ALL/?uri=CELEX:32015D0457.

as *"complex adaptive systems"* and a landmark classification was introduced to describe the different dimensions of infrastructure dependencies. Those were articulated into four categories, namely physical, cyber, geographic and logical interdependencies. In addition to such aspects, the authors observed the relevance of the environment in which infrastructures operate, including for instance the economic and business situation, policies, governmental actions, legal and technical aspects. A number of additional categorizations have been proposed in the literature for CI interdependencies, see [88] for a review.

Given this context, modern network science is offering opportunities for fertile interaction among a spectrum of disciplines, competencies and communities involved in NCIs. Supported by the developments of graph theory and cognate disciples such as statistical physics, discrete optimization and operations research, network theory provides a natural mathematical language to represent NCIs in terms of entities and relationships, able to echo the above-mentioned interdependence classifications. Indeed, [45] observes that *"complex networks are the skeletons of complex systems"*. Network-based approaches are included in [88] among the key modeling and simulation approaches to address infrastructure dependencies. Therein, a key way of exploiting network science in NCI analysis is outlined, wherein nodes are associated with CI components and edges express relations among them. Accordingly, analysis techniques are partitioned into topology-based and flow-based methods. In the first case, the system is described purely in terms of its connectivity plus discrete states (e.g. normal/failed) for each component. Topology-based methods are then further articulated into analytical and simulation methods, with node heterogeneity playing a key role in the selection of one or the other option. Instead, flow-based methods are oriented to service provision representation and, as such, they may be able to more accurately depict the mechanics of operations taking place over the network. As for the aspect of functional robustness of techno-social networks, [116] observes the following major features potentially leading to large-scale failures: the presence of *"multivariate and heterogeneous constraints"*, able to considerably influence the different operational modes of the system, and the *"multiscale nature"* of these networks, which may concentrate different processes involving various time and space scales.

In this chapter, we first trace some of the fundamental developments in network science that allowed, in time, to address the complexity emerging in many real-world systems, notably NCIs. The discussion starts with the case of standard networks, wherein relationships among nodes are assumed to be homogeneous in nature. In this context, the development of random networks theory was accompanied by an increasing interest in empirical networks. Accordingly, the classical theory was extended thanks to the investigation of the profound constitutive criteria determining the emerging topologies in various domains. In addition to the analysis of network topologies, processes taking place over them were studied, unveiling the role of connectivity in determining emerging phenomena. Besides, more and more attention was devoted to understanding networks from the higher perspective of multilayer systems, i.e. by taking into account how coupling strength and diversity can radically affect emerging properties and behaviors. Some key studies recently revealed radical modifications to the fragility properties of systems when observed from this viewpoint, with important implications on prospective developments in research and beyond. The insights provided by the above-mentioned studies, together with other factors such as data availability, are transforming our way of approaching the emerging complexity in systems such as NCIs. This transformation has long-reaching effects, as it calls for new ways of dealing with their design and operation, especially to cope with critical events.

As a second step, in this chapter we will address the relevance of network science from a NCI governance perspective, taking into account the advantage brought by a systemic view in monitoring a fast-evolving society, assessing risk and evaluating potential crisis scenarios towards impact mitigation. Stress tests are a significant example of approaches aiming to take into account the variable nature of CIs and their operation. In this respect, the banking sector is undoubtedly a most relevant example wherein macro-prudential policies emerged in recent years and aim at identifying systemic vulnerabilities. In time, stress tests in this field were subject to a systemic refocus, shifting the objectives from the analysis of the performance of single institutions to a broader scope, able to capture emerging high-order effects. To address this challenge, synergies are being found between stress tests and network theory-based approaches. In a similar direction, despite with different levels of maturity, we can observe the emergence of stress tests in other areas, such as towards the analysis of gas infrastructure resilience with respect to supply shocks. Recently, research initiatives such as FP7 project STREST[3] have made significant progress in translating stress testing principles into practical tools for understanding the complexity and level of resilience of different types of NCIs. A comprehensive exploitation of stress testing methodologies and network analysis techniques requires further efforts, yet it seems to open opportunities for an evolved approach to infrastructure governance.

The rest of this chapter is organized as follows: in Sect. 2 we deal with single-layer networks, including key theoretical developments, architectural aspects, and the study of processes and performance under perturbations; Sect. 3 is devoted to multilayer networks; in Sect. 4 we discuss banking stress tests, their current trends and the implications of network science applied in this domain. Finally, in Sect. 5 we briefly reflect on the relevance of these methods in the broad area of NCI analysis and resilience enhancement.

2 From Random Networks to Complex Networks

2.1 Theoretical Developments and Empirical Observations

One of the principal efforts displayed within modern network theory has been devoted to the construction of mathematical frameworks able to stylize architectures

[3]https://cordis.europa.eu/project/rcn/110339_en.html

of complex systems as observed in many different domains. A typical way of introducing network representations is by means of the underlying graph $G = (N, E)$, where N is the set of nodes (or vertexes) and $E \subseteq N \times N$ the set of edges (or links), with $|N| = n$ and $|E| = m$. To support our discussion, in the undirected case we define the degree distribution of G as (see [2])

$$P(k) = \frac{1}{n} \sum_{i=1}^{n} \delta(k_i - k)$$

where k_i is the degree of node i (i.e. the number of associated edges) and $\delta(\cdot)$ denotes Kronecker's delta function.

Fundamental to modern developments of network science was the theory of random graphs. In this domain, key were the contributions of Edgard Gilbert, Paul Erdős, and Alfréd Rényi. For a set of n nodes, Gilbert's random graph $G(n, p)$ is characterized by the fact that the presence of an edge occurs independently from the others and with probability $p \in (0, 1)$ [53]. Instead, Erdős-Rényi random graph $G(n, m)$ is obtained by assigning an equal probability of occurrence to graphs with m edges [43].

Throughout the years, random graph theory greatly evolved thanks to a constellation of mathematical results and characterizations [22, 23, 126]. At the same time, considerable scientific interest focused on the relationships between this theory and real-world complex systems [121], notably in terms of the balance between network regularity and disorder properties as empirically observed. Dorogovtsev and Mendes [42] highlight that *"the most important natural and artificial networks have a specific architecture based on a fat-tailed distribution of the number of connections of vertices"*. Among the relevant features found in real-world networks were for instance *hubs* (nodes with comparatively higher connectivity with respect to the others) and *motifs* (statistically recurrent structural patterns) [82]. See also [84] for an account about further properties detected in real networks, e.g. in terms of transitivity, clustering, and degree distributions. As a result of those observations, alternatives were proposed in the literature to both classical random graphs and lattice networks [34].

Seminal paper [125], starting from findings related to real-world networks and their commonalities, introduced the so-called Watts-Strogatz model, which encodes rules for the generation of graphs displaying intermediate characteristics in-between classical random graphs (small characteristic path lengths) and regular ring lattices (high clustering). These features were expressed in terms of a small-world property, characterized by the fact that the mean shortest path increases slowly enough with respect to the number of nodes. The reference also analyzes the significance of such class of graphs to represent the actual connectivity of dynamical systems.

Moreover, [10] observed, across a number of domains, the emergence of scale-free power-law degree distributions in large networks and the presence of considerable heterogeneity in node degrees, with the appearance of hubs. A power-law degree distribution can be expressed in terms of proportionality $P(k) \propto k^{-\zeta}$, where degree exponent ζ in many real-world networks takes values in the range $(2, 3)$. These considerations led to the detection of self-organizing criteria in networks and supported the formulation of the Barabási–Albert model, wherein network generation is scale-free (i.e. the asymptotic degree distribution follows a power-law) and based on a preferential attachment mechanism [2]. Such a mechanism was characterized analytically, e.g. in [24], and complemented by alternative scale-free criteria and extensions [85]. Small-world networks gather, indeed, a variety of particular network types; in this sense, [8] qualifies the three classes of scale-free networks, broad-scale networks, and single-scale networks, and notices that the emergence of a specific class depends on the constraints affecting a system's constitution. In the case of technological infrastructures, in particular, scale-free topologies are often observed, due to technological constraints limiting the growth of node degrees.

The mentioned references and a number of cognate contributions propelled extensive research developments in the area of network modeling and analysis. Progress was also enabled by dedicated research projects as well as the increasing availability of datasets and computational resources. Moreover, the transversality of network science to disciplines allowed to point out surprising aspects such as topological affinities among disparate complex systems. In recent years, different categorizations have been proposed for key real-world networks, such as in the case of the social, information, technological and biological networks presented in [84]. Also, considerable advancements have been registered in empirically assessing the proximity of particular categories to specific graph families, as well as their peculiarities and distinctive features. NCIs are also involved in this trend as some of the most appealing application areas; in Table 1, we report a non-exhaustive literature mapping related to this aspect.

An significant property of NCIs is that, in many cases, they belong to the class of spatial networks, which holds particular interest in a number of recent studies [14, 15]. Among the root causes determining the attributes of this kind of

Table 1 Assessing the proximity of empirical CI networks to particular graph families: a literature mapping of selected CIs mentioned in Section 3.1 of the *2004 EC-CIP Communication*

CI (2004 EC-CIP Communication)	Literature references
Energy installations and networks (e.g. electrical power, oil and gas production, storage facilities and refineries, transmission and distribution system)	[5, 35, 65, 90, 100]
Communications and Information Technology (e.g. telecommunications, broadcasting systems, software, hardware and networks including the Internet)	[3, 10, 46]
Finance (e.g. banking, securities and investment)	[20, 38, 60, 103, 110]
Water (e.g. dams, storage, treatment and networks)	[54, 127]
Transport (e.g. airports, ports, intermodal facilities, railway and mass transit networks, traffic control systems)	[12, 63, 72, 107, 113]
Government (e.g. critical services, facilities, information networks, assets and key national sites and monuments)	[39]

systems, we can count factors such as economy of wiring and geometric constraints. Observe that, in this context, transportation networks may be generally intended as *"structures that convey energy, matter, or information from one point to another"* [14]. As for the key theoretical models for spatial networks, the latter reference discusses in particular: geometric graphs (wherein the presence of edges depends on geometric criteria); spatial generalizations of Erdos-Rényi networks; variants of the Watts-Strogatz model; spatial extensions of the Barabasi-Albert growth model; optimal networks.

Networks should also be considered as time-evolving entities, whose scaling characteristics are worthy of investigation. Thanks to increasing dataset availability, this aspect is being studied for many systems including infrastructures. A corpus of literature in this domain relates to the evolution of urban systems and supporting services. For instance, [71] addresses the existence of scaling laws in urban supply systems, inferring from empirical data the presence of power-law distributions of relevant quantities as a function of population size; the authors also observe the presence of sublinear and superlinear scaling in some sectors, depending on criteria such as economy of scale. The temporal analysis of urban infrastructures also involves transportation networks. In [112], the evolution of the road mesh is scrutinized in the light of the two elementary processes of densification around existing urban centers and exploration leading to evolving urbanization, with different pace in time and spatial signatures. See also [55, 80, 81, 104] for further discussion. Other areas of empirical investigation from the evolution viewpoint involve, for instance, air transport systems [61, 122] and the internet [92]. Moreover, network transformation is being addressed also in terms of aging and survivability [76].

Alternative network representations may coexist and provide various meaningful abstractions of a given system's architecture. For instance, referring in particular to road networks, [14] makes reference to primal representations (wherein nodes and links portray intersections and road segments, respectively) and dual representations (wherein nodes are associated to roads and links express whether or not joining intersections exist) developed in the literature. In turn, the mentioned alternatives may be accompanied by different statistical properties. These observations shed light on the fact that, in general, transposing the system under analysis into a network representation may involve an array of choices, strategies, and assumptions.

2.2 *Network Analysis: Topology and Processes*

The investigation of constitutive criteria and models of real networks was also integrated by the development, in parallel with spectral graph theory, of network metrics and diagnostics. In addition to the mentioned aspect of degree distributions, features of interest include for instance centrality, betweenness, clustering, closeness, and efficiency. A vast number of references cover specific features and comparative aspects of the different tools developed in this domain, see for instance [85] and related literature for an introduction.

Analytical characterizations also supported the study of network structures under the effect of impacting perturbations. Interesting results, in this sense, have quantified the performance gaps between different classes of networks. For instance, it was observed in [4] that homogeneous Erdos-Rényi networks are very vulnerable to random failures, whilst heterogeneous scale-free networks display considerable robustness; at the same time, in the latter case error tolerance comes at the price of a reduced survivability to attacks targeting the most connected nodes. Among the key techniques used in network perturbation assessment appears percolation analysis, which borrows ideas from statistical physics and aims at exploring failure phenomena in terms of phase transition points leading to connectivity loss and other topological alterations, supporting fragility assessment of network architectures. Methods such as percolation analysis have also demonstrated their relevance in a risk analysis perspective. For instance, [74] exploits percolation theory for network reliability analysis, taking into account both random and real network models and considering different lifetime distributions for nodes and edges.

As a complement to the study of the topological principles of real-world networks, the research community intensified the investigation of dynamical processes taking place over them. The literature addressing this topic is quite extensive and spans different fields, see [13] for an introduction. Reaction-diffusion, contagion, cascading failures, search, dynamical decentralized decision, flow problems are just some examples of phenomena of interest. Vespignani [120] observes how the discipline developed in time, building on the study of macro-phenomena emerging from the specification of micro-processes and trending towards data-driven modeling approaches incorporating multiple degrees of granularity. Moreover, beyond process modeling over networks, a recent research thread is about the characterization of dynamic networks from a control perspective. This step involves a dialog between network science and systems theory, and one of the key concerns in the literature is about coping with limited-quality parameter specifications often found in applications. Some of the primary references in this area exploit ideas from structural control theory, notably the concept of structural controllability [75]. This allows to characterize the interplay of network architectures and dynamical features of nodes in the awareness of the above-mentioned constraints, for instance in order to define the minimal set of nodes allowing to fully control an assigned network and to address affine aspects, including controllability robustness, accessibility, and observability [77–79, 97].

Observe that, in the area of network-based process representation, a number of studies relevant to NCIs have addressed disaster spreading and response mechanisms. Notable examples include [30], where nodes are modeled as bistable dynamic components and edges induce delayed interactions. In [31], efficiency in recovery strategies is analyzed in terms of the effective distribution of resources across network elements, taking into account the topology of the system. Also, [89] expands on [30] by taking into account the role of redundancies in limiting the spread of perturbations. Overall, the literature addressing processes and resilience aspects over networks in the particular case of NCIs is under extensive development nowadays, see also [88] for a review.

3 Multilayer Networks

3.1 A Change in Perspective

In time, the need emerged to accommodate multiple kinds of entity-to-entity relationships in a given network representation. This led to the concept of network layering. In accordance with [21], we can express the graph structure underlying a multilayer network in terms of pair $\mathscr{M} = (\mathscr{G}, \mathscr{E})$, where $\mathscr{G} = \{G_\alpha, \alpha \in \{1, \ldots, l\}\}$ collects graphs $G_\alpha = (N_\alpha, E_\alpha)$ describing l layers, while $\mathscr{E} = \{E_{\alpha,\beta} \subseteq N_\alpha \times N_\beta, \alpha \neq \beta \in \{1, \ldots, l\}\}$ represents interlayer connections. The concept of multilayer network is quite broad and involves a variety of subcategories. According to the definitions proposed in [124], in particular,

- *multiplex networks* are characterized by layers sharing the same node set (i.e. $N_\alpha = N_\beta, \forall \alpha \neq \beta \in \{1, \ldots, l\}$), but displaying different link configurations among nodes;
- *interdependent networks* have disjoint node sets and dependency links relate nodes belonging to different layers;
- in *interconnected networks*, node sets are once again disjoint while the existing edges among nodes belonging to different layers are physical.

For a broader review of multilayer networks categories and definitions, including temporal networks, see also [21, 66].

In recent years, the theory of multilayer networks was boosted and a number of mathematical frameworks such as the tensor-based characterization proposed in [37] allowed the extension and generalization of some key concepts and results related to single-layer networks, as well as the definition of methods to extract information inherent to this kind of network representations. At the same time, many network-theoretical concepts mentioned in the previous section have been extended to the case of multilayer networks, e.g. network metrics [37], percolation analysis [21, 52, 109] and controllability results [96]. Observe that, in some areas, special aspects of interest emerge in relation to the multilayer nature of the system; this is the case, for instance, of percolation in antagonistic networks [21].

Growth criteria for multilayer networks were studied both from the theoretical point of view and in practical cases. Kim and Goh [64] addressed the coevolution of network layers, i.e. the entangled effects induced by the change of a layer on interdependent ones, together with the resulting effects on cascade dynamics. Wang et al. [123] studied coevolutive self-organization of multilayer networks towards optimality. Nicosia et al. [86] analyzed growth in multiplex networks as a function of interlayer couplings and node arrival times, while [87] introduced a nonlinear preferential attachment model able to stylize the appearance of different types of degree distributions and degree correlations among layers. Game-theoretical concepts were also recently proposed, such as in the case of evolutionary games [124]. In [105], the creation of new routes in multilayer transportation networks was interpreted in terms of a trade-off between efficiency and competition and supported with empirical observations on continental air transportation systems.

A change in perspective is called for when analyzing fragility from a multilayer networks viewpoint. Vespignani [119] observes that *"understanding the fragility induced by multiple interdependencies is one of the major challenges in the design of resilient infrastructures"*. The possibility that vulnerability may emerge from the coupling of layers has both analysis and design implications. In understanding multilayer systems fragility, in particular, significant was the observation that many statements true for single networks, e.g. in terms of failure tolerance, may become untrue when considering interdependent networks [29]. In this reference, percolation analysis is exploited to shed light on this issue by considering two interdependent networks. Moreover, [51] proposes an exact percolation law for a set of interdependent networks that generalizes the classical percolation theory devoted to the single-layer case. While the latter contribution focuses on random attacks, in [41] different types of targeted attacks are accounted for and vulnerability is related to layer centrality with respect to the overall system.

3.2 NCIs as Multilayer Networks

In the NCI domain, the concept of multilayer networks has been first of all approached from a risk analysis perspective. For instance, in [32] risk in interdependent infrastructures is studied in terms of events distribution probability as a function of coupling. Dependency risk paths in NCIs have been studied in [69, 70] through a risk-based methodology, allowing to determine higher-order dependency risks exceeding a threshold. Furthermore, [111] introduces criteria based on centrality metrics for the prioritization of risk mitigation measures. The topic is also contextualized, in [59], to the global scale with the associated globally networked risks, pointing out a number of potential drivers of systemic instabilities.

Interdependency modeling is at the core of today's CI analysis [93, 106]. Particularly relevant to the case of NCIs is the study of cascading events propagating across layers through dependency links. This topic is being addressed both from a static and a dynamic perspective by many scholars [88]. Also, depending on the type of infrastructures involved in the analysis, special factors such as the presence and relevance of buffering elements may determine the proper model choice [114].

One of the key applications is related to the power grid and related systems, e.g. the communication infrastructure. This is considered, for instance, in the above-mentioned reference [29]. In addition, in [101] the impacts of electrical grid failures on the telecommunication network is studied through a set of dynamical models, considering interlayer connections based on geographical proximity and a thresholding criterion depending on the number of active electrical nodes supporting the communication nodes; considerable amplification of effects is then pointed out even at moderate degrees of coupling. Distributed supervisory control architectures and resulting fragility properties are addressed in [83]. Bidirectional couplings are kept into account in [91], in order to determine the minimum set of node failures leading to the full blackout of both networks. In [50], power-ICT interdependencies are combined with an analysis of control strategies, while [115] addresses the related topic of controllability. Korkali et al. [68] warn against conclusions on cascading characteristics detected through simple topological models, comparing different interdependency models and observing how increased coupling between the power and communication layers may reduce vulnerability. A further aspect of interest is about cybersecurity, such as in the case of multilayer smart grid architectures [9, 128]. Other types of power grid-related analyses are about coupled power grids. For instance, [28] addresses cascading through a Bak-Tang-Wiesenfeld sandpile model observing compromises induced by the degree of connectivity and addressing cascade suppression, while [129] studies cascade-safe operating margins. Further aspects of interest relate the electrical grid to other infrastructures, also considering the economic perspective [47, 62].

Beyond electrical grid-related applications, we can observe huge interest in many other areas such as transportation and financial modeling [21]. The latter aspect, in particular, will be the object of the next section, where we consider network theory in its governance implications, referring in particular to its relevance in conjunction with stress testing methodologies in the banking sector.

4 Network Theory and Stress Tests in NCIs Governance: Lessons from the Banking Sector[4]

4.1 Micro and Macro Stress Testing

In economy, the use of stress testing methods is generally led by the idea of assessing the stability of financial institutions and instruments under shocks of various nature. In particular, analyzing specific stress scenarios might provide insights on adequate preventive measures to counteract potential future shocks displaying similar hallmarks. The interest in these methods partially stems from the fact that, as scholars suggest, *"while markets, asset types, players involved and the triggering events differ from one episode to another, risk accumulation cycles tend to be similar"* [98].

In [25], it is observed that stress tests in finance were originally instrumental *"to simulate the performance of individual portfolios and to gauge the stability of individual institutions"* (micro stress testing). This kind of stress tests served to banks as a micro-prudential complement to their internal models, helpful in order to enhance their own risk management capabilities. In this approach, stress tests are aiming at assessing the performance of single entities with relatively well-defined boundary conditions. However, the evolution of the banking system and of most economic and technological sectors in general dictated, in recent years, the

[4]The discussion proposed in Sects. 4 and 5 is partially based on [48, 49], which the reader is referred to for further discussion.

need to widen the scope of such tests. To this end, stress testing has been recently extended to address also the stability of sets of financial institutions impacting the overall economy (macro stress testing). As such, it represents a key tool in the general framework of macro-prudential analysis and it is strongly motivated by recent prudential bank regulation acts such as the Basel Accords. The objective is the assessment and monitoring of financial systems based on both quantitative and qualitative information about the institutional and regulatory framework [118]. From a network-theoretical perspective, this interpretation of macro stress tests suggests the need to analyze a multilayer network involving multiple governance components and financial dependencies.

Interesting insights have been provided in the literature about the concept of fragility and its relevance to stress testing. In [25], it is observed that *"the essence of financial instability is that normal-size shocks cause the system to break down"*. The authors also state that *"an unstable financial system is a fragile financial system; it is not one that would break down only if hit by severe macroeconomic shocks"*. Moreover, *"financial crises generally do not begin after output has collapsed, but before it contracts significantly"*. According to [98], the main components leading to a crisis are identified in the shock itself plus the contagion channels representing the propagation mechanisms for financial instability, see also [36]. The importance of stress testing is thus comprising elements of risk and fragility analysis of financial institutions. In particular, stress tests are most relevant for regulatory authorities, in order to improve awareness of the potential channels through which a systemic crisis may appear and propagate, ultimately allowing to identify the risk of distress and to evaluate countermeasures.

4.2 Merging Stress Testing Principles and Network Theory

In recent literature, considerable efforts are devoted to the analysis of the network architecture of financial institutions and the resulting properties, e.g. in terms of contagion risks and patterns [7, 108]. The necessity to bring a systemic viewpoint into risk analysis of the banking system was considerably emphasized during the last years, especially following the global financial crisis emerged in years 2007–2008. Haldane and May [57] discuss the interplay between complexity and stability in the system and the importance of adopting such a perspective, outlining fundamental implications towards the elaboration of public policies. These include, among other actions, the shaping of the financial network topology based on an assessment of its characteristics and involving the promotion of modular configurations able to mitigate potential cascades. Also, references [58, 59] point out how nowadays our highly interconnected societies exhibit complexity factors that are neither fully understood nor fully controlled. To a certain extent, this explains the underlying reason for which crises may begin even before the appearance of a significant disruptive event, as mentioned above. In turn, this may be the result of deep and sometimes overlooked interactions between elements of the system, which may lead to systemic instabilities and crises even in the event of small perturbations. This highly non-linear behavior is characteristic of complex multilayer systems, and applying control mechanisms seems a non-trivial task especially in the absence of an extensive knowledge of their architecture. The study of the correlation between interbank connectivity and contagion risks has revealed phase transition properties, characterized by enhanced stability against small and infrequent shocks in highly connected networks and inverse properties beyond certain thresholds [1].

Despite the importance of elaborating models describing the complexity of the banking system, the actual representation of interconnections by means of single or multilayer networks remains a challenge, also taking into account the continuous evolution and volatility of their topologies. One key research direction is about the formulation of generating criteria allowing the construction of random financial networks consistent with experimentally observed features, e.g. such as resulting from balance sheet information. Notable contributions in this direction include for instance [1, 26, 33, 56].

While many approaches focus on a single-layer representation of the system, see e.g. [26, 117], some recent studies move towards multilayer models. The necessity to grasp and quantify the strong interactions between layers is demonstrated, for instance, in [95]. Therein, the authors aim at quantifying the importance of an accurate representation of multilayer networks with a particular focus on systemic risk and its implications for financial crises. They provide substantial arguments on the importance of capturing the interplay of the various actors involved in the financial system who are linked in several different ways. In the proposed analysis of the Mexican banking system, the authors show how failing to capture the dynamics of the different layers can result in an underestimation of systemic risk up to 90%. Further relevant literature on multilayer modeling includes [6, 11, 27, 67].

Beyond network modeling, another key objective of the recent research is about the study of shock propagation and associated risks. In [16], a very interesting approach is presented on how to leverage graph theory and centrality metrics in order to quantify systemic risk. As an alternative to the standard notion of too-big-to-fail, the authors propose the concept of too-central-to-fail. This is associated to the introduction of a new network-based metric (DebtRank centrality) which aims at quantifying the importance of a node on the basis of the fraction of the total economic value in the network that is potentially affected by its distress or default. Controllability concepts affine to those discussed above in this chapter have also been exploited, in [40], to detect driver financial institutions.

More and more, the recent literature emphasizes the increasing importance of complexity analysis and awareness in financial regulation. This becomes primary in order to detect near-collapse situations, tipping points, and warning signals [18]. These aspects and the actual dynamics of recent crises seem to question the validity of the mentioned too-big-to-fail paradigm and network-theoretical concepts are holding the interest in order to evolve our approach to the problem, including stress testing. In this perspective, [73] aims at providing a methodology for stress testing rooted in graph theory. Most importantly, the proposed technique aims to capture systemic changes of the banking system on the basis of available datasets. A stress testing framework was also proposed in [17] by exploiting the DebtRank approach discussed above. The objective is allowing the analysis of various stress scenarios taking into account multi-order effects, particularly first-order effects (shocks to external assets), second-order effects (interbank distress) and third-order effects (fire sales). In this sense, the methodology extends traditional risk measures such as VaR and CVaR. Simulations based on realistic network representations suggest the dominance of high-order effects in determining systemic risk. Related applications involve, for instance, the evaluation of the implications of climate risks on the financial network [19]. Also, [94] proposes to combine network theory and credit risk techniques to perform bank ranking and determine resilience estimates. While the credit risk component brings in the assessment of default probabilities and of the tendency of institutions to fail together, the network component focuses on distress propagation.

In conclusion, bridging the gap between network theory and stress tests seems a game changer in the analysis of complex systems, such as in the banking sector. Firstly, this approach can allow capturing the interactions between the components of a complex system, specifically for the considered scenarios and possibly also against a number of other plausible shocks. Secondly, it can allow to track the time evolution of networks and make inferences on future trends. Thirdly, it can reveal vulnerabilities related to the scenario at hand which may not be adequately depicted in traditional representations. Ultimately, joining network theory to stress testing principles seems a very powerful technique to shed light on complex infrastructures, systemic risks and the emergence of cascade effects once a triggering event occurs.

5 Considerations on Stress Testing in NCIs

In the previous section, we outlined the links between network theory and stress tests in banking. Formulating stress testing methodologies and transforming them into a fully operational concept for a broader array of infrastructures remains a challenge, which recent contributions address with increasing interest [44]. At the European level, stress tests with respect to the security of supply are requested by the Member States in order to identify vulnerabilities, such as in the case of the European gas infrastructure.[5] The example provided by the mentioned sectors can help to identify elements to be considered for conducting stress tests able to reveal and test interdependencies among NCIs. In particular, the work of [25] identifies four components for the financial sector which can be extended to other categories of NCIs.

- **The set of risk exposures** These define stress extent and focus of the analysis. Often, in the case of banks, they basically refer to credit risk exposures; however,

[5] https://ec.europa.eu/energy/en/news/stress-tests-cooperation-key-coping-potential-gas-disruption

in advanced stress tests, they can also incorporate further risk components (e.g. market risk, liquidity risk, interbank contagion risk). In the case of NCIs, a robust analysis should start with a comprehensive list of risks to be considered in the scenario formulation.

- **The scenario** This component specifies the (exogenous) shocks affecting the exposures specified for the analysis. We can distinguish between the analysis of the impact of a single risk factor (sensitivity analysis) or of a multivariate scenario involving simultaneous changes in several risk factors (scenario analysis) [98]. The single-factor option usually allows faster computation while its usefulness may be narrow. On the other side, multi-factor tests can introduce both computational burdens and issues related to their compatibility with the stakeholders' assessment methods. In the definition of stress test scenarios, a common issue is about the choice of the severity of the stress level, which usually highly impacts the results of the analysis in view of the inherent nonlinearity of the system. As mentioned above, a common ground for stress test specification is that stress scenarios should be severe yet plausible [98]. In [118], in particular, a distinction is drawn, based on the level of shocks and scenario calibration criteria, between

  1. the worst case approach, based on searching for the most severe scenario while having a certain minimum degree of plausibility;
  2. the threshold approach, whose objective is to search for the largest shocks that let the system perform above a given threshold.

  While the second approach can provide advantages by eliminating the need for scenario calibration, it can be more computationally intensive when multiple risk factors are involved. The latter distinction has also implications on the way stress tests analysis results can be presented. The definition of the plausibility degree is in itself a matter of discussion. In practice, choices can vary according to circumstances specific to the moment when the analysis is performed and to the stress definition criteria (e.g. historical-data-based scenarios, hypothetical plausible and worst case scenarios). Typically, in current stress testing methodologies, institutions consider a set of possible scenarios with different degrees of severity for their analyses.
- **The model** This element maps the scenario (and the macroeconomic conditions it induces) to the outcome, and its choice can constitute a highly complex task, as usually confirmed in the case of macroeconomic analysis. We can summarize some of the main difficulties in constructing suitable macro stress testing models, see for instance [25]:

  1. taking into account the non-linearities affecting the system, especially during crisis periods which can strongly affect its dynamic behavior;
  2. incorporating a non-trivial description of feedback effects;
  3. detecting and modeling endogenous risk factors, which can produce crisis events even with moderate shocks and under otherwise favorable circumstances;
  4. taking into account how the endogenous vulnerabilities build up.

- **The outcome** It describes the impact of the considered scenario. The outcome is typically evaluated in terms of the impact of the scenarios on the level of functionality and service provision at the individual level, while it should help unravel complexity and hidden dependencies that lead to highly nonlinear behaviors. The outcome should also be mapped with respect to the expected output, in order to assess to which extent the current understanding of interdependencies reflects reality. At the same time, it should be also compared with past tests in order to identify trends with respect to the evolution of network complexity over time.

6 Conclusions

In this chapter, we addressed recent developments in network theory and stress testing applied to NCIs. Our discussion started with an overview of key methodological steps leading from the formulation of the classical random networks theory to the study of complex systems and the introduction of the concept of multilayer networks. The evolution of the discipline was complemented by the empirical observation of key real-world complex systems, which in many cases are indeed infrastructures. Moreover, methods for the analysis of both network architectures and associated processes allowed the study of phenomena, such as cascading events, that represent key aspects in today's protection- and resilience-oriented policies. At the same time, network theory is acquiring a role in supporting the governance of NCIs throughout their evolution in time. This aspect finds confirmation in emerging stress testing techniques, which aim at evaluating the operational boundaries of complex systems and have witnessed huge interest, in recent years, especially in banking and macro-prudential regulation. In this area, traditional risk analysis approaches are being more and more blended with network-based analysis. This trend can also provide input and guidance for the development of stress tests devoted to other families of high-complexity infrastructures.

References

1. Acemoglu D, Ozdaglar A, Tahbaz-Salehi A (2015) Systemic risk and stability in financial networks. Am Econ Rev 105(2):564–608
2. Albert R, Barabási AL (2002) Statistical mechanics of complex networks. Rev Mod Phys 74(1):47
3. Albert R, Jeong H, Barabási AL (1999) Internet: diameter of the world-wide web. Nature 401(6749):130
4. Albert R, Jeong H, Barabási AL (2000) Error and attack tolerance of complex networks. Nature 406(6794):378–382
5. Albert R, Albert I, Nakarado GL (2004) Structural vulnerability of the north american power grid. Phys Rev E 69(2):025103
6. Aldasoro I, Alves I (2018) Multiplex interbank networks and systemic importance: an application to European data. J Financ Stab 35:17–37

7. Allen F, Babus A (2009) Networks in finance. The network challenge: strategy, profit, and risk in an interlinked world, vol 367. Wharton School, Upper Saddle River
8. Amaral LAN, Scala A, Barthelemy M, Stanley HE (2000) Classes of small-world networks. Proc Natl Acad Sci 97(21):11149–11152
9. Ancillotti E, Bruno R, Conti M (2013) The role of communication systems in smart grids: architectures, technical solutions and research challenges. Comput Commun 36(17–18):1665–1697
10. Barabási AL, Albert R (1999) Emergence of scaling in random networks. Science 286(5439):509–512
11. Bargigli L, Di Iasio G, Infante L, Lillo F, Pierobon F (2015) The multiplex structure of interbank networks. Quant Finan 15(4):673–691
12. Barrat A, Barthelemy M, Vespignani A (2007) The architecture of complex weighted networks: measurements and models. In: Caldarelli G, Vespignani A (eds) Large scale structure and dynamics of complex networks: from information technology to finance and natural science, pp 67–92. World Scientific, Singapore
13. Barrat A, Barthelemy M, Vespignani A (2008) Dynamical processes on complex networks, vol 1. Cambridge University Press, Cambridge
14. Barthélemy M (2011) Spatial networks. Phys Rep 499(1–3):1–101
15. Barthelemy M (2017) Morphogenesis of spatial networks. Springer, Cham
16. Battiston S, Puliga M, Kaushik R, Tasca P, Caldarelli G (2012) Debtrank: too central to fail? Financial networks, the fed and systemic risk. Sci Rep 2:541
17. Battiston S, Caldarelli G, D'Errico M, Gurciullo S (2016) Leveraging the network: a stress-test framework based on debtrank. Stat Risk Model 33(3–4):117–138
18. Battiston S, Farmer JD, Flache A, Garlaschelli D, Haldane AG, Heesterbeek H, Hommes C, Jaeger C, May R, Scheffer M (2016) Complexity theory and financial regulation. Science 351(6275):818–819
19. Battiston S, Mandel A, Monasterolo I, Schütze F, Visentin G (2017) A climate stress-test of the financial system. Nat Clim Chang 7(4):283
20. Bech ML, Atalay E (2010) The topology of the federal funds market. Physica A Stat Mech Appl 389(22):5223–5246
21. Boccaletti S, Bianconi G, Criado R, Del Genio CI, Gómez-Gardenes J, Romance M, Sendina-Nadal I, Wang Z, Zanin M (2014) The structure and dynamics of multilayer networks. Phys Rep 544(1):1–122
22. Bollobás B (1998) Random graphs. In: Bollobás B (ed) Modern graph theory, pp 215–252. Springer, New York
23. Bollobás B (1998) Modern graph theory, vol 184. Springer Verlag, New York
24. Bollobás BE, Riordan O, Spencer J, Tusnády G (2001) The degree sequence of a scale-free random graph process. Random Struct Algoritm 18(3):279–290
25. Borio C, Drehmann M, Tsatsaronis K (2014) Stress-testing macro stress testing: does it live up to expectations? J Financ Stab 12:3–15
26. Boss M, Elsinger H, Summer M, Thurner 4 S (2004) Network topology of the interbank market. Quant Finan 4(6):677–684
27. Brummitt CD, Kobayashi T (2015) Cascades in multiplex financial networks with debts of different seniority. Phys Rev E 91(6):062813
28. Brummitt CD, D'Souza RM, Leicht EA (2012) Suppressing cascades of load in interdependent networks. Proc Natl Acad Sci 109(12):E680–E689
29. Buldyrev SV, Parshani R, Paul G, Stanley HE, Havlin S (2010) Catastrophic cascade of failures in interdependent networks. Nature 464(7291):1025–1028
30. Buzna L, Peters K, Helbing D (2006) Modelling the dynamics of disaster spreading in networks. Physica A Stat Mech Appl 363(1):132–140
31. Buzna L, Peters K, Ammoser H, Kühnert C, Helbing D (2007) Efficient response to cascading disaster spreading. Phys Rev E 75(5):056107

32. Carreras BA, Newman DE, Gradney P, Lynch VE, Dobson I (2007) Interdependent risk in interacting infrastructure systems. In: 40th Annual Hawaii International Conference on System Sciences (HICSS 2007), pp 112–112. IEEE, Los Alamitos
33. Cimini G, Squartini T, Garlaschelli D, Gabrielli A (2015) Systemic risk analysis on reconstructed economic and financial networks. Sci Rep 5:15758
34. Coxeter HSM, Ball WWR (1960) Mathematical recreations essays. Macmillan, New York
35. Crucitti P, Latora V, Marchiori M (2004) A topological analysis of the Italian electric power grid. Physica A Stat Mech Appl 338(1–2):92–97
36. Davis EP (1999) Financial data needs for macroprudential surveillance–what are the key indicators of risks to domestic financial stability? Lecture Series 2. Centre for Central Banking Studies, Bank of England
37. De Domenico M, Solé-Ribalta A, Cozzo E, Kivelä M, Moreno Y, Porter MA, Gómez S, Arenas A (2013) Mathematical formulation of multilayer networks. Phys Rev X 3(4):041022
38. De Masi G, Iori G, Caldarelli G (2006) Fitness model for the Italian interbank money market. Phys Rev E 74(6):066112
39. Del Vicario M, Bessi A, Zollo F, Petroni F, Scala A, Caldarelli G, Stanley HE, Quattrociocchi W (2016) The spreading of misinformation online. Proc Natl Acad Sci 113(3):554–559
40. Delpini D, Battiston S, Riccaboni M, Gabbi G, Pammolli F, Caldarelli G (2013) Evolution of controllability in interbank networks. Sci Rep 3:1626
41. Dong G, Gao J, Du R, Tian L, Stanley HE, Havlin S (2013) Robustness of network of networks under targeted attack. Phys Rev E 87(5):052804
42. Dorogovtsev SN, Mendes JF (2013) Evolution of networks: from biological nets to the internet and WWW. Oxford University Press, Oxford
43. Erdös P, Rényi A (1959) On random graphs, I. Publicationes Mathematicae (Debrecen) 6:290–297
44. Esposito S, Stojadinovic B, Babič A, Dolšek M, Iqbal S, Selva J, Giardini D (2017) Engineering risk-based methodology for stress testing of critical non-nuclear infrastructures (strest project). In: Proceedings of 16th World Conference on Earthquake, 16WCEE
45. Estrada E (2012) The structure of complex networks: theory and applications. Oxford University Press, New York
46. Faloutsos M, Faloutsos P, Faloutsos C (1999) On power-law relationships of the internet topology. In: ACM SIGCOMM Computer Communication Review, vol 29, pp 251–262. ACM
47. Galbusera L, Giannopoulos G (2018) On input-output economic models in disaster impact assessment. Int J Disaster Risk Reduct 30:186–198
48. Galbusera L, Ward D, Giannopoulos G (2014) Developing stress tests to improve the resilience of critical infrastructures: a feasibility analysis. Technical Report, JRC Science and Policy Reports JRC91129, European Commission
49. Galbusera L, Ward D, Giannopoulos G (2014) Stress tests and critical infrastructure protection-resilience. Technical Report, JRC Science and Policy Reports JRC93152, European Commission
50. Galbusera L, Theodoridis G, Giannopoulos G (2015) Intelligent energy systems: introducing power–ICT interdependency in modeling and control design. IEEE Trans Ind Electron 62(4):2468–2477
51. Gao J, Buldyrev SV, Havlin S, Stanley HE (2011) Robustness of a network of networks. Phys Rev Lett 107(19):195701
52. Gao J, Buldyrev SV, Stanley HE, Havlin S (2012) Networks formed from interdependent networks. Nat Phys 8(1):40
53. Gilbert EN (1959) Random graphs. Ann Math Stat 30(4):1141–1144
54. Giudicianni C, Di Nardo A, Di Natale M, Greco R, Santonastaso GF, Scala A (2018) Topological taxonomy of water distribution networks. Water 10(4):444
55. Gudmundsson A, Mohajeri N (2013) Entropy and order in urban street networks. Sci Rep 3:3324
56. Hałaj G, Kok C (2013) Assessing interbank contagion using simulated networks. Comput Manag Sci 10(2–3):157–186

57. Haldane AG, May RM (2011) Systemic risk in banking ecosystems. Nature 469(7330):351
58. Helbing D (2012) Systemic risks in society and economics. In: Social self-organization, pp 261–284. Springer, Berlin/New York
59. Helbing D (2013) Globally networked risks and how to respond. Nature 497(7447):51
60. Iori G, De Masi G, Precup OV, Gabbi G, Caldarelli G (2008) A network analysis of the Italian overnight money market. J Econ Dyn Control 32(1):259–278
61. Jia T, Qin K, Shan J (2014) An exploratory analysis on the evolution of the us airport network. Physica A Stat Mech Appl 413:266–279
62. Jonkeren O, Azzini I, Galbusera L, Ntalampiras S, Giannopoulos G (2015) Analysis of critical infrastructure network failure in the European union: a combined systems engineering and economic model. Netw Spat Econ 15(2):253–270
63. Kaluza P, Kölzsch A, Gastner MT, Blasius B (2010) The complex network of global cargo ship movements. J R Soc Interface 7(48):1093–1103
64. Kim JY, Goh KI (2013) Coevolution and correlated multiplexity in multiplex networks. Phys Rev Lett 111(5):058702
65. Kinney R, Crucitti P, Albert R, Latora V (2005) Modeling cascading failures in the north American power grid. Eur Phys J B Condens Matter Complex Syst 46(1):101–107
66. Kivelä M, Arenas A, Barthelemy M, Gleeson JP, Moreno Y, Porter MA (2014) Multilayer networks. J Complex Netw 2(3):203–271
67. Kok C, Montagna M (2016) Multi-layered interbank model for assessing systemic risk. Technical report, European Central Bank
68. Korkali M, Veneman JG, Tivnan BF, Bagrow JP, Hines PD (2017) Reducing cascading failure risk by increasing infrastructure network interdependence. Sci Rep 7:44499
69. Kotzanikolaou P, Theoharidou M, Gritzalis D (2013) Assessing n-order dependencies between critical infrastructures. Int J Crit Infrastruct 9(1–2):93–110
70. Kotzanikolaou P, Theoharidou M, Gritzalis D (2013) Cascading effects of common-cause failures in critical infrastructures. In: International Conference on Critical Infrastructure Protection, pp 171–182. Springer, Washington, DC
71. Kühnert C, Helbing D, West GB (2006) Scaling laws in urban supply networks. Physica A Stat Mech Appl 363(1):96–103
72. Latora V, Marchiori M (2002) Is the Boston subway a small-world network? Physica A Stat Mech Appl 314(1–4):109–113
73. Levy-Carciente S, Kenett DY, Avakian A, Stanley HE, Havlin S (2015) Dynamical macroprudential stress testing using network theory. J Bank Financ 59:164–181
74. Li D, Zhang Q, Zio E, Havlin S, Kang R (2015) Network reliability analysis based on percolation theory. Reliab Eng Syst Saf 142:556–562
75. Lin CT (1974) Structural controllability. IEEE Trans Autom Control 19(3):201–208
76. Lin Y, Patron A, Guo S, Kang R, Li D, Havlin S, Cohen R (2018) Design of survivable networks in the presence of aging. EPL (Europhysics Letters) 122(3):36003
77. Liu YY, Barabási AL (2016) Control principles of complex systems. Rev Mod Phys 88(3):035006
78. Liu YY, Slotine JJ, Barabási AL (2011) Controllability of complex networks. Nature 473(7346):167
79. Liu YY, Slotine JJ, Barabási AL (2012) Control centrality and hierarchical structure in complex networks. PLoS One 7(9):e44459
80. Louf R, Barthelemy M (2014) A typology of street patterns. J R Soc Interface 11(101):20140924
81. Masucci AP, Stanilov K, Batty M (2013) Limited urban growth: London's street network dynamics since the 18th century. PLoS One 8(8):e69469
82. Milo R, Shen-Orr S, Itzkovitz S, Kashtan N, Chklovskii D, Alon U (2002) Network motifs: simple building blocks of complex networks. Science 298(5594):824–827
83. Morris RG, Barthelemy M (2013) Interdependent networks: the fragility of control. Sci Rep 3:2764

84. Newman ME (2003) The structure and function of complex networks. SIAM Rev 45(2): 167–256
85. Newman M (2010) Networks: an introduction. Oxford University Press, Oxford
86. Nicosia V, Bianconi G, Latora V, Barthelemy M (2013) Growing multiplex networks. Phys Rev Lett 111(5):058701
87. Nicosia V, Bianconi G, Latora V, Barthelemy M (2014) Nonlinear growth and condensation in multiplex networks. Phys Rev E 90(4):042807
88. Ouyang M (2014) Review on modeling and simulation of interdependent critical infrastructure systems. Reliab Eng Syst Saf 121:43–60
89. Ouyang M, Fei Q, Yu MH, Wang GX, Luan EJ (2009) Effects of redundant systems on controlling the disaster spreading in networks. Simul Model Pract Theory 17(2):390–397
90. Pagani GA, Aiello M (2013) The power grid as a complex network: a survey. Physica A Stat Mech Appl 392(11):2688–2700
91. Parandehgheibi M, Modiano E (2013) Robustness of interdependent networks: the case of communication networks and the power grid. In: Global Communications Conference (GLOBECOM). IEEE, Piscataway, pp 2164–2169
92. Pastor-Satorras R, Vespignani A (2007) Evolution and structure of the internet: a statistical physics approach. Cambridge University Press, Cambridge
93. Pederson P, Dudenhoeffer D, Hartley S, Permann M (2006) Critical infrastructure interdependency modeling: a survey of us and international research. Idaho National Laboratory, pp 1–20
94. Petrone D, Latora V (2018) A dynamic approach merging network theory and credit risk techniques to assess systemic risk in financial networks. Sci Rep 8(1):5561
95. Poledna S, Molina-Borboa JL, Martínez-Jaramillo S, Van Der Leij M, Thurner S (2015) The multi-layer network nature of systemic risk and its implications for the costs of financial crises. J Financ Stab 20:70–81
96. Pósfai M, Gao J, Cornelius SP, Barabási AL, D'Souza RM (2016) Controllability of multiplex, multi-time-scale networks. Phys Rev E 94(3):032316
97. Pu CL, Pei WJ, Michaelson A (2012) Robustness analysis of network controllability. Physica A Stat Mech Appl 391(18):4420–4425
98. Quagliariello M (2009) Stress-testing the banking system: methodologies and applications. Cambridge University Press, Cambridge
99. Rinaldi SM, Peerenboom JP, Kelly TK (2001) Identifying, understanding, and analyzing critical infrastructure interdependencies. IEEE Control Syst 21(6):11–25
100. Rosato V, Bologna S, Tiriticco F (2007) Topological properties of high-voltage electrical transmission networks. Electr Power Syst Res 77(2):99–105
101. Rosato V, Issacharoff L, Tiriticco F, Meloni S, Porcellinis S, Setola R (2008) Modelling interdependent infrastructures using interacting dynamical models. Int J Crit Infrastruct 4(1–2):63–79
102. Rosato V, Meloni S, Simonsen I, Issacharoff L, Peters K, Von Festenberg N, Helbing D (2008) A complex system's view of critical infrastructures. In: Helbing D (ed) Managing complexity: insights, concepts, applications, pp 241–260. Springer, Berlin
103. Roukny T, Georg CP, Battiston S (2014) A network analysis of the evolution of the German interbank market. Technical report, Discussion Paper, Deutsche Bundesbank
104. Samaniego H, Moses ME (2008) Cities as organisms: allometric scaling of urban road networks. J Transp Land Use 1(1):21–39
105. Santoro A, Latora V, Nicosia G, Nicosia V (2017) Pareto optimality in multilayer network growth. arXiv preprint: 1710.01068
106. Satumtira G, Dueñas-Osorio L (2010) Synthesis of modeling and simulation methods on critical infrastructure interdependencies research. In: Sustainable and resilient critical infrastructure systems, pp 1–51. Springer, Berlin/Heidelberg
107. Scellato S, Cardillo A, Latora V, Porta S (2006) The backbone of a city. Eur Phys J B Condens Matter Complex Syst 50(1–2):221–225

108. Silva W, Kimura H, Sobreiro VA (2017) An analysis of the literature on systemic financial risk: a survey. J Financ Stab 28:91–114
109. Son SW, Bizhani G, Christensen C, Grassberger P, Paczuski M (2012) Percolation theory on interdependent networks based on epidemic spreading. EPL (Europhysics Letters) 97(1):16006
110. Soramäki K, Bech ML, Arnold J, Glass RJ, Beyeler WE (2007) The topology of interbank payment flows. Physica A Stat Mech Appl 379(1):317–333
111. Stergiopoulos G, Kotzanikolaou P, Theocharidou M, Gritzalis D (2015) Risk mitigation strategies for critical infrastructures based on graph centrality analysis. Int J Crit Infrastruct Prot 10:34–44
112. Strano E, Nicosia V, Latora V, Porta S, Barthélemy M (2012) Elementary processes governing the evolution of road networks. Sci Rep 2:296
113. Strano E, Viana M, da Fontoura Costa L, Cardillo A, Porta S, Latora, V (2013) Urban street networks, a comparative analysis of ten European cities. Environ Plann B Plann Des 40(6):1071–1086
114. Svendsen NK, Wolthusen SD (2007) Connectivity models of interdependency in mixed-type critical infrastructure networks. Inf Secur Tech Rep 12(1):44–55
115. Theodoridis G, Galbusera L, Giannopoulos G (2015) Controllability assessment for cascade effects in ICT-enabled power grids. In: International Conference on Critical Information Infrastructures Security, pp 147–158. Springer, Berlin
116. Toroczkai Z, Vespignani A (2016) Understanding the fundamental principles underlying the survival and efficient recovery of multi-scale techno-social networks following a WMD event (a). Technical report, University of Notre Dame Du Lac Notre Dame United States
117. Upper C, Worms A (2004) Estimating bilateral exposures in the German interbank market: is there a danger of contagion? Eur Econ Rev 48(4):827–849
118. Čihák M (2004) Stress testing: a review of key concepts. Research and Policy Notes 2004/02, Czech National Bank, Research Department. http://ideas.repec.org/p/cnb/rpnrpn/2004-02.html
119. Vespignani A (2010) Complex networks: the fragility of interdependency. Nature 464(7291):984–985
120. Vespignani A (2012) Modelling dynamical processes in complex socio-technical systems. Nat Phys 8(1):32
121. Vespignani A (2018) Twenty years of network science. Nature 558:528–529
122. Wandelt S, Sun X, Zhang J (2017) Evolution of domestic airport networks: a review and comparative analysis. Transportmetrica B Trans Dyn 13:1–17
123. Wang Z, Szolnoki A, Perc M (2014) Self-organization towards optimally interdependent networks by means of co evolution. New J Phys 16(3):033041
124. Wang Z, Wang L, Szolnoki A, Perc M (2015) Evolutionary games on multilayer networks: a colloquium. Eur Phys J B 88(5):124
125. Watts DJ, Strogatz SH (1998) Collective dynamics of "small-world" networks. Nature 393:440–442
126. West DB et al (2001) Introduction to graph theory, vol 2. Prentice Hall, Upper Saddle River
127. Yazdani A, Jeffrey P (2011) Complex network analysis of water distribution systems. Chaos Interdisciplinary J Nonlinear Sci 21(1):016111
128. Zhang Y, Wang L, Sun W, Green II RC, Alam M (2011) Distributed intrusion detection system in a multi-layer network architecture of smart grids. IEEE Trans Smart Grid 2(4):796–808
129. Zio E, Sansavini G (2011) Modeling interdependent network systems for identifying cascade-safe operating margins. IEEE Trans Reliab 60(1):94–101

Part III
Industrial & Automation Control Systems

Micro-Grid Control Security Analysis: Analysis of Current and Emerging Vulnerabilities

Peter Beaumont and Stephen Wolthusen

Abstract Micro-grids (MG) enable autonomous operation of remote or islanded power networks such as critical infrastructure assets, but also the integration of Distributed Energy Resources (DER) into power distribution networks coupled to a transmission system, enhancing robustness of and reducing strain on the power network. Micro-grid control technology relies heavily on networked or distributed control techniques which are exposed to cyber-security threats than conventional power networks. Consequences of security violations could manifest as loss of electrical power to critical loads or dangerous operating states; given the severity of the risks every effort should be taken to reduce exposure to vulnerabilities. This chapter reviews common MG architectures and selected system elements feeding into a structured high-level security analysis based on the Systems Theoretic Process Analysis for Security (STPA-Sec) framework to generate causal scenarios for security violations in MGs and, accordingly, to identify the priority areas for future security research in the MG domain.

P. Beaumont (✉)
School of Mathematics and Information Security, Royal Holloway, University of London, Egham, UK
e-mail: Peter.Beaumont.2014@live.rhul.ac.uk

S. Wolthusen
School of Mathematics and Information Security, Royal Holloway, University of London, Egham, UK

Department of Information Security and Communication Technology, Norwegian University of Science, Gjøvik, Norway
e-mail: stephen.wolthusen@rhul.ac.uk

D. Gritzalis et al. (eds.), *Critical Infrastructure Security and Resilience*, Advanced Sciences and Technologies for Security Applications,
https://doi.org/10.1007/978-3-030-00024-0_9

1 Introduction

Access to affordable and reliable electrical power is necessary to support economic growth [24] and improvements to quality of life. Consequently, a reliable source of power is an expectation in higher income countries and an aspiration in those regions without it. Delivering against these expectations and aspirations requires power network operators to innovate and to evolve their infrastructure so that networks operate as efficiently as possible whilst delivering sufficient power of a high enough quality to meet demand where-ever it may be needed. Micro-Grids (MG) are a recent product of this innovation and are defined as "*a group of interconnected loads and DER with clearly defined electrical boundaries that act as a single controllable entity with respect to the grid [the utility] and can connect and disconnect from the grid to enable it to operate in both grid-connected or island modes*" [26].

MGs were born out of the requirement to connect increasing numbers of Distributed Energy Resources (DER) to Medium Voltage (MV) electrical distribution networks without compromising the operation of the utility [14], but may also operate in isolation. The non-security benefits of installing DER at the utility level can be summarised as follows:

- Installing generation close to loads (on the edges of the network) increases the installed capacity of national power grids without the need to reinforce transmission networks.
- Increased use of DER reduces the electrical transmission and distribution losses which would be incurred if the same capacity were to be installed centrally.
- Aggregating DER into MGs provides consumers with a means to reduce their dependency on utility companies and to increase their security of supply through diversification [10].
- Enabling DER to operate in parallel with the utility enables the increased penetration of Renewable Energy Sources (RES), thereby helping to contribute toward emissions reduction targets.

Moreover, both security of supply in case of disasters or contingencies are aided by distributed, redundant resources, and as will be shown later, so can the ability to sustain an isolated network such as for critical infrastructures in case of cyber attacks where conventional emergency generation capabilities are insufficient.

DER have many positive characteristics and have the potential to help resolve some of today's most pressing power engineering problems. However, connecting DER to the utility presents technical challenges relating to system management, stability and protection [21]. Meeting these challenges would not be feasible if the utility was required to directly control each individual DER. MGs overcome these problems by aggregating DER and loads into MG systems which are viewed by the utility as a single controllable entity. This process of aggregation significantly eases the control burden on the utility [7, 13]. One corollary also important for cyber security considerations is that the MG assumes responsibility for the control of its DER and loads; ultimately this control is achieved via controllable Power Electronics (PE) which are used to couple DER to the MG.

PE provide the means by which power generated by DC (Direct Current) or variable frequency AC (Alternating Current) sources can be converted so as to be able to operate in parallel with other DER and with the AC supply provided by the utility [4]. Inverters are used as the final stage of DER power conversion in AC MGs and they covert a DC input to an AC output. Inverters are able to control the voltage (V) and frequency (ω) of their terminal values; and can use a changes in ω to modify their phase angle (δ) relative to a reference voltage. Manipulating these terminal values is required to enable parallel operation with other sources. A large part of the successful control of a MG can be described as manipulating the terminal values of each inverter on the MG so that the system as a whole is *stable* and operating *optimally*. This is a key observation: Much of the control effort in a MG is dedicated to regulating the output ω, V and δ of multiple inverters so that they integrate in a manner that is stable and optimal. MG control architectures are designed to achieve this task reliably and are structured hierarchically [2] to enable them to do so.

The hierarchical structure consists of 3 levels, each of which has its own objectives. Data is processed at each level and communication occurs within and between the levels. These actions are enabled via a networked Industrial Control System (ICS) which governs the operation of the MG.

The amalgamation of sources, loads and ICS form a Cyber-Physical System (CPS). Unfortunately, recent attacks [3, 12] against CPS have proven that it is feasible for adversaries to influence the operation of physical systems by corrupting the integrity and/or reducing the availability of data on the ICS at several levels from supply chain to data manipulations. This chapter will explore the opportunities for, and impact of, adversaries attacking a MG, concentrating on civilian MG that may not be fully isolated from external communication.

The remainder of this chapter is structured as follows: Sect. 2 provides an introduction to MGs and discusses architectures, modes of operation and control hierarchies whilst Sect. 3 explains the purpose and operation of the major MG elements. The aim of Sects. 2 and 3 is to elaborate consequences of MG element failure to operate according to their design specification as could occur if they were to be attacked. Based on this, Sect. 4 undertakes a STPA-Sec analysis of a typical MG. The output of the analysis is a set of causal scenarios that describe events that could lead to losses. Causal scenarios are valuable artefacts that can help analysts to better understand the manner in which attackers could achieve their objectives. Improved understanding should then lead to the development of techniques and processes that improve the security of CPS such as MG.

2 Micro-Grid Architecture, Modes of Operation, and Control Hierarchy

A MG is a CPS which employs networked communications to enable the control of physical components, themselves interconnected by a power network. Figure 1 represents a MG as the sum of its cyber and physical components. The figure illustrates a typical MG architecture consisting of multiple electrical feeders, a mix of DER types (including Energy Storage Systems (ESS)), loads and a Micro-Grid Central Controller (MGCC) that is networked to each node in the MG and to the Distribution Management System (DMS) of the utility.

A MG has the primary objective of supplying power of sufficient quality to local loads and to the utility in a manner that is optimal. Power quality refers to power that is reliably supplied within acceptable voltage and frequency bounds—these bounds may be defined by local grid-codes or by the more onerous requirements of specific sensitive loads on the MG (e.g. a micro-grid serving a hospital complex). Optimality is a function of the objectives of the MG operator and the utility which generally relate to cost effective operation and availability of the supply. A MG may also have the secondary objective of providing ancillary services such as voltage and frequency support to the utility [20].

MG architectures are developed in order to meet the design objectives of the specific system [5]. The term architecture refers both to the topology, mix of the

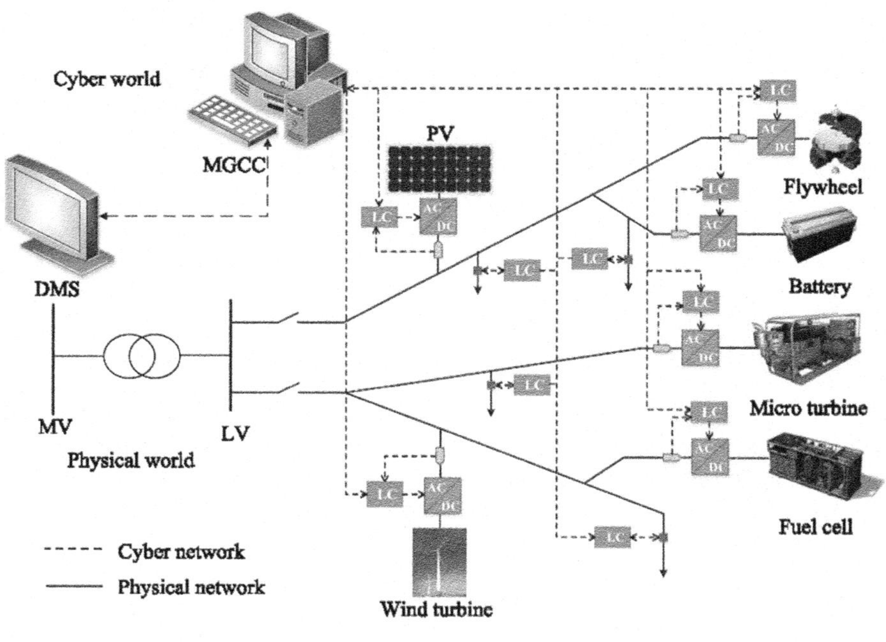

Fig. 1 Representation of a MG as a Cyber Physical System [18]

electrical elements [9] in the MG, and to the control methodologies employed to integrate them. Architectures are highly significant in security assessments as they determine the nature and extent of system vulnerabilities. For example, the mix of DER and load types and the control methodologies employed will dictate system stability margins [8], whilst the use of decentralised or centralised control architectures will dictate the amount and distribution of data traversing the system and the access to it that an attacker may be able obtain. Hence Sects. 2 and 3 will describe the modes of operation, control hierarchy and system elements employed in MGs and the security implications thereof, acknowledging the strong dependence on architecture that may result in some security objectives not being possible to satisfy for a given architecture. The result is that the MG must balance supply and demand locally in order to maintain power quality, which may imply selective load shedding. Transition between modes can occur as the result of either planned or unplanned events. The ability to successfully transition due to an unplanned event enables the MG to continue to supply local loads when there are faults or outages on the utility—this capability is a significant advantage for MG connected loads and enhances reliability of supply. As the two modes of operation are very different, so are the nature and complexity of the control challenges associated with them.

Operating in parallel with the utility in grid-connected mode enables DER to regulate their power output against a reference provided by the utility, thereby removing the requirement for the MG to generate its own voltage reference. It also allows the MG to operate non-dispatchable DER—such as wind-turbines and photovoltaic (PV) arrays—at their maximum power points without the need to consider local power imbalances. This means that in grid-connected mode the MG control problem is largely reduced to that of ensuring that the MG operates economically [17].

In contrast, operating in islanded mode presents a much more challenging control problem. The islanded MG must have mechanisms to:

- establish a local voltage reference
- ensure supply/demand balance of active and reactive power
- share active and reactive load across the DER in order to avoid overloading individual DER

Establishing reference values requires that at least one inverter on the MG fulfils a grid-forming role whilst other inverters either contribute toward grid-forming or to follow the reference [11]. Achieving load/supply balance requires that sufficient DER capacity is available (known as Unit Commitment (UC)), that DER set points are appropriate (known as Economic Dispatch (ED)) for PQ controlled VSI, and that there is sufficient energy stored in ESS to enable the MG to respond to load changes. Achieving accurate load sharing without introducing damaging circulating currents requires that load is shared between DER in proportion to their capacity, whilst voltage control loops accurately track set-points. The control mechanisms that achieve these tasks in islanded mode all present vulnerabilities that could threaten the MG if they were to be exploited.

2.1 Security Implications of Grid-Connected Mode

A MG is less vulnerable to cyber-attack whilst in grid connected mode than it is in islanded mode, i.e. it is more difficult for an attacker to compromise the objectives of the MG in grid-connected mode. This is a consequence of the stabilising effect of the utility. The most significant cyber-physical security implications of a MG operating in grid-connected mode relate to an adversary manipulating data flows and control signals between the tertiary controller/MGCC and the DMS [6]. Modifying these data flows and control signals could change the power flow across the PCC from levels requested by the DMS to levels that are dictated by an attacker. The result could be the reduction of power quality on the utility, in the area of the PCC, as unexpected reactive power flows from the MG counteract the voltage regulation measures employed by the utility, which may result in safe operating paramenters of both loads and the MG being breached.

A further consideration is that adversaries may try to find means of forcing a MG from grid-connected to islanded mode, in order to launch further attacks in the more vulnerable islanded mode; this transition is typically formulated in terms of power quality and hence a consequence of the aforementioned attacks.

2.2 Security Implications of Islanded Mode

MGs are more vulnerable to cyber-attack in islanded mode as stability margins are reduced (due to the absence of the utility) and medium term supply and demand is more finely balanced due to the reliance on stochastic weather and load forecasts that are used to optimise the system and determine UC and ED states (see Sect. 2.6).

The exact vulnerabilities associated with islanded mode operation differ according to system architecture. However, generic vulnerabilities at secondary and tertiary levels include but are not limited to the following examples:

- **Secondary Control: Loss of Integrity or Availability of Secondary Control Variables** MG operation in islanded mode often relies upon droop control methods [25] at the primary level to ensure accurate load sharing. However, droop controls introduce offsets to nominal voltage and frequency values that require correction. These corrections are calculated at the secondary control level and communicated to DER by the MGCC (assuming w.l.o.g. a centralised architecture). Malicious modification of these corrections could introduce voltage profiles across the MG that lead to circulating currents, lead to power oscillations that cause system instability, or result in inadvertent discharge of ESS.
- **Tertiary Control: Loss of Integrity or Availability of Load and Source Forecasts** Inaccurate load and source forecasts would lead to poor system optimisation decisions which may, in turn, result in the capacity of on-line DER (in the case of dispatchable DER) and the State of Charge (SOC) of ESS being insufficient to support loads.

Table 1 Micro-grid control hierarchy and control objectives

Control hierarchy	Response time	Control objectives
Primary	1–10 ms	Voltage and current control and power sharing control
Secondary	100 ms–1 s	Power quality control
Tertiary	s—hours	Micro-grid supervision, utility grid observation, unit commitment, economic dispatch

The key observation with regard to the security implications of the modes of operation is that in islanded mode stability margins are lower and there is a greater flow of data required to ensure stability and optimality. This results in a larger number system vulnerabilities and more opportunities for adversaries to maliciously modify data.

2.3 *Control Hierarchy*

A MG is best considered a system of systems, and implementing effective control is a demanding task. Control of the system as a whole involves the control of each discrete sub-system and then measures to ensure that the sub-systems work together collaboratively to achieve a global objective. In the face of such complexity a divide and conquer approach has been developed to simplify the analysis of the control problem. The result is a conceptual framework which partitions a MG's control objectives into three levels: primary, secondary, and tertiary control [2]. Table 1 summarises the control objectives of each of the 3 levels and details indicative response times. The three-level control hierarchy allows MG control to be analysed in modules, this is particularly useful in security analysis as each control module has different characteristics which lead to different security vulnerabilities.

2.4 *Primary Control*

Primary level control consists of the mechanisms that ensure stable power quality throughout the MG and accurate load sharing between DER. Stability entails fast-acting tracking of V and ω references such that these variables remain within acceptable limits in the event of both small-signal and transient disturbances. Accurate load sharing requires a means of modulating the P and Q power outputs of each DER in response to changes in system demand. Tracking of voltage and frequency references is accomplished by current and voltage loops operating at zero level (a sub-set of the primary control level) and load sharing is most commonly achieved by droop control mechanisms at the primary level in islanded mode.

In grid-connected mode the objective of the MG is to maximise the harvest of RES, this involves injecting the maximum power possible into the MG for the given environmental conditions (wind levels and irradiance). To accomplish this the inverters of RES are operated as Current Controlled Voltage Source Inverters (CCVSI) and power outputs are determined by Maximum Power Point Trackers (MPPT) configured to each DER. This mode of inverter operation is a zero level control function and is referred to as grid-feeding [23]. Primary control for CCVSIs is achieved by providing P and Q set points to each inverter and these set points are communicated to the inverters by the MGCC.

In grid-connected mode the correct operation of the inverters relies upon the accurate calculation and transmission of P and Q set-points. This in turn relies upon the network connections used for communication and upon the Proportional Integral and Differential (PID) gain parameters used in the P, Q control loops at the zero level. False data injection or modification attacks against P, Q set points (e.g. via man-in-the-middle attacks (MITM)) would result in unfavourable import/export conditions at the PCC and possibly to voltage instability up to tripping to ensure equipment safety. Unauthorised modification of gain terms—which would require an attacker to gain access to the settings of the CCVSI controller—could lead to voltage and frequency instability within the MG.

In islanded mode the MG must generate its own voltage reference, which this is achieved at the zero level of control by operating one or several inverters as Voltage Controlled Voltage Source Inverters (VCVSI). This mode ensures that the VSI's terminal values are maintained close to nominal values, thereby allowing the inverter's output current to vary according to load demand. The phase angle of VCVSIs must be synchronised and are treated as the reference phase angle ($\angle 0$) in the system.

Primary control for VCVSIs involves determining the reference voltage for the inverter to track. The reference voltage is determined via droop control and calculates the desired V and ω values of the inverter as a function of the active and reactive power being drawn by the load. Adjusting terminal values (very marginally) in this way re-balances power flow in the MG so as to ensure that the load is shared between DER in proportion to their capacities. Droop control does not require data exchange between controllers and the MGCC to achieve load balancing. Droop control relies only upon data collected locally—at the inverters terminals—and locally performed control calculations to achieve system wide load sharing. This fact is significant from both a security and a reliability perspective as it means that there is no dependency on communications links and the system has a reduced attack surface as compared to active (communication based) load sharing techniques.

In islanded mode the correct operation of the inverters relies upon the droop gains employed in the droop controllers and the PID terms used in the current and voltage control loops at the zero level. Unauthorised modification of droop gains has the effect of modifying system dynamics and would change the eigenvalues of the system. Sufficiently large changes to droop gain terms would lead to system instability [8] as would modification of PID controller gains at the zero level.

2.5 Secondary Control

Secondary control has the objective of maximising power quality across the grid when it is operating in islanded mode. A consequence of the use of droop techniques at the primary control level is that MG voltage and frequency are likely to reach steady-state conditions that are off-set from nominal values. Secondary control techniques are employed to correct these off-set steady-state values so that they return to nominal values. The correction is calculated via feedback loops typically employing a PID controller.

In islanded mode the correct operation of secondary control relies upon the integrity of the V_{MG} and ω_{MG} measurements received by the secondary controller and upon the correct tuning of the PID (or equivalent control law) gains. An attacker who is able to modify either of these variables, by modifying sensor readings or gaining unauthorised access to PID gain settings, could cause permanent voltage and frequency offsets or instability throughout the MG.

2.6 Tertiary Control

The tertiary control level is responsible for optimising the MG and the operational decisions made at this level are the most computationally intensive in the hierarchy—hence the control loops are the slowest. Operational decisions relate to optimal generator scheduling and dispatch, Demand Response (DR), ESS charge management and management of ancillary services to the utility. Control at the tertiary level is generally referred to as energy management and the apparatus used to implement it—the MGCC and the control software—are referred to as the Energy Management System (EMS).

The objectives of the EMS change according to the MG operating mode. In grid-connected mode the EMS will tend to make operational decisions in order to minimise operating costs; this will involve maximising revenue from exports to the utility (known as energy arbitrage), minimising the operation of local synchronous generators, and avoiding excessive cycling of ESS, which reduces the life of ESS units. In islanded mode the EMS will make decisions based upon ensuring a continuous supply to critical loads over what could be extended periods. These very different strategies involve different decision making logic [27] in the form of different objective functions. An attacker who was able to interfere with the operation of the EMS would be able to engineer increased MG operating costs, excessive cycling of system components, or reduce the preparedness of a grid-connected MG to operate in islanded mode.

A typical EMS operates by continually minimising multi-dimensional objective functions. Dimensionality increases with the size of the MG and the number of

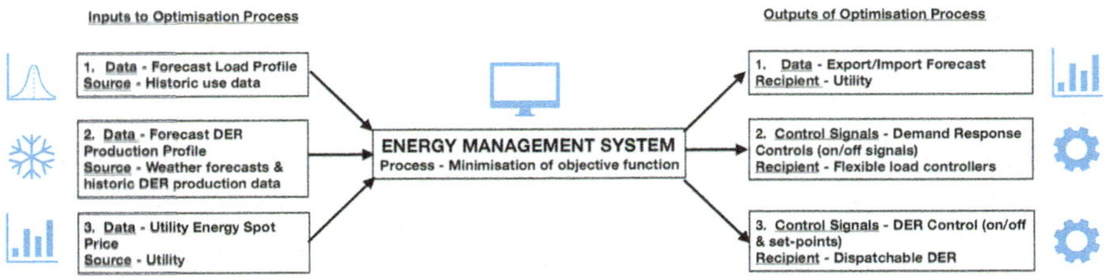

Fig. 2 Role of the MGCC as a Data, Processing and Signals Hub

controllable variables. When successfully minimised the objective function will return an optimal system state for each time-interval over which the system has been optimised. The MGCC can then issue control instructions such as on/off commands to DER and to flexible loads and set points to dispatchable DER in order to enforce the optimal state. Minimisation of the objective function relies on inputs such as customer load forecasts, generation forecasts for variable sources and spot prices for power. Figure 2 illustrates inputs and outputs of a generic EMS.

As can be seen from Fig. 2, the correct operation of the EMS relies on accurate inputs, the correct formulation and weighting of each term of the objective function, and correct as well as timely transmission of control signals to enforce the required system state. An attacker able to compromise any of those elements could negatively influence the optimality of the MG, thereby increasing operating costs or impairing operation in islanded mode.

This section has described how differing MG architectures and modes of operation can impact on MG security and also how attacks that influence the different levels of the control hierarchy can achieve differing effects on MG operation. The implication is that an intelligent attacker can design attacks to achieve specific effects or to take advantage of specific system vulnerabilities dependent on a specific architecture and control laws. Attacks on MG need not be simple naive attempts to cause power loss to loads, but can take the form of highly engineered attacks aiming to cause varying degrees of operational or economic impact or causing failures at a deliberately chosen time and location.

3 Micro-Grid Elements

This section continues the technical review of MGs by investigating a selection of different element types that are employed in systems; these are distinguished by their functions and logical dependencies; consequently, vulnerabilities in elements also will impact the system in different ways.

3.1 Distributed Generation Units

Distributed Generation (DG) units are the micro-sources that are directly connected to the MG. The capacity of an individual DG unit is likely to be in the 10 s of KW—10 MW range, although definitions of size differ throughout the literature [1]. The aggregate generating capacity installed on a single MG will be limited by the relevant grid codes; as an example the US Department of Energy specifies that the total generating capacity of a MG must not exceed 10 MW [26]. The size of MGs is relevant to security analysis as the larger the MG, relative to the capacity of the utility connection, the greater the potential of the MG to adversely impact power quality on the utility if it is poorly controlled.

The DG technologies that are integrated into a MG can take numerous forms and the mix of forms will vary dependent upon the system's objectives and the local climatic conditions. In MGs where the system operator's aim is to reduce energy costs and emissions there is likely to be a heavy reliance on RES. In MGs in which availability of supply is more heavily valued (such as when the primary purpose of the MG is to provide an Uninterpretable Power Source (UPS)) the system may consist entirely of ESS and conventional DG such as diesel generators or micro-turbines. Frequently, a mix of RES and conventional sources will be employed in order to achieve a balance between capital expenditure, operational expenditure, availability of supply and system stability [9].

DG are classed as either dispatchable or non-dispatchable sources. Dispatchable sources such as diesel generators are fully controllable within their capacity range and operating constraints such as ramp rates and minimum and maximum operating periods. RES are typically non-dispatchable sources and will be controlled simply in order to operate at their maximum power point. The maximum power point is a function of environmental conditions and denotes maximum power output for given climatic-conditions. Dispatchable sources will respond to control signals from the tertiary control layer of the MGCC to run up or close down (on/off) (known as UC) and can also modulate their outputs in accordance with set-points (known as ED).

The most significant vulnerabilities regarding DG concern the operation of the PE that are used to electronically couple the DG to the MG as this provides an attacker with the greatest range of options to influence the MG. However, it is possible for DGs to be attacked directly. This could occur by attackers spoofing the on/off instructions sent between the MGCC and the DG or locally attacking the DG controls. Attacks of this nature would negatively impact the economic operation of the MG in grid-connected mode and could lead to equipment tripping, uncontrolled loss of supply to loads in islanded mode or at least to load-shedding.

3.2 Power Electronics

Electronic coupling through PE is required to connect any non-synchronous DER to the MG. This conversion process involves a multi-stage PE package, the final stage of which uses an inverter to convert a DC input to an AC output and involves close control of the V, ω and δ at the inverter's terminals. The manner in which the inverter is controlled to feed power into the MG play a large role in determining the stability and stability margins on the MG; for the purposes of this chapter we do not consider internal operation of the DG resource in scope.

Inverters function by sequencing the operation of internal switches in order to generate an output voltage that matches a reference voltage. In a common model, a signal representing the reference voltage is fed to a Pulse Width Modulation (PWM) controller which determines the correct operation of the internal switches of the inverter. Consequently, the operation of the inverter is driven by the reference voltage provided to the PWM. In a MG the reference signal fed to the PWM is a product of the primary, secondary and tertiary control loops and the correct operation of the inverter is entirely dependent upon the correct operation of the control loops that generate the reference signal.

Inverters are sensitive electronic devices and are liable to damage as a result of circulating currents caused by voltage and frequency imbalances between neighbouring devices [22]. An attacker who is able to manipulate the secondary control loops of MGs is likely to be able to induce sufficiently high circulating currents to cause permanent damage to inverters.

Inverters are particularly vulnerable during the transition from grid-connected to islanded modes, if the control system can be induced to close the PCC CBs when the grid and utility are in an asynchronous state, there would be a large in-rush current across the PCC [28]. The size of the in-rush current would be dependent on the extent to which the utility and MG were out of phase, but could exceed the fault current ratings of inverters and cause permanent damage to electronic switches.

3.3 Energy Storage Systems

ESS are a critical component of MGs and satisfy two key functions:

- They enable rapid load following by allowing the MG to maintain load/supply balance during instances of load or supply fluctuations. ESS are required in this role as the time constants of DER such as diesel generators and fuel-cells are too long to enable them to meet rapid demand changes.
- They enable the MG to continue to supply loads over extended periods when demand exceeds supply.

MGs that employ a large proportion of PE coupled DER lack physical inertia. In conventional power grids that are supplied by synchronous generators, the rotors of the machines provide a source and a sink for energy when there is an imbalance between supply and demand on the grid. Rotor speeds increase when supply exceeds demand and decrease when demand exceeds supply. Generator governors are able to detect these changes in rotor speed and will adjust fuel flow to re-establish the supply/demand balance. MGs lack the inertia to allow them to regulate voltage in this manner and so they rely on fast acting ESS systems to provide the same service. These ESS provide the initial source/sink of energy to absorb the imbalance whilst DG units ramp up or down to meet the change in load in the longer term. ESS come in a variety of forms such as flywheels, super-capacitors and batteries, each with differing speeds of response and storage capacity.

The most severe instance of load/supply imbalance occurs when the MG transitions from grid-connected to islanded mode. At this instant there is likely to be a large imbalance between supply and demand within the MG; this will occur if the MG had been either importing or exporting from the utility immediately prior to the islanding event. The size of this imbalance is likely to constitute a transient disturbance, the ability of the MG to maintain stability throughout this transient phase depends upon the stability margin of the system which in turn depends upon the state of charge and the maximum discharge current of the ESS.

ESS also allow MG to overcome longer term energy imbalances in supply and demand when operating in islanded mode [16]. An example of such would be in a system that relies heavily on solar PV sources that are unable to generate overnight. In this case excess energy can be stored in ESS during the day and released to loads overnight. Decisions as to when to store and release energy in this manner are made at the tertiary control level as part of the MG optimisation process.

Cyber-physical security considerations relating the use of ESS must take account of how an attacker could disrupt the functions of the ESS, as described above, so as to disrupt the MG as a whole. When operating in island mode the ESS will respond to over and under frequency conditions (in respect to nominal values) on the MG by absorbing or injecting active power and will respond in the same way to over and under voltage conditions by injecting reactive power. Measurements of absolute and rate of change values of MG frequency and voltage measurements will be taken locally to enable the ESS to respond appropriately. An attacker who was able to manipulate secondary control operation in order to introduce a permanent negative offset to MG frequency and voltage values could force ESS into a state of constant discharge. If the ESS has discharged sufficiently, it is no longer able to compensate for short or long term frequency imbalances in the MG. which will lead to instability. Alternatively, an attacker who is able to modify the MG state such that ESS are forced to cycle (charge and discharge) will be able to reduce the life expectancy of the ESS and increase operational costs of the MG (as ESS will need to be replaced more frequently). Moreover, rapid discharging will typically impose thermal stresses that also limit the ability to rapidly re-charge.

3.4 Controllable Loads

MGs exist to feed power to loads and, if the MG is to be successful and efficient, it must be designed around the specific characteristics of its loads. The nature of these loads differ depending upon the MG's use case, however, loads will typically be cyclic and follow daily and seasonal patterns; they will also be prioritised differently by consumers. Some loads will be considered critical—the consumer places the highest priority on maintaining a power supply to these loads such as operating theatres or ICU in a hospital or local air traffic control dispatching—whilst others will be flexible.

Flexible loads can be further categorised as being either curtailable or deferrable. Curtailable loads are those loads that can accept a loss of supply and can be isolated by the MGCC without further consideration. The advantage of curtailable loads is that they can be immediately isolated as part of frequency support strategies, thereby contributing toward power quality on the MG in islanded mode. This technique is referred to as peak lopping. Deferrable loads, such as heat storage loads, are more complex in that they can tolerate a loss of supply so long as, in aggregate, they receive a supply in any given time period that exceeds a lower bound. The advantage of deferrable loads is that they can be scheduled by a controller so that they draw power during periods when total system load is predicted to be low. This technique is referred to as load shifting. Peak lopping and load shifting are both Demand Response (DR) techniques and allow the MG to pro-actively shape its load profile in order to best match its predicted forecast schedule.

The process of deferring and curtailing loads is part of the tertiary control process. During this process the cost, or inconvenience, of employing DR techniques is measured against the cost of achieving other system objectives. When the global benefit of reducing MG load outweighs the local inconvenience cost that results from peak lopping or load shifting, decisions will be made to curtail or defer controllable loads. Control signals will then be sent from the MGCC to static switches which will isolate individual loads or entire non-critical feeders dependent upon the MG topology.

The risks associated with the use of DSM and controllable loads are self-evident. If an attacker is able to spoof control signals to switches they will be able to override legitimate controls and isolate or energise flexible loads, which may in the case of deferrable loads also impose limits on flexibility at a later time.

Section 3 has described the major element types of MGs in outline and concentrated on effects on power network stability and equipment safety in the event of attacks.

4 STPA-Sec and Causal Scenarios

This section builds upon the previously documented MG control architectures and system elements to construct a set of generic causal scenarios that describe how intentional disruptions to a MG control system such as those caused by an attacker could lead to system losses. System-losses are those events in which the high level objectives of the MG, such as continuous supply of critical loads, are disrupted and the MG ceases to be effective.

This work uses the *Systems-Theoretic Process Analysis–Security* (STPA-Sec) framework [29] to build these causal scenarios. STPA-Sec is a method of assessing security vulnerabilities that draws heavily on the work of the industrial systems safety community to better understand and control the risks associated with complex systems of systems. The complexity of modern systems has led to a recognition in industrial safety thinking that hazards can be caused not only by component failure, the traditional focus of industrial safety, but also as a result of the unintended interaction of control loops. This recognition led to the development of Systems-Theoretic Accident Modeling and Processes (STAMP) and latterly to STPA [15]. These unintended interactions result in emergent behaviours that were not anticipated by system designers and are potentially hazardous. Emergent behaviours are a function of the system as a whole rather than any single element in isolation.

STPA-Sec develops the systems approach used in STPA in recognition that the same complexity that can lead unintentional disruptions to systems can also be exploited by intelligent attackers to deliberately engineer system losses. STPA-Sec conducts top-down analysis of systems with the aim of designing control systems that are constrained so as to prevent emergent behaviours resulting in vulnerable states. These constraints safeguard system objectives by preventing the possibility of vulnerable states regardless of whether systems are being influenced by adversarial action, component failure or unintended control interaction. The ability of STPA-Sec to take account of the complexities of heavily interconnected and interdependent systems make it an ideal tool for analysing the security of MGs.

The STPA-Sec methodology employs the following steps, elaborated below:

- Step 1—Establish the systems engineering foundation.
- Step 2—Identify unsafe/insecure control actions
- Step 3—Develop security requirements and constraints
- Step 4—Identify causal scenarios

4.1 Step 1: Establish the Systems Engineering Foundation

The systems engineering foundation step describes the system to be analysed and the associated losses and vulnerabilities. It is important to appreciate that the losses and vulnerabilities are determined not only by the technical aspects of the MG but also by the specific system objectives.

The reference MG that will be used in this analysis has been selected so as to generate generic causal scenarios that are applicable to the widest range of MGs. Consequently, the system specification is deliberately high level and incorporates the most common MG topologies and technologies. The reference topology is illustrated in Fig. 1 and employs the following control methodologies and elements: a centralised tertiary and secondary control architecture, a communications infras-

Table 2 Loss events and system vulnerabilities

Loss events	System vulnerabilities			
	V1: instability	V2: poor optimisation	V3: dangerous operating states	V4: poor power quality
L1: loss of supply to MG loads	✓	✓	✓	✓
L2: damage to MG equipment	✓		✓	
L3: high system operating costs		✓	✓	

tructure based on WAN employing unencrypted ICS protocols, a distributed primary control system employing droop control, and a communications link between the DMS and the MGCC in order to allow the utility to regulate ancillary services. The MG is installed on a campus site and loads consist of a mix of critical and flexible loads. The MG operator wishes to reduce electricity costs (as compared to a standard utility feed) and increase the reliability of supply to its critical loads.

The first step of the analysis involves identifying the loss events that threaten the objectives of the MG and the vulnerabilites that could precipitate them. Vulnerabilities can be thought of as those system states that expose the system to the possibility of loss. Table 2 identifies 3 categories of MG loss L1-3 and 4 vulnerabilities that could lead to these losses V1-4. The potential for specific losses to be caused by specific system vulnerabilities are indicated with a checkmark. Explanations of system vulnerabilities are given in the following sub-sections.

4.1.1 Vulnerabilites: Instability

Instability in a MG refers to the situation in which the system variables (V, ω and δ) fail to converge to a steady-state after the occurrence of a disturbance. The task of achieving steady-state operation is the function of the MG control system and instability is a result of the failure of the control system to operate correctly or of the system being subjected to disturbances which exceed the ability of the control system to regulate them, in which case the disturbances are said to exceed the stability margins of the system. The potential for instability poses a security vulnerability to MGs as, if an attacker can engineer the conditions necessary to achieve instability, the consequences are likely to be loss of supply to MG loads and, dependent upon protection settings, damage to equipment.

Small-signal stability requires the MG to respond in a stable manner to routine system disturbances such as changes in load. Ineffective operation of droop controllers is the predominant cause of small-signal instability [19] and poorly designed or tuned primary controllers could lead to voltage spikes (MG voltage temporarily exceeding acceptable bounds) and frequency oscillations. Both situations could lead to individual loads or DER tripping or a blackout of the entire MG.

Transient stability requires the MG to respond in a stable manner to large disturbances of the type that would be caused by the MG changing operating mode from grid connected to islanded, unanticipated loss of a DER, or a fault. As explained in Sect. 3.3 ESS are a critical component in maintaining system stability in the event of transient conditions.

4.1.2 Vulnerabilities: Poor Optimisation

Table 2 states that poor system optimisation can lead to high system operating costs and potentially to a loss of supply of loads. MG optimisation is briefly discussed in Sect. 2.6 and it was explained that optimisation decisions are made at the tertiary control level in order to ensure that the MG is able to maximise revenue through energy arbitrage whilst remaining well poised to meet system objectives given forecasts of future operating conditions. Any actions that an attacker is able to take so as to undermine the correctness of these decisions may compromise the MG's objectives.

The following strategies are a selection of viable means by which an attacker could impair the ability of the MG to prepare itself for future operating conditions:

- Modification of the load, supply and cost data that are used as inputs to the optimisation process at the MGCC as part of tertiary control.
- Modification of the weighting terms used to encode the relative importance of each term of the system's objective function.
- Modification of the control signals sent from the MGCC to the dispatchable DER and controllable loads.

4.1.3 Vulnerabilities: Dangerous Operating States

Table 2 states that dangerous operating states can lead to loss of supply to loads, damage to MG equipment and high operating costs as a result of the requirement to replace damaged components. Dangerous operating states refer to those states in which the MG could cause physical damage to itself. The transition between islanded and grid-connected mode presents the potential for a MG to enter a dangerous state if the control systems fail to ensure sufficiently accurate synchronisation prior to the closure of the PCC CBs. Whilst operating in islanded mode voltage reference of the MG is likely to deviate from that of the utility, hence it is necessary to resynchronise prior to returning to grid-connected mode, which involves re-closure of circuit breakers at the PCC. Re-synchronisation involves the comparison of a reference voltage from the utility, which may also be communicated from a remote Phasor Measurement Unit (PMU) via a network connection, with the MG voltage and control measures to align the MG voltage to the utility. If the communication channels or control loops involved in this process are disrupted then circuit breakers may be instructed to close when the voltages remain out of phase [6].

Failure to ensure accurate synchronisation of voltage, frequency and phase angle would result in high torques being applied to the rotors of synchronous generators on the MG and very high in-rush currents in the PE serving non-synchronous DER. Both of these conditions could seriously damage equipment as well as jeopardise loads.

4.1.4 Vulnerabilities: Poor Power Quality

Table 2 identifies that poor power quality can result in loss of supply to loads. Poor power quality in the context of this analysis refers to steady-state deviations in the voltage waveform from nominal values. MGs are equipped with isolation devices that will react to poor power quality by isolating DER and loads from the supply in order to avoid damage. The implication is that an attacker could achieve a blackout in a MG by influencing the MG control system in such a way that power quality is reduced below acceptable limits (but whilst the MG remains stable); this will cause protection devices to trigger and loads to be isolated.

4.1.5 High Level Control Structure Model

Production of a High Level Control Structure (HLCS) is the final stage of step 1 of STPA-Sec. The purpose of producing this model is to aid the analyst in identifying each control loop in the system so they can be assessed for security vulnerabilities. Figure 3 illustrates the High Level Control Structure (HLCS) that enables the MG

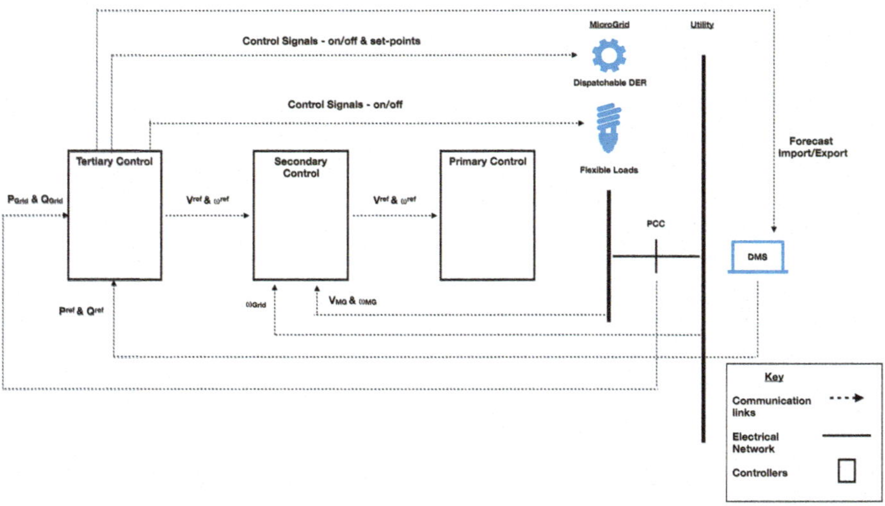

Fig. 3 High level control structure model

to regulate frequency and voltage in grid connected and islanded modes. The figure also shows the feedback loops involved in regulating the primary, secondary and tertiary control objectives and also shows the control channels required for the purpose of controlling dispatchable DER and DR techniques.

4.2 Step 2: Identify Unsafe/Unsecure Control Actions

The process of identifying unsafe or insecure control actions aims to determine the circumstances in which the control loops identified in the HLCS could lead to system vulnerabilities. The STPA-Sec methodology specifies the following 4 circumstances in which Unsafe Control Actions (UCA) may lead to vulnerabilities:

- UCA 1—Not providing a control action leads to a hazard or exploits a vulnerability
- UCA 2—Providing a control action leads to a hazard or exploits a vulnerability
- UCA 3—Providing control actions too late, too early, or in the wrong order leads to a hazard or exploits a vulnerability
- UCA 4—Stopping a control action too soon or continuing it too long leads to a hazard or exploits a vulnerability

Table 3 provides a summary of the most significant UCA relating to a MG matching the specification given in Sect. 4 above. Relevant control actions have been drawn from assessment of the HLCS model and the potential for unsafe consequences arising from these control actions is based on the MG review discussed in Sects. 2 and 3.

4.3 Step 3: Develop Security Requirements and Constraints

Security requirements and constraints are simple statements that negate the possibility of the occurrence of any of the unsafe control actions listed in Table 3. For instance, in relation to PCC closure, an appropriate control constraint is that the PCC should not close when the utility and MG are operating out of phase; we illustrate this case and omit the full list of constraints relating to Table 3.

4.4 Step 4: Identify Causal Scenarios and Priority Areas for Future Research

The final step in the STPA-Sec is to develop the causal scenarios that could lead to the unsafe control actions identified in Table 3. The casual scenarios describe how,

Table 3 Unsafe/Unsecure control actions

Control Action	UCA 1	UCA 2	UCA 3	UCA 4
1. Modification of V, ω set points @ primary control	1. V, ω set points are not modified when P, Q values change at inverter terminals. Leads to poor load balancing and circulating currents [V2] [V3]	1. V, ω set points are modified when P, Q values are static at inverter terminals. Leads to poor load balancing and circulating currents [V2] [V3]	1. V, ω set points are modulated Leads to poor load balancing, circulating currents and instability [V2][V3][V1}	N/A
2. Modification of V, ω set points @ secondary control	1. V, ω set points are not modified when MG variables deviate from nominal. Leads to poor power quality and possible discharge of ESS [V4] [V2]	1. V, ω set points are modified when MG variables are at nominal. Leads to poor power quality and possible discharge of ESS [V4] [V2]	1. V and ω set points are modulated. Leads to poor power quality, instability and possible cycling of ESS [V4][V1]	N/A
3. Modification of V, ω set points @ tertiary control	1. MG fails to respond to instructions from DMS to adjust power flow across PCC. Leads to disruption of voltage regulation on utility [V4]	1. MG fails to appropriately respond to instructions from DMS to adjust power flow across PCC. Leads to disruption of voltage regulation on utility. [V4]	1. As per UCA 1, 2	N/A
4. PCC closure	N/A	N/A	1. PCC CBs operate when utility and MG are asynchronous. Leads to large in-rush currents, transient disturbance, and damage to hardware [V1][V3][V4]	N/A
5. Isolation of flexible loads.	1. Loads fail to disconnect in response a requirement to load shed on the MG. Leads to overloaded DER and poor power quality [V3][V4]	1. Loads are disconnected when there is no requirement to load shed. Leads to unnecessary loss of supply [V4]	1. Loads are rapidly connected and disconnected. Leads to unnecessary loss of supply and to transients causing instability [V1][V4]	N/A
6. Isolation of dispatchable DER	1. Dispatchable DER fail to disconnect when there is too much capacity on MG. Leads to poor optimisation of the MG and higher operating costs. [V2]	1. Dispatchable DER are disconnected when there is insufficient capacity on MG. Leads to poor power quality and to loss of supplies to loads [V4] [V2]	1. DER are rapidly connected and disconnected. Leads to large transients causing instability. [V1]	N/A

given the physical and logical infrastructure that constitutes the MG, constraints can be violated which lead to the possibility of losses. Table 4 describes these scenarios.

Examination of Table 4 reveals a number of recurring themes in terms of the requirements of MG security. These themes are:

- **Requirement for Protected Data Channels** Data channels used to communicate set-points between the levels of the control hierarchy need to be secure. An intelligent adversary who is able to conduct MiTM attacks against the data on these channels will be able to engineer attacks leading to loss of supply of loads or, potentially, damage to equipment. Examination of Tables 3 and 4 suggests that the data links that present the greatest vulnerability to a MG are the links between the secondary controller and the DER. These links present the greatest range of opportunities for an attacker and are subject to the most diverse range of causal scenarios.
- **Requirement for Secure Controller Parameters** Controller parameters must to be secure as, once again, an intelligent attacker who is able to modify these parameters will be able to engineer attacks leading to loss of load, or damage to equipment. The MG is equally vulnerable to parameter modifications of controllers at each of the hierarchical layers. Consequently strong access control measures are required at all controllers.

The problem of securing data channels can be achieved partially cryptographic primitives for many issues identified in Table 4. However, some requirements also are time-sensitive and may create novel attack vectors resulting from the use of cryptographic mechanisms; moreover, key management and related overheads must also be resolved while particularly in deployment in lower income countries may limit the ability to maintain systems at desirable level for requirements that at least initially do not appear to be functional.

Consequently, this analysis implies that supplementary measures in addition to cryptographic primitives should be considered a research priority for securing the data channels between the secondary control level and individual DER. Such research should focus on methods to improve:

- **Detection** The detection of anomalous data streams that indicate cyber-attacks
- Enhancing **resilience** The enhancement of MG control schemes such that the MG can continue to operate, potentially in an *emergency* state, in the event of an attack by a limited adversary.

Detection schemes are likely to involve the use of bad data detection techniques that are sensitive to bad data as a result of unintentional causes (sensor failures or communication errors) and intentional causes (errors introduced deliberately by an adversary). Such schemes are likely to incorporate "physics aware" techniques such as the use of state estimators as security monitors [19] or may consist of simple statistical checks.

Table 4 Causal scenarios

Control action		Causal scenarios
1. Modification of V and ω set points generated by primary controller	UCA 1	The inverter terminal values used as feedback to the primary controller are measured and communicated locally and, as such, can be considered to be reliable. Therefore, an attacker must rely on attacking the controllers droop control parameter and setting it to a zero gradient to prevent the controller from adjusting set points in accordance with the droop scheme. Such an attack depends upon an attacker gaining unauthorised access to the primary controller, remotely or locally, in order to effect a parameter change
	UCA 2	No feasible scenario
	UCA 3	V and ω set points could be forced to modulate by an attacker who was able to decrease the dynamic response time of the secondary controller (speed up the control loop), this could be achieved by increasing secondary controller PID gains to in order to reduce rise times. This would achieve the effect of coupling the primary and secondary loops, whereby the dynamics of the secondary loop would impact on the dynamics of the primary. The 2 control loops would interact and primary control set points would modulate
2. Modification of V and ω set points generated by the secondary controller	UCA 1	MG V and ω set points could be prevented from tracking nominal values by an attacker who is able to achieve either of the following: (a) Denial of Service (DOS) against channels carrying the V and ω values received from the tertiary control level. (b) DOS against the channels carrying the V and ω values sent to the primary controller
	UCA 2	MG V and ω set points could be forced away from nominal values by an attacker who is able to achieve either of the following: (a) Use MiTM techniques to modify V and ω references received from the tertiary control level. (b) Use MiTM techniques to modify V and ω references sent to the primary controller
	UCA 3	V and ω set points could be forced to modulate by an attacker who was able to: (a) Decrease the dynamic response time of the tertiary controller. The 2 control loops would then interact and secondary set points would modulate
3. Modification of V and ω set points generated by tertiary controller	UCA 1	The MG could be prevented from responding to utility requests for ancillary services by an attacker who is able to achieve either of the following: (a) DOS against the P Q set points communicated from the DMS (b) DOS against the V and ω references sent to the secondary controller
	UCA 2	The MG could be induced to establish inappropriate (undesirable from the perspective of the utility) power flows across the PCC. This could be achieved by an attacker by: (a) Using MiTM techniques to manipulate the P Q set points communicated from the DMS (b) Using MiTM techniques to manipulate the V and ω references sent to the secondary controller
	UCA 3	As per 3.UCA1 and 3.UCA2

4. PCC Closure	UCA 1	N/A
	UCA2	N/A
	UCA3	Circuit breakers at the PCC could be induced to close when the MG and utility are in an asynchronous state if the Phasor Measurement Unit (PMU) data (which provides very accurate readings relating to measured voltage waveforms) received from the utility are incorrect. This would result in an error between the actual utility phase angle and the phase angle that the MG believes the utility to be operating at. This could be achieved by an attacker who is able to: (a) Conduct DOS against the channel carrying PMU data. (b) Use MiTM techniques to modify the time stamps or waverform data in the communication packets.
5. Isolation of flexible loads	UCA 1	An attacker could force a load to remain connected, despite a requirement for it to be shed, by employing DOS techniques against the controller of the flexible load or the MGCC
	UCA 2	An attacker could force a load to disconnect, despite there being no requirement to do so, by spoofing control signals to the controller of the flexible load
	UCA 3	An attacker could force a load, or several loads, to modulate by rapidly switching them on and off by spoofing control signals to the controller of the flexible load
6. Isolation of dispatchable DER	UCA 1	No feasible scenario
	UCA 2	An attacker could force a DER to disconnect from the MG, despite there being no requirement to do so, by spoofing control signals to the controllers of dispatchable DER
	UCA 3	An attacker could force a DER, or several DER, to connect and disconnect rapidly (particularly fast acting DER such as ESS and solar PV). This could be achieved by spoofing the control signals to the controllers of DERs

5 Conclusions

The chapter has sought to conduct a high level security analysis of the MG concept in order identify possible causal scenarios for security violations in MGs leading to the identification of priority areas for security research in the MG domain as this has thus far received limited attention.

To this end we have reviewed core concepts in Sects. 2 and 3 including microgrid modes of operation, control methodologies, and major system elements. The aim of the review was to highlight the security implications associated with the structures of MGs in order to provide the foundation required for the conduct of a security analysis. The review identified that the data links that act as the interface between the levels of the control hierarchy present access points through which attackers could employ e.g. MitM techniques to manipulate data and control signals. It was also shown how data inputs to the tertiary control layer could be manipulated in order to increase MG operating costs.

Section 4 conducted a high level security analysis of a MG, using a reference model based on common MG architectural choices. The intention was to generate casual scenarios that would be applicable to the broadest range of MGs. The analysis was based on the STPA-Sec framework and showed that the data links carrying controller set points and controller gain parameters present critical vulnerabilities to MG operations and need to be protected accordingly. It was also found that the data links carrying secondary control set points from the MGCC to the DER controllers presented the most serious vulnerability and should be considered as a priority for future MG security research.

The analysis also suggested that supplementary measures beyond cryptographic protection mechanisms should be considered for securing data streams. This is because of the difficulty of maintaining encryption schemes in the medium to long term, a problem which is particularly acute in Lower Income Countries where many MGs are likely to be deployed. We conclude with the recommendation that future MG security research needs to be directed towards designing control algorithms that are able to detect adversarial action and are robust in the presence of such activities and faults.

References

1. Ackermann T, Andersson G, Söder L (2001) Distributed generation: a definition. Electr Power Syst Res 57(3):195–204
2. Bidram A, Davoudi A (2012) Hierarchical structure of microgrids control system. IEEE Trans Smart Grid 3(4):1963–1976
3. Case DU (2016) Analysis of the cyber attack on the ukrainian power grid. Electricity Information Sharing and Analysis Center (E-ISAC)

4. Chandorkar MC, Divan DM, Adapa R (1993) Control of parallel connected inverters in standalone ac supply systems. IEEE Trans Ind Appl 29(1):136–143
5. Domenech B, Ranaboldo M, Ferrer-Martí L, Pastor R, Flynn D (2017) Local and regional microgrid models to optimise the design of isolated electrification projects. Renew Energ 119:795–808
6. Friedberg I, Laverty D, McLaughlin K, Smith P (2015) A cyber-physical security analysis of synchronous-islanded microgrid operation. In: Proceedings of the 3rd International Symposium for ICS & SCADA Cyber Security Research. British Computer Society, Swindon, pp 52–62
7. Guerrero JM, Chandorkar M, Lee TL, Loh PC (2013) Advanced control architectures for intelligent microgrids—part I: decentralized and hierarchical control. IEEE Trans Ind Electron 60(4):1254–1262
8. Guo X, Lu Z, Wang B, Sun X, Wang L, Guerrero JM (2014) Dynamic phasors-based modeling and stability analysis of droop-controlled inverters for microgrid applications. IEEE Trans Smart Grid 5(6):2980–2987
9. Hafez O, Bhattacharya K (2012) Optimal planning and design of a renewable energy based supply system for microgrids. Renew Energy 45:7–15
10. Hatziargyriou N, Asano H, Iravani R, Marnay C (2007) Microgrids. IEEE Power Energ Mag 5(4):78–94
11. Katiraei F, Iravani R, Hatziargyriou N, Dimeas A (2008) Microgrids management. IEEE Power Energ Mag 6(3):54–65
12. Langner R (2011) Stuxnet: dissecting a cyberwarfare weapon. IEEE Secur Priv 9(3):49–51
13. Lasseter RH (2011) Smart distribution: coupled microgrids. Proc IEEE 99(6):1074–1082
14. Lasseter R, Akhil A, Marnay C, Stephens J, Dagle J, Guttromson R, Meliopoulous A, Yinger R, Eto J (2002) The CERTS microgrid concept. White paper for Transmission Reliability Program, Office of Power Technologies, US Department of Energy 2(3), 30
15. Leveson N (2011) Engineering a safer world: systems thinking applied to safety. MIT Press, Cambridge/London
16. Levron Y, Guerrero JM, Beck Y (2013) Optimal power flow in microgrids with energy storage. IEEE Trans Power Syst 28(3):3226–3234
17. Li C, Cao C, Cao Y, Kuang Y, Zeng L, Fang B (2014) A review of islanding detection methods for microgrid. Renew Sust Energ Rev 35:211–220
18. Li H, Zang C, Zeng P, Yu H, Li Z (2015) A stochastic programming strategy in microgrid cyber physical energy system for energy optimal operation. IEEE/CAA J Automat Sin 2(3):296–303
19. Majumder R (2013) Some aspects of stability in microgrids. IEEE Trans Power Syst 28(3):3243–3252
20. Meng L, Sanseverino ER, Luna A, Dragicevic T, Vasquez JC, Guerrero JM (2016) Microgrid supervisory controllers and energy management systems: a literature review. Renew Sust Energ Rev 60:1263–1273
21. Olivares DE, Mehrizi-Sani A, Etemadi AH, Cañizares CA, Iravani R, Kazerani M, Hajimiragha AH, Gomis-Bellmunt O, Saeedifard M, Palma-Behnke R et al (2014) Trends in microgrid control. IEEE Trans Smart Grid 5(4):1905–1919
22. Piagi P, Lasseter RH (2006) Autonomous control of microgrids. In: Power engineering society general meeting, 2006. IEEE, Piscataway, pp 1–8
23. Rocabert J, Luna A, Blaabjerg F, Rodriguez P (2012) Control of power converters in AC microgrids. IEEE Trans Power Electron 27(11):4734–4749
24. Stern DI, Burke PJ, Bruns SB (2017) The impact of electricity on economic development: a macroeconomic perspective. EEG state-of-knowledge paper series. Oxford: Oxford policy management center for effective glocal action
25. Tayab UB, Roslan MAB, Hwai LJ, Kashif M (2017) A review of droop control techniques for microgrid. Renew Sust Energ Rev 76:717–727

26. Ton DT, Smith MA (2012) The us department of energy's microgrid initiative. Electr J 25(8):84–94
27. Tsikalakis AG, Hatziargyriou ND (2011) Centralized control for optimizing microgrids operation. In: Power and energy society general meeting, 2011. IEEE, Piscataway, pp 1–8
28. Williston D, Finney D (2011) Consequences of out-of-phase reclosing on feeders with distributed generators. IEEE SCC21 standards coordinating committee on fuel cells, photovoltaics, dispersed generation, and energy storage, pp 1–8
29. Young W, Leveson N (2013) Systems thinking for safety and security. In: Proceedings of the 29th Annual Computer Security Applications Conference. ACM, New York, pp 1–8

Engineering Edge Security in Industrial Control Systems

Piroska Haller, Béla Genge, and Adrian-Vasile Duka

Abstract Industrial Controllers (e.g., Programmable Logical Controllers – PLCs, and Remote Terminal Units – RTUs) have been specialized to deliver robust control strategies. However, little has been done towards the integration of security strategies within their application-layer. This chapter investigates the integration of security solutions within the industrial control system's "edge" devices – the Industrial Controller (IC). As a specific case study it demonstrates the implementation of a simple anomaly detection engine traditional in control applications. The approach shows that the scheduling rate of control applications is significantly affected by various events, such as a change in the number of network packets, configuration interventions, etc. Implementations realized on a Phoenix Contact ILC 350-PN controller demonstrate the feasibility and applicability of the proposed methodology.

1 Introduction

Supervisory Control and Data Acquisition (SCADA) systems provide essential functions for the operation of industrial processes (e.g., oil and gas pipelines, water distribution systems, smart energy systems, air and gas transportation systems). Their architecture combines traditional Information and Communication Technology (ICT) hardware and software with industrial controllers (e.g., Programmable Logical Controllers – PLCs, and Remote Terminal Units – RTUs) in order to facilitate a cost-effective implementation of local actuation strategies.

Industrial controllers (IC) are, in most cases, embedded devices that are specialized for real-time applications in manufacturing and process control. These are available in a wide variety of configurations running a diverse palette of operating systems together with dedicated real-time schedulers. While these controllers

P. Haller (✉) · B. Genge · A.-V. Duka
Petru Maior University of Tîrgu Mureş, Tîrgu Mureş, Mureş, Romania
e-mail: phaller@upm.ro; bela.genge@ing.upm.ro; adrian.duka@ing.upm.ro

D. Gritzalis et al. (eds.), *Critical Infrastructure Security and Resilience*, Advanced Sciences and Technologies for Security Applications,
https://doi.org/10.1007/978-3-030-00024-0_10

have been specialized to deliver robust and effective control strategies, little has been done towards the integration of security strategies within the application of industrial controllers. In fact, in the vast majority of cases, control applications do not account for the security and the inner monitoring of their behavior. This immense responsibility has been transferred to external devices such as "bump-in-the-wire" monitoring devices (Intrusion/Anomaly Detection Systems, process monitoring systems), and cryptographic devices providing the secure tunneling of legacy industrial protocols (e.g., IPSec). However, security operations, if carefully engineered, may be integrated into the edge devices found in Industrial Control Systems (ICS), namely, critical devices such as PLCs.

To this end, a considerable amount of research has been focused on the development of Intrusion Detection Systems (IDS) [1, 3, 21–23] for the industrial realm. Different strategies have been developed by leveraging diverse techniques such as classification [16, 25, 26], multivariate statistical analysis including principal component analysis [13, 20], and data fusion [5, 6, 10]. Nevertheless, we observe that the practical implementation of previous methodologies within the industrial realm would require major software/hardware changes. Furthermore, in most cases, the complexity of the suggested detection strategy does not permit their integration within the application of industrial controllers. This is owed to the time constraints imposed to the scheduling of real-time control applications, where complex computations may significantly affect the schedulability of such applications.

Based on the aforementioned issues, this chapter investigates the integration of security solutions within the ICS's "edge" devices – the IC. As a specific case study it demonstrates the implementation of a simple anomaly detection engine traditional in control applications. The approach leverages the scheduling rates found in control applications, and their deviation from the "normal" behavior. As it turns out, the approach is well-suited for scenarios in which the encountered behavior is sufficiently narrow to allow meaningful detection from the "normal". Furthermore, we show that the scheduling rate of control applications is significantly affected by various events, including a change in the number of network packets, administrator's interventions, configuration changes, etc. Implementations realized on a Phoenix Contact ILC 350-PN controller demonstrate the feasibility and applicability of the proposed methodology.

2 Related Studies

The dramatic impact of traditional computer system attacks on industrial processes has been first demonstrated by the Stuxnet malware [14, 24]. In light of such threats, there have been several proposals towards enhancing the security of existing and future industrial installations [7, 11]. In fact, a considerable amount of research has been allocated to understanding the design of comprehensive anomaly detection systems for industrial systems. Nonetheless, the practical integration of such schemes was not sufficiently explored. In the remaining of this section we provide an overview of the main anomaly detection techniques for SCADA systems available in the literature.

We start by mentioning techniques that leverage different parameters exposed by the cyber and/or the physical dimension of SCADA systems. The work of Kiss et al. [16], explored the applicability of data clustering techniques for anomaly detection. The Gaussian mixture model was compared to the K-means clustering, and the superior performance of the former was demonstrated. The work of Wan et al. [25], integrated the monitoring of network characteristics with the process-specific parameters. In their work, a weighted mixed Kernel function and parameter optimization methods were developed to improve the performance of the classification. A similar attempt for the classification of different events was undertaken by Wang and Mao in [26]. Here, an ensemble of two models of one-class classification were developed, and the performance of the approach was demonstrated in the context of two industrial installations and several public datasets.

In the direction of multivariate statistical analysis we find the work of Daegeun Ha et al. [13]. Here, the multi-mode principal component analysis (PCA) was used together with k-nearest neighbor algorithm for process monitoring and data classification. The approach was demonstrated in the context of a mixed refrigeration physical process. In a similar direction, a weighted adaptive recursive PCA approach was developed by Portnoy et al. in [20], while the works mentioned in [5, 6, 10] adopted the data fusion strategy as an attempt to develop a multivariate reasoning methodology.

Starting with their work in [4], Cárdenas et al. demonstrated that, by incorporating knowledge of the physical process in the control system, it is possible to detect traditional computer attacks that change the behavior of the underlying control system. By leveraging the knowledge of the physical process, the detection can focus on the attack target, rather than on the various strategies that an adversary may undertake. More recently, in [12], Giraldo et al. stepped further and designed an attack-resilient controller. In terms of detection, [12] adopted the non-parametric cumulative sum model.

Lastly, for completeness, we mention that besides the aforementioned techniques, a wide variety of other strategies have been designed for abnormal behavior in industrial settings [3, 8, 9, 15]. Nevertheless, our analysis focused on the techniques leveraging network and process-specific parameters, that may be integrated in the application of industrial controllers.

According to the analysis above we emphasize that, while a wide variety of techniques have been proposed, few cases addressed the practical integration of detection strategies within the industrial realm. Therefore, the complexity of detection schemes based on computation-intensive operations such as the ones proposed in [5, 6, 10, 13, 16, 20, 25, 26] may require external hardware. Conversely, the integration of process-specific information in the design of the control system, as proposed in [4] and [12], may represent a feasible solution. However, the authors do not provide specific details on the practical implementation of such control strategies. Furthermore, their proposal focuses on the monitoring of physical process, while in the paper at hand, our aim is to monitor the operation of the control hardware from the application layer, and to detect abnormal behavior accordingly.

2.1 Architecture of Industrial Controllers

The architecture of modern industrial controllers varies among different vendors. Accordingly, on top of the hardware we may find a classical operating system (e.g., Linux Version 2.6 or later, FreeRTOS OPENRTOS, RTX, Windows CE 6.0, Windows Embedded Compact 7, etc.) together with a dedicated real-time operating system (e.g., ProconOS), or a real-time scheduler [19]. Irrespective on the underlying solution, control code is usually organized in several distinct *user* tasks. These are scheduled for periodic execution by the real-time preemptive scheduler (or the real-time operating system), which governs the temporal determinism of user tasks.

Besides user tasks, ICs also host *system* tasks that implement the typical functions for handling remote connections, processing of network packets. The scheduler uses the task priority levels in the task scheduling algorithm. To this end, the processor time is always assigned in the favor of user tasks in order to ensure that critical control functions are executed in the expected deadline. Consequently, in the case of increased load (e.g., due to application state changes, or network traffic-based attacks), the tasks that are in charge of communications (e.g., via Modbus/TCP, OPC) are among the first to exhibit a change in their scheduling behavior. This is owed to the fact that, underneath, communications are handled by interrupt service routines (ISR) that, in case of disturbances (e.g., increasing number of packets), are overloaded with processing requests. As a result, the execution of the ISR will require additional computational resources, which can delay the scheduling of user tasks.

User tasks can be classified as cyclic tasks, event tasks, and default tasks. Cyclic tasks are activated periodically (i.e., they change their state to "ready to run"), but they are actually scheduled to run according to their configured priority and deadline. The default task has the lowest priority and is executed when no other task is active. Conversely, event-based tasks are activated by the OS when the ISR finishes its execution. The actual scheduling time of the event tasks depends on the priority of each task.

The time between two consecutive runs of the same periodic task (difference between the start time) differs from the task period even for the highest priority task. This is mainly owed to the hardware interrupts and to the execution time of the interrupt service routines. The start and end time of low priority cyclic tasks are difficult to predict as they depend on the execution time of high priority tasks, the number of interrupts, and the overhead in the OS kernel. However, the finishing time for a task should be less than a predefined watchdog time, otherwise an error event is generated at run-time.

An example task scheduling scenario is illustrated in Fig. 1. Here, we depicted both circular and event-based tasks, together with the monitoring task (i.e. our

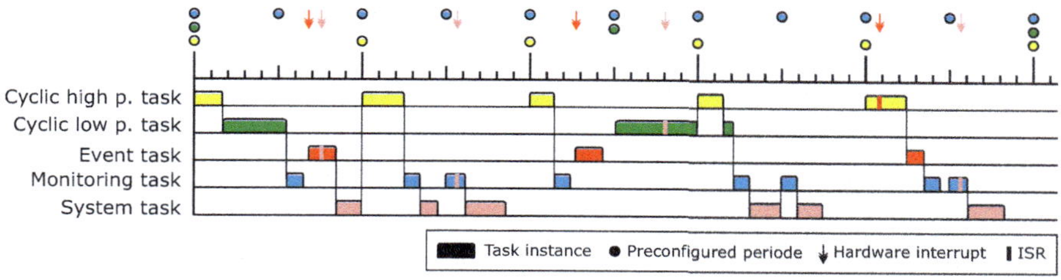

Fig. 1 Example of task scheduling

proposal) and the system task. It can be seen that high priority tasks are scheduled with a higher accuracy, while the execution of low priority tasks is interrupted and delayed by the high priority tasks. We further observe that task execution is also interrupted by ISRs, which are called as a response to external events. A particularly interesting aspect is that, an increasing number of executed ISRs can delay the planning of user tasks (even high priority tasks). Therefore, the careful monitoring of task scheduling delays can yield an effective instrument in the design of an anomaly detection system for ICs.

2.2 Design Considerations

The design of an anomaly detection system for industrial controllers needs to account for several restrictions. While apparently the hardware of modern ICs may provide sufficient computing power to run complex detection algorithms (e.g., neural networks, clustering techniques) as a distinct OS process (i.e., on top of the OS), this strategy brings a significant competitor to the real-time scheduler for the available processing time. Subsequently, it can cause significant delays in the scheduling of critical tasks, and, ultimately, it can halt the scheduler (a behavior that is triggered automatically when a task cannot be scheduled within a specific time period – configured with the help of a "watchdog timer"). Besides these aspects, normally, control application developers do not have access to the underlying OS. This restriction stems from the immense risk of altering the correct behavior of the real-time scheduler when uncontrolled changes are performed to the OS. Consequently, developers must use the available programming models and languages, which significantly reduce the size and complexity of the applicable instruction set.

According to these observations, it is clear that any practical proposal for a detection system needs to be positioned in the user task and it must account for the resources available at this application layer. Therefore, in order to ensure its suitability for existing control applications, we identify two fundamental requirements: (i) the detection algorithm's implementation must be modular and it needs to be isolated from the existing control logic in order to minimize the required changes and to reduce the complexity of maintaining its code (e.g., debugging, updating);

and (ii) the anomaly detection algorithm must account for the limited programming features available at this level (e.g., adopt simple arithmetic operations), while ensuring that the schedulability of the user tasks is not affected.

2.3 Developed Anomaly Detection System

The architecture of the proposed anomaly detection system addresses the aforementioned requirements as follows.

Firstly, the detection algorithm is implemented as a separate user task, hereinafter called the *monitoring* task. This is a periodic task configured with the lowest priority, which repeatedly records the task's start time and runs a lightweight anomaly detection algorithm. The period of the monitoring task is chosen as low as possible in order to ensure high sensitivity to scheduling delays caused by an increasing number of interrupts (e.g., input events, increased number of packets received on a networking interface), or by the abnormal behavior (e.g., change of load) of other user tasks. An important requirement is to configure the cyclic monitoring task to skip execution cycles if time overrun occurs, due to the preemption caused by higher priority tasks. An example execution of the proposed monitoring task has been included as part of Fig. 1. Here we observe that the monitoring task is configured with the lowest priority (among user tasks) in order to ensure that its scheduling time is influenced by all the planned tasks (through subsequent interruptions and delays). As a result, a careful recording of the delays occurring in the scheduling time of the monitoring task can indicate the change of behavior in the application's execution. This can further indicate the presence of internal/external disturbances.

Secondly, the monitoring task implements a detection algorithm that leverages the sensitivity of the scheduling time of user tasks to disturbances (e.g., internal software execution state changes, external cyber attacks), to detect abnormal application behavior. The detection algorithm is formulated in terms of simple arithmetic operations in order to minimize the execution overhead and to guarantee that schedulability is not affected.

The detection algorithm is implemented as a separate (monitoring) task. Its operation builds exclusively on the information available to this task. It leverages the measured start time of the task, which denotes the time at which the task begins its execution. Let t_i be the i-th start time of the monitoring task, and $T_i = t_i - t_{i-1}$ the i-th measured task period, that is, the time elapsed from the previous run. We note that for a high priority task the measured period has a mean value close to the configured task period, while exhibiting a slight variation (i.e., jitter) due to the system interrupts. Conversely, a task with the lowest priority has a significantly higher jitter for the measured task period.

Note that the PLC runs a set of periodic tasks, which are repeated after each *hyper-period*. The *hyper-period* is the least common multiple of all task periods, and is large enough such that all tasks are run at least once. Using a moving average filter on the monitoring task, we measure its average period over the hyper-period (over L

consecutive samples), in order to reduce the jitter of different tasks scheduled across different monitoring intervals. In the following, the filtered measured period $\widetilde{T}_j$ is used in the anomaly detection algorithm, computed as follows:

$$\widetilde{T}_j = \sum_{i=(j-1)L}^{jL} T_i/L. \tag{1}$$

In order to minimize the computational overhead and to ensure the proposal's practical integration with existing (legacy) controllers, the proposed detection algorithm is the statistical cumulative sum. A similar technique has been widely used by previous studies in the construction of efficient detection strategies [12]. Briefly, based on the work firstly proposed by Page [18], and Montgomery [17], a two-sided cumulative sum is computed in order to detect the increase and decrease of the mean value of the monitoring task period. The upper (C_j^+) and lower (C_j^-) cumulative sums are computed based on the mean shift value (K), the measured task period (denoted by $\widetilde{T}_j$), the expected task period (denoted by $\hat{T}$), and the deviation of the measured task period σ_T, according to the following two equations:

$$C_j^+ = max\Big(0, C_{j-1}^+ + \widetilde{T}_j - (\hat{T} + K)\Big), \tag{2}$$

$$C_j^- = max\Big(0, C_j^- + (\hat{T} - K) - \widetilde{T}_j\Big), \tag{3}$$

where K is usually expressed through the standard deviation unit $K = \frac{k}{2}\sigma_T$.

According to this detection strategy, a change point is detected when $(C_j^+ \geq h) \cup (C_j^- \geq h)$, where h is the detection threshold. Obviously, the selection of k and h has a major impact on the detection sensitivity and accuracy. Several threshold values can be chosen (e.g., $h_1, h_2, \ldots$) in order to define different levels of criticality, which would also permit to define a different set of actions for each distinct level. Consequently, several distinct detection levels can be chosen to send notifications, stop noncritical system tasks, and to send critical alarm status, reducing thus the negative effects of false alarms to the industrial process.

In case multiple change points need to be detected (i.e., in case a disturbance is persistent), in the above-mentioned method the next starting point is configured as the cumulative sum of the previously measured change point. In this case, the new expected value needs to be computed once again. According to [17], this is computed as follows:

$$\hat{T^1} = \begin{cases} \hat{T} + K + C_j^+/N^+, & \text{for}(C_j^+ \geq h), \\ \hat{T} - K - C_j^-/N^-, & \text{for}(C_j^- \geq h), \end{cases} \tag{4}$$

where N^+, and N^- denote the number of consecutive periods in which $C_j^- \neq 0$ and $C_j^+ \neq 0$.

3 Experimental Assessment

3.1 Test Infrastructure

The implementation has been tested in the context of a real SCADA system operating in a Romanian gas transportation network. The system builds on a primary controller (PLC^P) produced by Phoenix Contact, model ILC 350-PN. PLC^P runs the necessary control logic and handles the communication (OPC, Modbus TCP, Modbus RTU) with the other components which include the secondary controllers (PLC^S), the Modbus RTU slaves, HMIs, which are all typically found in the automation solution of a gas distribution node in the Romanian gas transportation network. A simplified view of the automation system's architecture is shown in Fig. 2.

Given the significance of the primary controller, the following tests focus on this controller. Each task running on PLC^P is described as a tuple of (period, deadline, [best case worst case execution time], priority). Accordingly, PLC^P runs the following periodic tasks: a control task (100 ms, 200 ms, [10 30], 1); one task for reading inputs (50 ms, 100 ms, [1 10], 1); a Modbus/RTU task (10 ms, 100 ms, [1 2], 0); a Modbus/TCP task (100 ms, 500 ms, [1 2], 2), and monitoring task (10, 200, [1 1], 3). The average filter window length (L) has 100 samples, which permits every task to run at least once.

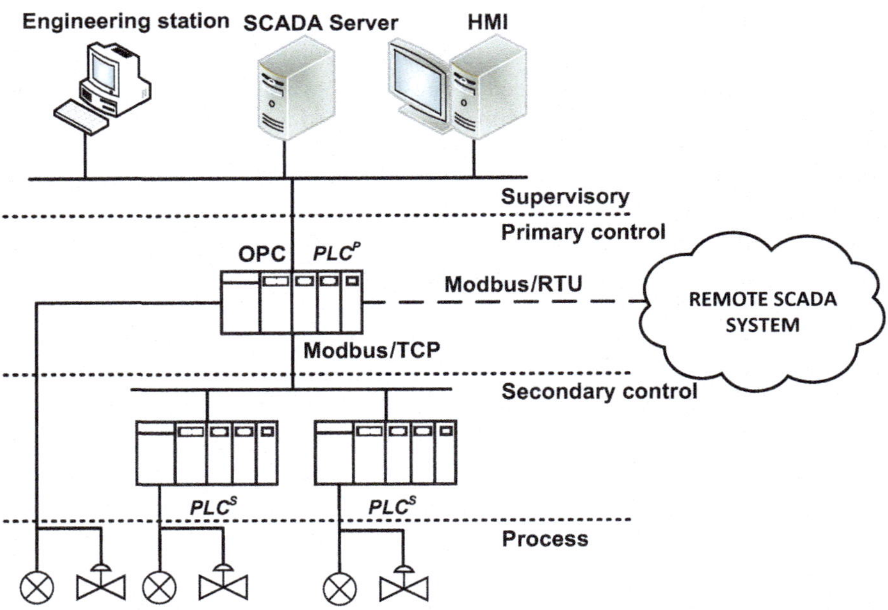

Fig. 2 Simplified architecture of a gas distribution node in a Romanian gas transportation network

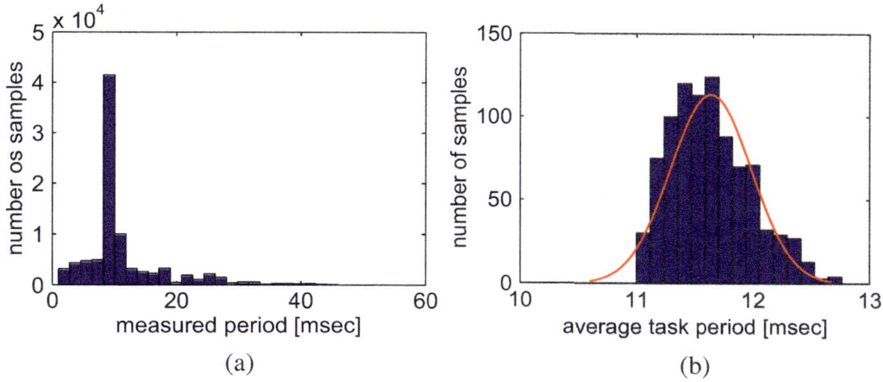

Fig. 3 The distribution of task periods. (**a**) T_i, (**b**) $\widetilde{T}_j$

The implementation and tuning of the proposed detection engine needs to start with an off-line analysis of the recorded attack-free run of the system and in the same time use the existing methods [2] for estimation of the upper bounds on worst case response times based on the task parameters. This is needed to determine the monitoring task period's mean value and deviation. In control systems the asynchronous events have a minimal inter-arrival time, and can be either measured a priori based on the physical process analysis, or can be determined using statistical analysis on the recorded task execution time (Fig. 3a).

Accordingly, the filtered monitoring task period shows a normal distribution (Fig. 3b) with a mean value of 11.63 ms, a standard deviation of 0.346 ms, and a maximum value of 12.76 ms. Even in the disturbance-free run, the expected value of the monitoring task period is larger than 10 ms, since the monitoring task has the lowest priority. The selected parameters for the anomaly detector include half of a deviation shift in both directions ($k = 1/2$), an alert threshold on three standard deviations ($h_1 = 3$), and an error threshold equal to four standard deviations ($h_2 = 4$).

3.2 Measured Results

In order to test the developed anomaly detection system, the network traffic generated by the `nmap` tool (full host scan) was replayed at various packet rates (i.e., the attack traffic) against PLC^P. Figure 4a, b show the behavior and output of the developed anomaly detection system (ADS) in the attack-free scenario. Here, the two threshold levels have been highlighted with horizontal red lines.

Next, in the first set of experiments two 60 s attacks were launched at a 30 s time intervals. The first attack issued 200 packets/s, while the second attack issued 500 packets/s. The detection algorithm was configured to run the two sided

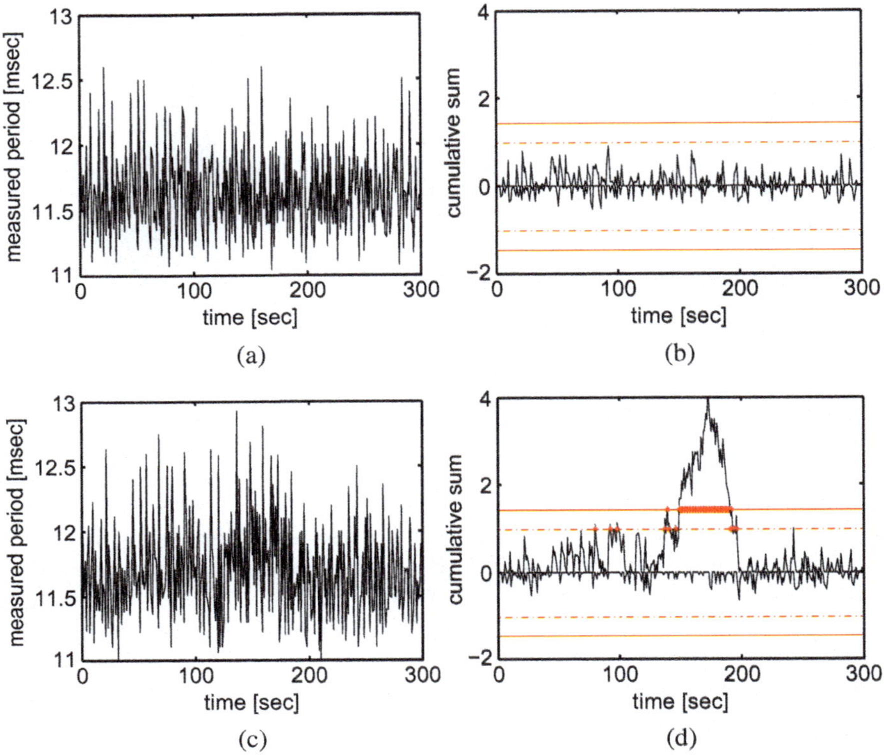

Fig. 4 Monitoring task period and output in the first experimental scenario. (**a**) Task period without attack. (**b**) Detection without attack. (**c**) Task period with attack. (**d**) Detection with attack

cumulative sum. As shown in Fig. 4c, the change in the measured task period is not trivial to notice. However, the anomaly detection algorithm (see Fig. 4d) signals an alert at 100 s, and later on it signals an error level, when the value of the detection algorithm reaches h_2. We observe that the error signal persists even after the attack is stopped (at 170 s). This is a direct consequence of the cumulative sum, which is designed to detect a single point of change. However, if a human intervention is needed when errors are signaled, the identification of the single point of change may not be sufficient.

Next, we launched a second set of experiments with the proposed multiple change point detection algorithm, and with two 120 s attacks launched at 30 s time differences. The first attack replayed the network scan traffic with a 200 packets/s rate, while the second attack increased the replayed packets rate to 700 packets/s. The first rate was chosen to be comparable in the number of packets to the total number of packets processed by the primary controller (PLC^P), while the second rate is equal to the total throughput of PLC^P as shown in Fig. 5.

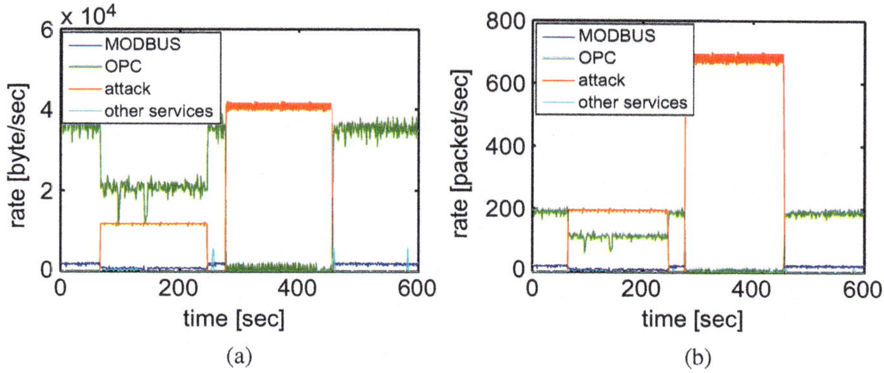

Fig. 5 Network traffic with attack. (**a**) Throughput. (**b**) Packet count

In order to showcase the significance of leveraging multiple change points, two distinct experiment instances have been considered: a first instance that leverages the single change point strategy, and a second instance that leverages the multiple change points strategy. The measured task period for both instances is shown in Fig. 6a.

The results in Fig. 6b show that the magnitude of the attack triggers a significant increase in the value of the cumulative sum. This yields an error state that persists almost up to 600 s, while the attack ended at 450 s. Conversely, in the case of the multiple change point detection strategy, once a threshold is exceeded, the controller estimates the new task period mean (see Fig. 6c) and resets the value of the cumulative sum. Figure 6d shows that the multiple change point detects both the presence and the absence of the attack in a few samples.

The second experiment was repeated under different system load, by changing the control task algorithm including the best case and worst case execution time. The system utilization rate was computed based only on the recorded execution time for the user tasks. The communication tasks and the network traffic have been left unchanged, while the throughput was identical to the one presented in Fig. 5. The mean system utilization rate was chosen between 20% and 50%, but the maximal load did not exceed the schedulability condition. The variation of the system utilization rate is shown in Fig. 7, and the number of detected signals (alert and error) are included in Table 1. As expected, if the system utilization rate is low, the monitoring task scheduling is not delayed by other task or system interrupts generated by the incoming packets. In this case the filtered monitoring task period has a mean value of 10.11 ms, a standard deviation of 0.11 ms for 20% system utilization, a mean value of 12.47 ms, and a standard deviation of 0.527 ms for 50% system utilization. In order to increase the sensitivity on the low utilization rate, the period of the monitoring task can be reduced accordingly.

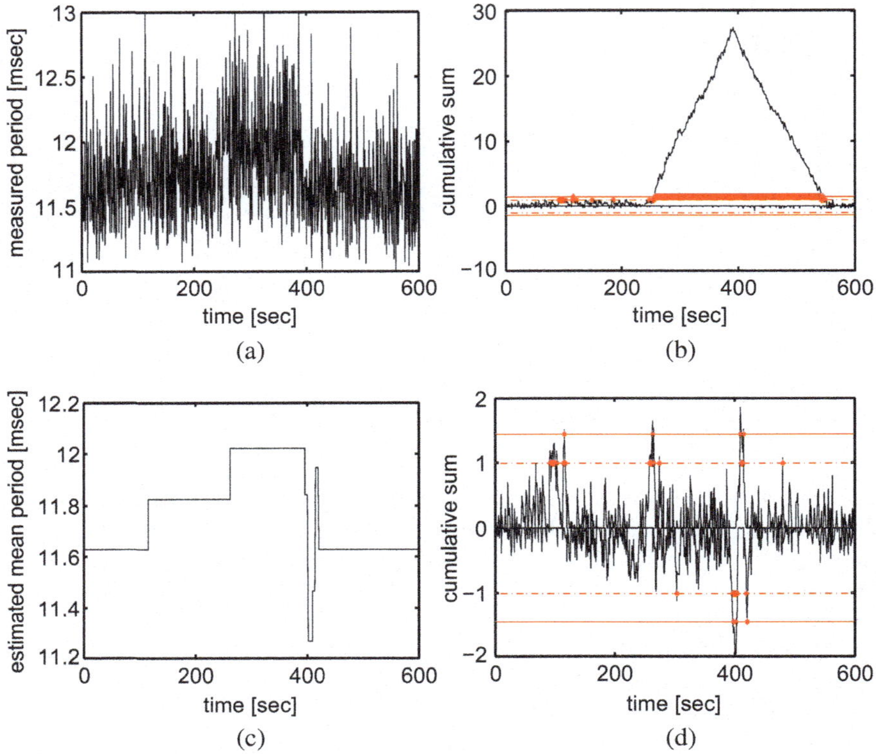

Fig. 6 Monitoring task period and output in the second experimental scenario. (**a**) Task period with attack. (**b**) Detection with single change point. (**c**) Estimated task period mean. (**d**) Detection with multiple change points

Based on the network traffic analysis a significant reduction in OPC network traffic can be observed in the presence of the attack traffic. The throughput on the OPC communication channel was reduced proportionally with the attack traffic and eventually stopped if number of attack packets are increased. Besides this, monitoring only the network traffic is not enough to identify the system anomalies. The third set of experiments modify the OPC traffic rate, by downloading data and code from the engineering station through the OPC communication protocol (see Fig. 8). We observe that when the download/upload operation is initiated, the traffic rate on OPC increases from 40 to 250 kbyte, while the packet rate is adjusted accordingly by the OS running on the PLC. This is needed in order to preserve the packet processing time near a constant value without causing a major impact on

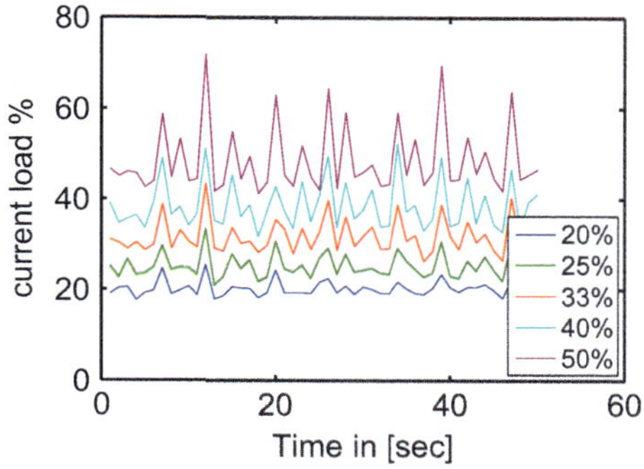

Fig. 7 The variation of the system utilization rate

Table 1 Number of detected alerts for different system utilization rates

	Attack rate	System utilization rate				
Detection level	packet/s	20%	25%	33%	40%	50%
Alert	200	2	3	12	28	164
	700	4	164	179	207	280
Error	200	0	0	4	8	155
	700	1	159	170	200	280

the scheduling of user tasks. As the monitoring task period is not influenced by the regular interventions from the engineering station, the proposed anomaly detection system doesn't generate false alarms.

4 Conclusions

We presented a methodology for enabling edge security in Industrial Control System. In essence, the approach shows that a proper tailoring of operations, together with a careful analysis of industrial controllers (i.e., the edge devices), an effective security module can be integrated into these critical devices as well. More specifically, our proposal is twofold: (i) a monitoring task that repeatedly records the task's start time; and (ii) a lightweight anomaly detection engine based on the computation of the cumulative sum. Intrinsic details have been presented with respect to the functioning of IC, and the implementation of the proposed approach. Experimental results on a real industrial controller confirmed the feasibility and applicability of the approach. As future work, we intend to evaluate its applicability to other industrial controllers as well.

Acknowledgements This work was supported by a grant of the Romanian National Authority for Scientific Research and Innovation, CNCS/CCCDI-UEFISCDI, project number PN-III-P2-2.1-BG-2016-0013, within PNCDI III.

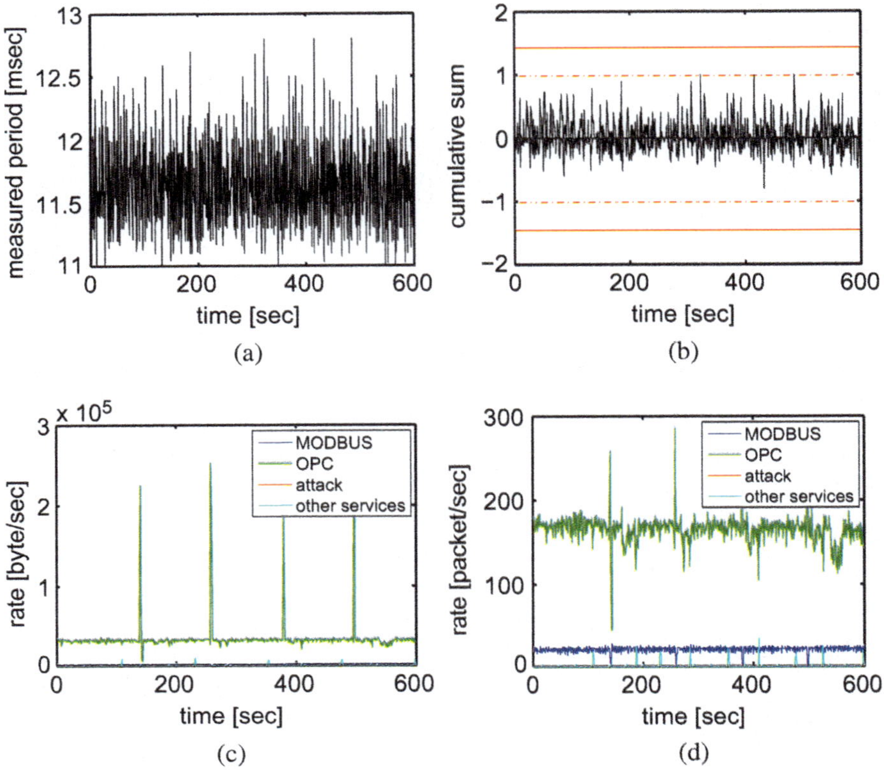

Fig. 8 Monitoring task period, output and network traffic in the third experimental scenario. (**a**) Task period. (**b**) Detection with single change point. (**c**) Throughput. (**d**) Packet count

References

1. Almalawi A, Fahad A, Tari Z, Alamri A, AlGhamdi R, Zomaya AY (2016) An efficient data-driven clustering technique to detect attacks in scada systems. IEEE Trans Inf Forensics Secur 11(5):893–906. https://doi.org/10.1109/TIFS.2015.2512522
2. Bini E, Nguyen THC, Richard P, Baruah SK (2009) A response-time bound in fixed-priority scheduling with arbitrary deadlines. IEEE Trans Comput 58(2):279–286
3. Carcano A, Coletta A, Guglielmi M, Masera M, Fovino IN, Trombetta A (2011) A multidimensional critical state analysis for detecting intrusions in SCADA systems. IEEE Trans Ind Inf 7(2):179–186. https://doi.org/10.1109/TII.2010.2099234
4. Cárdenas AA, Amin S, Lin ZS, Huang YL, Huang CY, Sastry S (2011) Attacks against process control systems: risk assessment, detection, and response. In: Proceedings of the 6th ACM Symposium on Information, Computer and Communications Security, ASIACCS'11. ACM, New York, pp 355–366. https://doi.org/10.1145/1966913.1966959
5. Chen B, Ho DWC, Zhang WA, Yu L (2017) Distributed dimensionality reduction fusion estimation for cyber-physical systems under dos attacks. IEEE Trans Syst Man Cybern Syst PP(99):1–14. https://doi.org/10.1109/TSMC.2017.2697450

6. Di Pietro A, Panzieri S, Gasparri A (2015) Situational awareness using distributed data fusion with evidence discounting. In: Rice M, Shenoi S (eds) Critical infrastructure protection IX. Springer, Cham, pp 281–296
7. Filippini R, Silva A (2014) A modeling framework for the resilience analysis of networked systems-of-systems based on functional dependencies. Reliab Eng Syst Saf 125:82–91. https://doi.org/10.1016/j.ress.2013.09.010, http://www.sciencedirect.com/science/article/pii/S0951832013002676
8. Fovino IN, Coletta A, Carcano A, Masera M (2012) Critical state-based filtering system for securing SCADA network protocols. IEEE Trans Ind Electron 59(10):3943–3950. https://doi.org/10.1109/TIE.2011.2181132
9. Genge B, Rusu DA, Haller P (2014) A connection pattern-based approach to detect network traffic anomalies in critical infrastructures. In: Proceedings of the Seventh European Workshop on System Security, EuroSec'14. ACM, New York, pp 1:1–1:6. https://doi.org/10.1145/2592791.2592792
10. Genge B, Siaterlis C, Karopoulos G (2013) Data fusion-base anomay detection in networked critical infrastructures. In: 2013 43rd Annual IEEE/IFIP Conference on Dependable Systems and Networks Workshop (DSN-W), pp 1–8. https://doi.org/10.1109/DSNW.2013.6615505
11. Giani A, Bent R, Pan F (2014) Phasor measurement unit selection for unobservable electric power data integrity attack detection. Int J Crit Infrastruct Prot 7(3):155–164. https://doi.org/10.1016/j.ijcip.2014.06.001, http://www.sciencedirect.com/science/article/pii/S1874548214000407
12. Giraldo J, Cardenas A, Quijano N (2017) Integrity attacks on real-time pricing in smart grids: impact and countermeasures. IEEE Trans Smart Grid 8(5):2249–2257. https://doi.org/10.1109/TSG.2016.2521339
13. Ha D, Ahmed U, Pyun H, Lee CJ, Baek KH, Han C (2017) Multi-mode operation of principal component analysis with k-nearest neighbor algorithm to monitor compressors for liquefied natural gas mixed refrigerant processes. Comput Chem Eng 106:96–105. https://doi.org/10.1016/j.compchemeng.2017.05.029, http://www.sciencedirect.com/science/article/pii/S0098135417302466. ESCAPE-26
14. Hagerott M (2014) Stuxnet and the vital role of critical infrastructure operators and engineers. Int J Crit Infrastruct Prot 7(4):244–246
15. Haller P, Genge B (2017) Using sensitivity analysis and cross-association for the design of intrusion detection systems in industrial cyber-physical systems. IEEE Access 5:9336–9347. https://doi.org/10.1109/ACCESS.2017.2703906
16. Kiss I, Genge B, Haller P, Sebestyén G (2014) Data clustering-based anomaly detection in industrial control systems. In: 2014 IEEE 10th International Conference on Intelligent Computer Communication and Processing (ICCP), pp 275–281. https://doi.org/10.1109/ICCP.2014.6937009
17. Montgomery DC (2013) Introduction to statistical quality control. Wiley, New York
18. Page ES (1954) Continuous inspection schemes. Biometrika 41(1/2):100–115
19. Phoenix Contact GmbH Co. K (2010) PC WORX 6 IEC 61131-Programming
20. Portnoy I, Melendez K, Pinzon H, Sanjuan M (2016) An improved weighted recursive PCA algorithm for adaptive fault detection. Control Eng Pract 50:69–83. https://doi.org/10.1016/j.conengprac.2016.02.010, http://www.sciencedirect.com/science/article/pii/S0967066116300326
21. Rubio JE, Alcaraz C, Roman R, Lopez J (2017) Analysis of intrusion detection systems in industrial ecosystems. In: Proceedings of the 14th International Joint Conference on E-Business and Telecommunications (ICETE 2017) – vol 4: SECRYPT, Madrid, 24–26 July 2017, pp 116–128. https://doi.org/10.5220/0006426301160128
22. Shitharth S, Prince Winston D (2017) An enhanced optimization based algorithm for intrusion detection in SCADA network. Comput Secur 70(Supplement C):16–26. https://doi.org/10.1016/j.cose.2017.04.012, http://www.sciencedirect.com/science/article/pii/S0167404817300901

23. Stone S, Temple M (2012) Radio-frequency-based anomaly detection for programmable logic controllers in the critical infrastructure. Int J Crit Infrastruct Prot 5(2):66–73. https://doi.org/10.1016/j.ijcip.2012.05.001, http://www.sciencedirect.com/science/article/pii/S1874548212000200
24. Symantec (2014) Dragonfly: cyberespionage attacks against energy suppliers. Symantec Security Response
25. Wan M, Shang W, Zeng P (2017) Double behavior characteristics for one-class classification anomaly detection in networked control systems. IEEE Trans Inf Forensics Secur 12(12):3011–3023. https://doi.org/10.1109/TIFS.2017.2730581
26. Wang B, Mao Z (2018) One-class classifiers ensemble based anomaly detection scheme for process control systems. Trans Inst Meas Control 40(12):3466–3476

Secure Interconnection of IT-OT Networks in Industry 4.0

Cristina Alcaraz

Abstract Increasingly, the society is witnessing how today's industry is adapting the new technologies and communication protocols to offer more optimal and reliable services to end-users, with support for inter-domain communication belonging to diverse critical infrastructures. As a consequence of this technological revolution, interconnection mechanisms are required to offer transparency in the connections and protection in the different application domains, without this implying a significant degradation of the control requirements. Therefore, this book chapter presents a reference architecture for Industry 4.0 where the interconnection core is mainly concentrated in the Policy Decision Points (PDP), which can be deployed in high volume data processing and storage technologies such as cloud and fog servers. Each PDP authorizes actions in the field/plant according to a set of factors (entities, context and risks) computed through the existing access control measures, such as RBAC+ABAC+Risk-BAC (Role/Attribute/Risk-Based Access Control, respectively), to establish coordinated and constrained accesses in extreme situations. Part of these actions also includes proactive risk assessment measures to respond to anomalies or intrusive threats in time.

1 Introduction

Industry, in general, is accepting the incorporation of the new technologies, networks and communication protocols to modernize its systems and allow a wider connection from anywhere, at any time and in anyhow. There are already several related works reflecting this progress [16, 19, 30, 65, 66], in which multiple cyber-physical devices interact with control processes and manufacturing chains for greater production, distribution and quality of service. This technological confluence is mainly based on the new paradigms of the Internet of Things (IoT), such as

---

C. Alcaraz (✉)
Computer Science Department, University of Malaga, Malaga, Spain
e-mail: alcaraz@lcc.uma.es

D. Gritzalis et al. (eds.), *Critical Infrastructure Security and Resilience*,
Advanced Sciences and Technologies for Security Applications,
https://doi.org/10.1007/978-3-030-00024-0_11

the Industrial IoT (IIoT), and the new edge computing infrastructures, such as cloud and fog computing [34]; all of them working as part of a heterogeneous network where Information Technologies (IT) merge with the Operational Technologies (OT). In this way, it is possible to maximize, optimize and customize the production tasks, and offer a greater range of functional services for a better industrial sector, economy and society [21].

But when different IT-OT domains have to coexist to collaborate each other, interconnection mechanisms have to be extensively considered as mentioned in our previous works [10, 11]. In both works, different entities and application domains of the smart grid interconnect to provide a rapid and effective action in the field. Now, we expand the concept to include the Policy Decision Points (PDP) in the edge computing (i.e. in the cloud as a centralized component and in the fogs as part of each application domain) to not only simplify computational costs involved in the interconnection processes, but also take advantage and benefits of the new technologies of Industry 4.0. In this sense, we provide a reference architecture for any “smart” scenario (e.g. smart factories [16, 54], smart cities [39], smart health-care, etc.) of Industry 4.0 together with its influence sectors, ensuring at all time operational and control performance, dependability, survivability, sustainability and safety-critical [7].

Through the PDP nodes, different stakeholders and industrial domains can converge in the connections and cooperate in a same common environment, offering a federated network composed of multi-domains. However, this type of collaboration and the need to modernize control and operational processes may also bring about numerous classes of anomalies that may, in turn, lead to subsequent and drastic threats [50]. For this reason, the access to our domains is strictly restricted to: The type of roles assigned to each entity (either IT-OT devices, software processes or physical entities) that wishes to take access to the different resources of the system, the real state of the context (e.g. severity level of a threat, criticality level of the context, number of isolated controllers, segmented and uncontrolled areas, etc.) and the risks associated with that context. To orchestrate all these actions, our approach contemplates the traditional authorization mechanisms [41] based on RBAC+ABAC+Risk-BAC (Role/Attribute/Risk-Based Access Control, respectively), as well as the IEC-62351-8 standard [35].

The standard IEC-62351 [33] comprises specific eleven parts to manage critical environments, such as power grids and their substations. Concretely, these parts include the specification of security profiles for IEC 60870-5 objects [32], XML files and communication channels, as well as the definition of security architectures and roles. But from these eleven parts, we especially focus on the IEC-62351-8 [35] by encompassing a useful set of particular entities, such as human operators, security administrators and engineers, together with their roles and rights. Apart from considering this standard as part of our approach, the architecture proposed also addresses aspects related to risk management from a proactive perspective, so as to offer an imminent response before major and serious disruptions arise within the system or between systems.

In either case, all these functional aspects are widely described in this chapter, which is organized as follows: Sect. 2 presents the interconnection architecture taking into account the restrictions of the context and the characteristics of the new industry. In this section, a set of assumptions are establihed to simplify the design and the scope of the approach. Each component of the architecture is widely described in Sects. 3.1 and 3.2, in which we consider the inclusion of the new edge computing infrastructures to address the policy decision points. The feasibility of the approach is, to the contrary, analyzed in Sect. 3.3 so as to show the effectiveness of the components and guarantee protection to each of the industrial areas and their final services. Lastly, Sect. 4 concludes the book chapter and presents future work.

2 Interconnection Architecture for Industry 4.0 Scenarios

When different application domains need to be interconnected each other, it is commonly applied interconnection frameworks based on Policy Enforcement Points (PEP) and PDP [60]. Through PEP, entities (i.e. physical members, IT-OT devices or software processes) can request access to the different resources of the system. In this case, the PEP intercepts and forwards the request to the PDP so that this latter can manage the authorization policies and determine the access level to the different sections of the system according to a set of factors: The type of entity, the resources and the context. Once the decision is taken by the PDP, the PEP processes it to permit or deny access to the interested entity, thereby protecting the critical resources of the system.

This way of connecting systems can also allow today's industry to interconnect industrial multi-domains, at which the creation of a cooperative environments is generally required to transparently connect providers, customers and other industrial networks [66]. In this sense, our architecture should follow a collaborative interconnection model where interconnection components (i.e. PDP) should maintain certain information of the own federated network. The architectures presented in [11, 25] and [10] are clear federation examples. The former is a patent where users and domains are able to transparently connect each other. The patent characterizes the inter-domain communication through an additional Meta Policy Decision Point (MPDP) to manage authentication and authorization processes between domains. The works [11] and [10], to the contrary, assign all the authentication process in the respective domains and concentrate all the authorization process in intermediaries PDP working like proxies.

If we unify both ideas and adapt them to our architecture, we can find a way to connect different industrial domains together with their application sub-domains, at which different protocols and technologies can coexist. To do this, we assume the following structural conditions, technologies and stakeholders:

Structural conditions Today the new industrial revolution accepts the inclusion of the new IT to manage, manipulate and store operational data and processes. This also means that industrial networks have to protect IT-OT connections through perimeter protection elements such as industrial firewalls and/or Virtual LAN (VLAN) for segmentation, Intrusion Detection/Prevention Systems (IDS/IPS) and Virtual Private Networks (VPN) for a secure tunneling through IPSec.

Technologies Apart from the technological diversity in control terms (e.g. sensors, actuators, controllers – remote terminal units or programmable logic controllers –, robot units, etc.) and the proliferation of industrial communication protocols (e.g. OPC-UA, 6LowPAN, IO-Link, EtherNet/IP and EtherCAT, WirelessHART, ISA100.11a or ZigBee PRO) [6, 16], there is an important need to integrate IT services to render large industrial data streams and processes. Among these IT services, we stress the cloud and the fog computing [34], which can compute contextual information for future administrative or operational actions, and benefit the control (per domain) and the processes related to context management, predictive maintenance, detection of anomalies and equipment failures, performance monitoring, governance, auditing or forensic.

As for security, it is widely assumed that all sections of the interconnection system, including the Machine-to-Machine (M2M) communication between devices, are protected through the existing security mechanisms and standards [33]. Beyond the perimeter protection, cryptography, key management systems, identity management, access control and traditional security protocols such as Secure Sockets Layer (SSL) or Transport Layer Security (TLS) are also essential for processing, storing and transferring critical data from a secure perspective [24]; without ruling out high-level security services such as privacy, trust or quality of service [12].

Stakeholders As stated in [16], customers and providers may also be part of the operational procedures to accelerate, customize and optimize the manufacturing and logistic processes, maximizing operational performance and costs in the plant/field. This also means that the model proposed should allow the influence of external connections with access to IT networks, such as the cloud or the fog. From the set of entities specified in [35], we also identify, among others, the participation of engineers, auditors and security administrators since they can interact with the system to offer essential actions for the production and distribution of minimal services to end-users, such as energy, water or food.

All these assumptions are also illustrated in Fig. 1. This figure clearly represents the technological confluence of Industry 4.0 composed of diverse operational and control areas, and multiple types of stakeholders. As can be observed, each domain comprises a set of OT devices working with different communication protocols and interacting with IT networks, such as industrial wireless sensor networks, RFID (Radio-Frequency Identification) or fog-computing. The role of the fog-computing is to locally provide a mean of processing and storage of large volumes of data, the information of which can also be compiled by a federated cloud infrastructure, common for all the application domains. The cloud technology, to the contrary, serves as a holistic environment capable of managing data related to users, control and context belonging to the different "smart world" scenarios (e.g. smart factories, smart grid, smart cities, smart health-care, etc.), the services of which are fundamental for social and economic well-being.

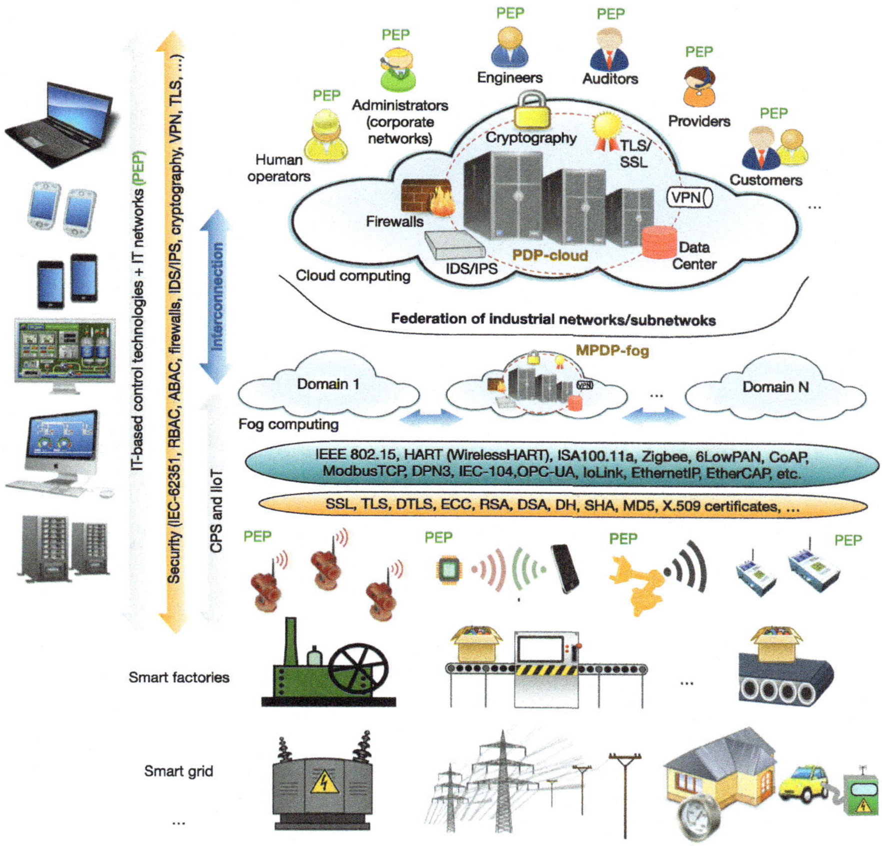

Fig. 1 Secure interconnection architecture for Industry 4.0 scenarios

To articulate all these connections, the architecture accommodates two classes of PDP: One global to the entire system and another local to each application domain. The global PDP is shaped in the cloud to (i) receive information of the context from each local PDP deployed in the fog and (ii) offer an overview of the state of the entire system and its correct performance. The PDP in the cloud is denoted here as PDP-cloud and the PDP in the fog is called MPDP-fog in relation to the MPDP described in [25]. The access to each one of these two kinds of policy decision points relies on the type of entity (human operators, providers, customers, administrators, auditors, engineers, processes or IT-OT devices). Local entities linked to local operational actions in the field or in the process plant should consider the access through its corresponding MPDP-fog; whilst remote entities (administrators, engineers, operators located at SCADA (Supervisory Control And Data Acquisition) centers, providers, auditors, etc.) to the different local domains should access through the PDP-cloud. This functional characteristic is also illustrated in Fig. 2.

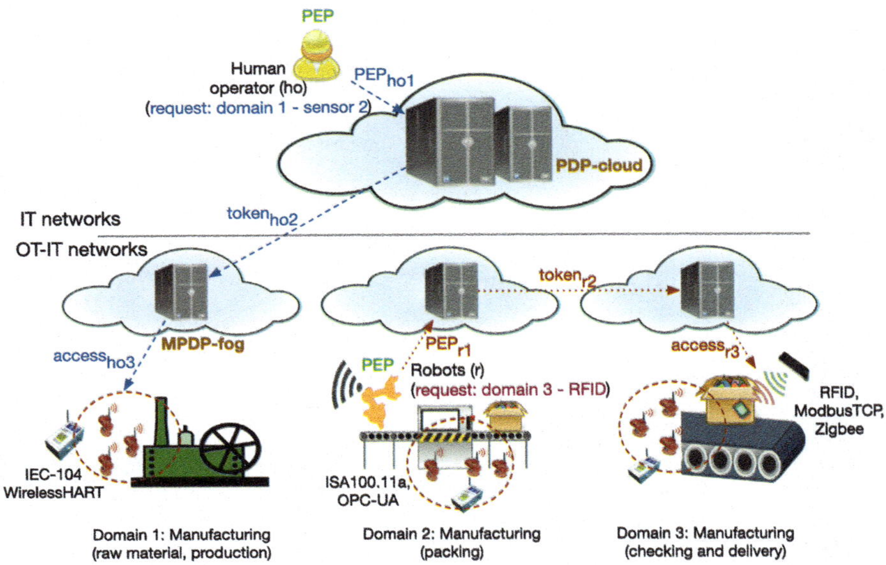

Fig. 2 Connection between: entities, cloud server and fog servers

Figure 2 is an example of how remote stakeholders are able to gain access through PEP instances to the PDP-cloud. However, the secure interoperability between IT-OT networks, the devices of which generally present performance limitations [7, 10], adds the need to locally delegate all the authorization process and translation actions of security policies and communication protocols to the MPDP-fog nodes. This condition endorses that the PDP-cloud is only able to authenticate external entities and validate the access according to the context, leaving all access responsibility to the meta PDP. In this way, the architecture simplifies the centralized actions in the cloud and any bottleneck occurrence. Note that this restriction is also subject to M2M communications of each domain. In this case, the authentication procedure is concentrated in each MPDP to locally handle PEP calls between domains and unburden the cloud of these operations.

3 Interconnection Components for Industry 4.0 Connections

Both the architecture of the PDP-cloud and the MPDP-fog are described in detail in this section together with those components that these include. More specifically, the actions taken by the PDP-cloud are firstly addressed to show how external connections are managed from an independent infrastructure to each domain, and later the specific components of the meta PDP are analyzed.

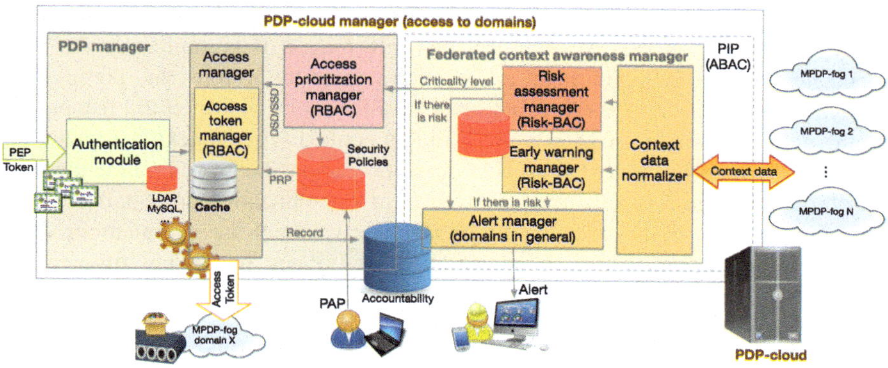

Fig. 3 PDP-cloud: architecture and functional components

3.1 *PDP-Cloud: Modules and Functionality*

Figure 3 represents the architectonic design of the modules that integrate the PDP operations required between entities and domains. Particularly, the architecture adds two chief components: The *PDP manager* and the *context awareness manager*. The former is in charge of validating the authentication tokens provided by each entity. This means that each entity must authenticate by itself from its own organization, delegating all the authorization process in the policy decision points.

Authentication is a procedure required to validate the identity of an entity and favor legitimate access to resources of the system. If the authentication is made from the entity premise and the access through the cloud, then it is required to consider the solutions described in [13]. This survey classifies the methods according to the location of the authentication modules, where the methods implemented in the "entity side" are mainly based on identity and context schemes. Chow et al. for example, define in [52] an identity-based authentication scheme, the core of wich is focused on the zero-knowledge authentication, the digital signature, and the fuzzy method. In contrast, Schwab and Yang specify in [18] a federated authentication framework, known as TrustCube with similar features to the OpenID technology, managing multiples types of policies related to the platform, devices and users. This way of authenticating in the entity side would not only reduce maintenance costs of databases in the cloud side, but it would also benefit the user's mobility. Human operators, engineers or even customers using mobile devices within a specific application scenario, such as manufacturing plants in smart factories or smart grid substations, can request PEP instances from any where, at any time and in any how, thereby promoting the new paradigms of the IoT; i.e. the IIoT.

But despite this local procedure, any validated identity in its premise also has to show its authenticity and legitimacy in the PDP-cloud through the use of authentication tokens. These tokens should add certain information about the previous authentication process and specific information about the PEP request, such as: The identity of the resource and the domain, and the type of action to be performed on the resource. All this information is compiled by the PDP manager together with additional information related to the roles and permissions assigned to the entity, the criticality level of the context in which the resource is being deployed and the risks associated to that context. The context information is obtained through the context awareness manager, responsible of computing the level of observation and controllability received from the application domain itself. This information is generally associated with attribute values that explain among other things: Which sensors, actuators or controllers are isolated, how many sub-areas are segregated, which nodes are working and which are not, status of communication links, operating systems or network parameters, etc.

Apart from the *authentication module* of tokens, the PDP manager is also composed of two further components: The *access token manager* and the *access prioritization manager*. These two components are based on the Role-Based Access Control (RBAC) strategy as recommended by the standard IEC-62351-part 8 [35]. Concretely, the standard defines seven specific roles for engineering and control scenarios managing different types of rights, such as the human operator with the capacity for viewing, reading, reporting and controlling operational objects and processes, or the engineer with the ability for viewing, reading, reporting, configuring and managing objects, databases and file systems. In addition to these roles, the standard reserves until 32.767 roles for private use, allowing to allocate new Industry 4.0 stakeholders as identified in Sect. 2. In our case, we could define capacities for viewing, reading and reporting operational objects assigned to auditors and customers, adding configuration support to providers.

This way of orchestrating permissions together with the dynamic capacity of RBAC for separation of duties, commonly known as Dynamic Separation of Duties (DSD), permits the system to redistribute security controls according to the security policies of each organization and the contextual conditions, adding versatility in the approach and dynamism in the protection process. To do this, the *risk assessment manager*, included as part the context awareness manager, has to compile all the information from the domains and contrast the existence or the persistence of possible risks [49] in the domains demanded where the control should prevail in extreme situations. This means that each entity should support at least two roles, one working as primary and other as secondary; and in this way, when control areas lack of enough connectivity, only authorized entities with determined roles could gain access to the affected area and take the control of this one. This propriety of DSD is widely described and implemented in [10, 11].

The context can also be managed by the *early warning manager* to estimate in optimal times and from a local or global perspective, the real state of the system for the next stage; and in the worst case, to prepare and activate the protection mechanisms related to location and alerting of human operators, as well as establish the prioritization levels taking into account the DSD properties. Any estimation must be loaded to the database for future risk assessments, in which a set of parameters should be evaluated, such as: The frequency, the relevance and the severity of the threat in the different domain/s, the criticality of the scenarios and their resources, the degree of devastation and the consequences (e.g. in social or economic terms), etc. The computation of all of these inputs will allow to compute and estimate any cascading effect between subsystems or systems, track and visualize in real time the threat in order to tackle the problem, and improve the regulatory procedures related to governance, auditing and forensic.

All this context information is part of the Policy Information Point (PIP) as specified in the RFC-2904 [60] for the interconnection of systems. A PIP refers to the management point where a set of attribute values related to resources, subjects and environment is compiled and normalized, to later determine the severity degree of the area and permit or not access to the area. This features also allows us to adapt the methods Attribute-Based Access Control (ABAC) and Risk-Based Access Control (Risk-BAC), and combine them with RBAC, in order to further restrict access conditions. Through ABAC+Risk-BAC, it is possible to take more stringent decisions established by the real attributes of the context and the risks associated with that context [57], further delimiting the access conditions by dynamically managing roles. In the literature, there several related works for IoT and IIoT environments [14, 29, 38], which can be considered for future implementations.

Finally, the *access manager*, integrated in the PDP manager, computes not only the information received from the respective modules but also verifies the legitimacy of the permissions to be executed in the field. For this action, it is necessary to contrast the information with the security policies stored in the databases, which are managed by technical administrators, installers or engineers through Policy Administration Points (PAP). Once the information is processed, the manager generates an access token to later validate the entity and the access itself in the destination domain. To accelerate the management of future related PEP instances or detect possible abuses in the requests (i.e. replay attacks), the access manager also needs to keep a temporal copy of each instance managed through a cache memory.

3.2 MPDP-Fog: Modules and Functionality

This section presents the architectonic model of the meta policy decision points configured in the respective fog infrastructures installed in each of the application domains (see Fig. 4). Similar to the PDP-cloud architecture, each MPDP-fog includes two chief modules: The *access manager* and the *domain awareness manager*. The first module contains an *authentication component* capable of addressing two types of actions depending on the origin and the class of token: (i) Verify the authenticity of the access tokens received from the cloud and (ii) validate the identity of those PEP instances established from other domains.

In this state, the technical capacities of the technologies are also keys to determine the authentication mode. For example, M2M communication based on IIoT devices and manufacturing machines (e.g. sensors, actuators, controllers or robots) are not generally tamper-resistant to attacks and they are based on constrained hardware components [4], working by themselves at remote locations such as substations or operational plants [40]. To reduce computational and communication overheads, the use of lightweight authentication schemes at the application layer

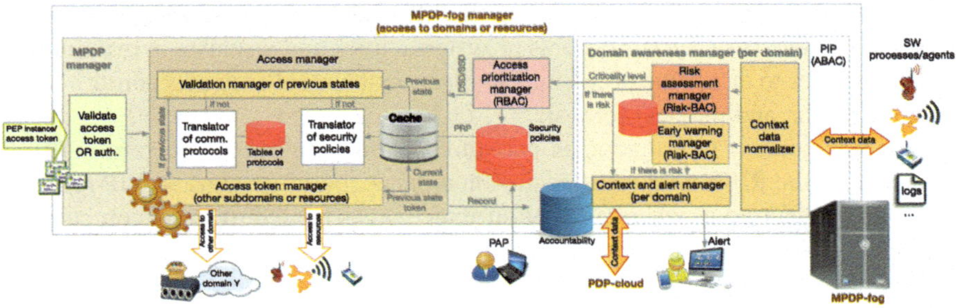

Fig. 4 MPDP-fog: architecture and functional components

and security protocols at the transport layer (TLS or Datagram TLS (DTLS)) are extensively considered in the literature [1, 28]. However, the design of lightweight solutions (at the application layer) for certain paradigms like IIoT, is still a great challenge for the scientific community [67]. In this case, we stress some works related to cyber-physical systems and IIoT such as [26, 53] and [17]. Esfahani et al. propose in [26] a mutual authentication mechanism for M2M communication using simple primitives and mathematical operations (hashes and XOR), thereby simplifying the authentication processes. In [53], the authors, to the contrary, offer an authentication framework to validate the identity of each object in the IIoT according to the device-specific information; and in [17], Chin et al. similarly propose M2M two-layer-based authentication framework for smart grid scenarios where smart meters are authenticated by a public key infrastructure and digital signature.

At the transport layer, there are already available several communication protocols for IoT, such as [58, 64]: 6LowPAN (IPV6 over Low power wireless Personal Area Network) [47], MQTT (Message Queue Telemetry Transport) [44, 58], AMQP (Advanced Message Queuing Protocol) [58], XMPP (Extensible Messaging and Presence Protocol) [61], DDS (Data Distribution Service) [45], and CoAP (Constrained Application Protocol) [48]; all of them supporting authentication measures through SSL and DTLS sessions. Namely, all the protocols except CoAP are based on TLS, whilst CoAP is focused on DTLS [1, 26]. Moreover, XMPP and AMQP can also use the Simple Authentication and Security Layer (SASL) protocol to authenticate devices [42, 43]. However, for all these protocols and the existing works related to the IoT field [31, 37, 62, 63] is recommendable to verify the suitability of the approach taking into account the technical restrictions of the IIoT devices together with control requirements as specified in [7].

Continuing with the actions of the access manager, the system has to validate all the previous states before computing any new access request. The goal is to reduce any computational cost involved in the context evaluation and translation of security policies and communication protocols. As stated in Sect. 2, the operational performance is critical at this interconnection point since multiple and concurrent access requests are generally demanded in this stage; either from the cloud or from any application area (through a new PEP request). To ensure this performance level, the system needs to temporarily cache all the actions performed by the access manager to avoid passing through the translators of communication protocols and security policies. Normally, both modules demand computation and time to address translation tasks considering the management and updating of specific tables for the matching of protocols (including ports and IP addresses) and policies. Nonetheless, this computational consumption is heavily dependent on the type of implementation designed for the translation engine. For example, the work [23] proposes a protocol translator for industrial communication based on a service-oriented architecture, translating on-demand and at a low-latency cost; whilst [36] traduces the communication according to algebraic specifications and [10, 11] are based on a rule-based expert translation system.

In either case, these translations benefit interoperability tasks in such a way that IIoT entities in general, can connect with each other transparently as stated in [10, 11]. Both works reflect similar goals to the proposed approach, in which different interfaces can establish connectivity without need to follow an equivalent security policy criterion for all parties and taking into account the natural conditions of the context to activate the DSD mechanisms if they are necessary. To go beyond these two works, our meta PDP nodes are not only able to handle the access according to the RBAC+ABAC properties, but they are also able to proactively determine the accessibility level according to the risks of the context. At this point, the risk management is critical to locally determine the severity degree of a threat and assess the consequences to establish much more restrictive conditions per area instead of only processing it in a centralized node as outlined in [10, 11].

Therefore, all our policy decision points, pertinent to the PDP-cloud and MPDP-fog, manage the access taking into account the capacities provided by RBAC+ABAC+Risk-BAC [41]. In particular, the access prioritization is under the restrictions given by the RBAC-based *access prioritization manager* as specified in Sect. 3.1. This manager activates the DSD mechanism according to the risk evaluation given by the *domain awareness manager*, which includes four similar components to the context manager of the PDP-cloud. The main difference that keeps the awareness manager of the MPDP-fog regarding the PDP-cloud is that the domain awareness manager is mainly focused on locally computing the context at which the application scenario is being developed. The information processed by this module can be very versatile, the data of which can belong to the physical world (e.g. humidity, temperature, pressure, etc.) and/or the virtual world through software processes, software agents (e.g. through opinion dynamics [49, 51]) or logs.

3.3 Suitability of the Architecture for Industry 4.0 Scenarios

Considering the control requirements specified in [7], this section analyses the suitability of the architecture proposed in Sect. 2 and its functions for future Industry 4.0 scenarios. In [7] five requirements for industrial control systems are identified: Real-time performance, dependability, sustainability, survivability and safety-critical; and for each of these requirements, the impact on the different elements and services of the system (information, resources, control, minimal services) is assessed. To adapt these five control requirements to our architecture, the analysis will primarily be focused on evaluating the five control requirements taking into account the primary needs of Industry 4.0 and the interconnection requirements defined in [9], such as rapid access, transparency in the connections, communication in real time, availability and reliability, also adding protection of devices and security in the multi-domain connections.

Real-time performance One of the main goals of including policy decision points in edge computing infrastructures is precisely to decrease the number of connections to the different application domains. Entities connecting from the cloud, first need to validate their access. If the access is not proper, then the system denies the entry in the field/plant, thereby reducing the number of connections in the domains and unnecessary overloads. This feature is also contemplated in each domain individually where entities first has to locally authenticate in their premise, so as to later gain access to the resources of other domain, thereby protecting the access to constrained resources. On the other hand, the use of cache memories and different authorization mechanisms, in which access privileges are restricted according to roles, contextual conditions of each domain and risks associated with these domains, also avoid serious overheads that may hamper the operational and control processes and cause significant delays.

Dependability and survivability The possibility of managing risks from a proactive and reactive perspective, allows the system to detect anomalies and response accordingly, ensuring availability of resources at all time and reliability of their services. Many of the anomalies come from the malfunctions or unsuitable configurations of systems or networks, or deficiencies in the coexistence of multiple systems [8], which may consequently bring about numerous security problems [15, 50]. Moreover, this manner of offering automatic fault detection also adds a significant reduction of maintenance costs and benefits the future Industry 4.0 services allocated in the cloud, such as predictive maintenance and the optimization of operational services and equipment. In this case, our risk assessment and early warning managers should connect with external services to feed up any suspicious of threat, risk or anomaly, or could even connect with specialized cyber-security centers (e.g. computer emergency response teams such as the CNN-CERT [20] or the ICS-CERT [22]) to alert of extreme situations. Also related to cyber-defense, the use of cache memories aids to detect replay attacks by simply tracking the last PEP requests, IP addresses and timestamps as specified in [5]. And though the advanced security services are not extensively considered in this chapter, such as privacy and trust, they are also essential as part the M2M communications and particularly between cloud/fog-IIoT devices [46, 55].

Sustainability The abilities of the system to manage risks and supply accountability capacities (see Figs. 3 and 4) allow the system to provide a more reliable governance, auditing and forensic services. The records in each one of the incoming points of the system can determine the type of access in the field/plant, the actions carried out in the resources, the entities or organizations responsible for these actions, the access periods and abuses in the connections. These inputs can even feed up the risk assessment and early warning managers to estimate inappropriate actions, anomalies or threats, and this can also help the system to review its security policies and any regulation framework required to respond accordingly. Evidently, if this process is rigorously considered, the system can comply with the interconnection requirements at all time and be sustainable for the control; i.e. maintain control services at all times and for a long period, at an acceptable level for the protection of resources and critical infrastructures [7]. This sustainability feature is also supported by the abilities of each MPDP to translate protocols and security policies, and if, in addition, the corresponding modules are regularly updated, the system also ensures a tenable interconnection.

Safety-critical In this aspect, we highlight the capacity of the system to protect the critical resources from external accesses, and especially when the domain hosting the resources present extreme crisis situations. Under these critical circunstances, it is always recommendable to recover and return the control [2, 3] to the affected area, and to avoid, as much as possible, expanding the effect of the threat to the rest of interconnected domains, known as cascading effect. In addition to this, the management of proactive responses also aims to reduce possible secondary effects in the system or between systems, reducing the risk levels in advance [56] and any threatening effect that may entail a drastic cascading effect.

Taking into account all these control and interconnection principles, we consider that our architecture is suitable for the new control industry, in which a set of (IT-OT) technologies, protocols and networks have to coexist for a long period of time. From these technologies, we particularly focus on the cloud and fog infrastructures to accommodate the approach and reduce computational and communication costs, as well as enhance their resources to add additional capacities related to interconnection and protection in different terms and levels; all of them necessary for Industry 4.0 scenarios.

4 Conclusions and Future Work

A multi-domain interconnection architecture is proposed in this book chapter to connect multiple federated areas belonging to critical infrastructures (e.g. manufacturing industry and supply chains, food production plants, power grids and smart cities [39], and water treatment plants) without breaking the control requirements that generally these infrastructures demand. Typical domains are, for example, the generation, transmission and distribution substations configured as part of smart grid, or the different manufacturing sections corresponding to smart factories or supply chains. To do this, the architecture is based on a two layer interconnection system composed of two kinds of policy decision points; one located at a centralized system and another distributed throughout the different application domains. The centralized node corresponds to a cloud server capable of managing any entry belonging to external entities of the system or subsystems, such as customers, providers, auditors, etc.; whilst the distributed PDP are in charge of controlling any access coming from other domains or from the cloud.

This architecture based on two-layers incorporates in each PDP a set of functional modules with the ability to handle the access according to the characteristics and intentions of each entity together with their roles, the real state of the context and its resources, as well as the risks associated to this context. Therefore, the approach includes components capable of orchestrating aspects related to RBAC+ABAC+Risk-BAC with support for proactive solutions before serious interruptions may arise within the entire system. For the future, we intend to implement all these components in our laboratory [59] to later include them as part of the goals of the European SealGRID project [27]. And with this, show all the functionalities of the architecture for the new control industry, further considering the incorporation of specific services related to protection of communication channels (entities-cloud/fog, cloud-fog, M2M), privacy and trust.

Acknowledgements This work has been mainly supported by the EU H2020 project SealGRID (8.06.UE/47.8035), with partial support of the project DISS-IIoT financed by the University of Malaga (UMA) by means of the "I Plan Propio de Investigación y Transferencia" of UMA where specific knowledge about assembly and configuration of IIoT and control components has been widely received.

References

1. Al-Fuqaha A, Guizani M, Mohammadi M, Aledhari M, Ayyash M (2015) Internet of things: a survey on enabling technologies, protocols, and applications. IEEE Commun Surv Tutorials 17(4):2347–2376
2. Alcaraz C (2017) Resilient industrial control systems based on multiple redundancy. Int J Crit Infrastruct (IJCIS) 13(2/3):278–295
3. Alcaraz C (2018) Cloud-assisted dynamic resilience for cyber-physical control systems. IEEE Wirel Commun 25(1):76–82
4. Alcaraz C, Cazorla L, Fernandez G (2015) Context-awareness using anomaly-based detectors for smart grid domains. In: 9th International Conference on Risks and Security of Internet and Systems, vol 8924. Springer, Trento, pp 17–34
5. Alcaraz C, Fernandez-Gago C, Lopez J (2011) An early warning system based on reputation for energy control systems. IEEE Trans Smart Grid 2(4):827–834
6. Alcaraz C, Lopez J (2010) A security analysis for wireless sensor mesh networks in highly critical systems. IEEE Trans Syst Man Cybern Part C Appl Rev 40(4):419–428
7. Alcaraz C, Lopez J (2012) Analysis of requirements for critical control systems. Int J Crit Infrastruct Prot (IJCIP) 5:137–145

8. Alcaraz C, Lopez J (2013) Wide-area situational awareness for critical infrastructure protection. IEEE Comput 46(4):30–37
9. Alcaraz C, Lopez J (2017) Secure interoperability in cyber-physical systems. In: Security solutions and applied cryptography in smart grid communications, chap 8. IGI Global, USA, pp 137–158
10. Alcaraz C, Lopez J, Choo KKR (2017) Resilient interconnection in cyber-physical control systems. Comput Secur 71:2–14
11. Alcaraz C, Lopez J, Wolthusen S (2016) Policy enforcement system for secure interoperable control in distributed smart grid systems. J Netw Comput Appl 59:301–314
12. Alcaraz C, Zeadally S (2013) Critical control system protection in the 21st century: threats and solutions. IEEE Comput 46(10):74–83. https://doi.org/10.1109/MC.2013.69
13. Alizadeh M, Abolfazli S, Zamani M, Baharun S, Sakurai K (2016) Authentication in mobile cloud computing: a survey. J Netw Comput Appl 61:59–80
14. H. F. Atlam, A. Alenezi, R. J. Walters, G. B. Wills and J. Daniel, "Developing an Adaptive Risk-Based Access Control Model for the Internet of Things," 2017 IEEE International Conference on Internet of Things (iThings) and IEEE Green Computing and Communications (GreenCom) and IEEE Cyber, Physical and Social Computing (CPSCom) and IEEE Smart Data (SmartData), Exeter, 2017, pp. 655–661.
15. Cazorla L, Alcaraz C, Lopez J (2018) Cyber stealth attacks in critical information infrastructures. IEEE Syst J 12:1778–1792
16. Chen B, Wan J, Shu L, Li P, Mukherjee M, Yin B (2018) Smart factory of industry 4.0: key technologies, application case, and challenges. IEEE Access 6:6505–6519
17. Chin WL, Lin YH, Chen HH (2016) A framework of machine-to-machine authentication in smart grid: a two-layer approach. IEEE Commun Mag 54(12):102–107
18. Chow R, Jakobsson M, Masuoka R, Molina J, Niu Y, Shi E, Song Z (2010) Authentication in the clouds: a framework and its application to mobile users. In: Proceedings of the 2010 ACM Workshop on Cloud Computing Security Workshop, CCSW'10. ACM, New York, pp 1–6
19. Cisneros-Cabrera S, Ramzan A, Sampaio P, Mehandjiev N (2017) Digital marketplaces for industry 4.0: a survey and gap analysis. In: Camarinha-Matos LM, Afsarmanesh H, Fornasiero R (eds) Collaboration in a data-rich world. Springer, Cham, pp 18–27
20. CNN-CERT (2006) Centro Cripotológico Nacional. https://www.ccn-cert.cni.es. Last retrieved in June 2018
21. Dar KS, Taherkordi A, Eliassen F (2016) Enhancing dependability of cloud-based IoT services through virtualization. In: 2016 IEEE First International Conference on Internet-of-Things Design and Implementation (IoTDI). IEEE, Berlin, pp 106–116
22. Department of Homeland Security (2004) Industrial control systems cyber emergency response team (ICS-CERT). https://ics-cert.us-cert.gov. Last retrieved in June 2018
23. Derhamy H, Eliasson J, Delsing J (2017) Iot interoperability on-demand and low latency transparent multiprotocol translator. IEEE Internet Things J 4(5):1754–1763. https://doi.org/10.1109/JIOT.2017.2697718
24. Dzung D, Naedele M, Von Hoff TP, Crevatin M (2005) Security for industrial communication systems. Proc IEEE 93(6):1152–1177
25. Edwards NJ, Rouault J (2008) Multi-domain authorization and authentication. US 7.444,666B2
26. Esfahani A, Mantas G, Matischek R, Saghezchi FB, Rodriguez J, Bicaku A, Maksuti S, Tauber M, Schmittner C, Bastos J (2017) A lightweight authentication mechanism for m2m communications in industrial IoT environment. IEEE Internet Things J 1–1. https://ieeexplore.ieee.org/abstract/document/8006209/
27. European Commision (2018) SealGRID: scalable, trustEd, and interoperAble pLatform for sEcureD smart GRID. http://www.sgrid.eu/. Last retrieved in June 2018
28. Ferrag MA, Maglaras LA, Janicke H, Jiang J (2016) Authentication protocols for internet of things: a comprehensive survey. CoRR abs/1612.07206
29. Fraile F, Tagawa T, Poler R, Ortiz A (2018) Trustworthy industrial IoT gateways for interoperability platforms and ecosystems. IEEE Internet Things J 1–1. https://ieeexplore.ieee.org/document/8353121/

30. Grangel-González I, Baptista P, Halilaj L, Lohmann S, Vidal ME, Mader C, Auer S (2017) The industry 4.0 standards landscape from a semantic integration perspective. In: 2017 22nd IEEE International Conference on Emerging Technologies and Factory Automation (ETFA), pp 1–8
31. Hernández-Ramos JL, Pawlowski MP, Jara AJ, Skarmeta AF, Ladid L (2015) Toward a lightweight authentication and authorization framework for smart objects. IEEE J Sel Areas Commun 33(4):690–702
32. IEC-61850 (2003) Power utility automation – communication networks and systems in substations – parts 1–10. TC 57 – Power systems management and associated information exchange
33. IEC-62351 (2007–2011) IEC-62351 parts 1-8: information security for power system control operations, international electrotechnical commission. http://www.iec.ch/smartgrid/standards/. Last retrieved in June 2018
34. Industrial Internet Consortium, Edge Computing Task Group (2018) Introduction to edge computing in IIoT. An Industrial Internet Consortium White Paper, IIC:WHT:IN24:V1.0:PB:20180618. https://www.iiconsortium.org. Last retrieved in June 2018
35. International Electrotechnical Commission (2011) IEC-62351-8, Power systems management and associated information exchange – data and communications security – part 8: role-based access control. http://www.iec.ch/smartgrid/standards/. Last retrieved in June 2018
36. Ishihara Y, Seki H, Kasami T (1993) A translation method from natural language specifications into formal specifications using contextual dependencies. In: Proceedings of the IEEE International Symposium on Requirements Engineering, pp 232–239
37. Lee JY, Lin WC, Huang YH (2014) A lightweight authentication protocol for internet of things. In: International Symposium on Next-Generation Electronics (ISNE), pp 1–2
38. Liu Q, Zhang H, Wan J, Chen X (2017) An access control model for resource sharing based on the role-based access control intended for multi-domain manufacturing internet of things. IEEE Access 5:7001–7011
39. Lom M, Pribyl O, Svitek M (2016) Industry 4.0 as a part of smart cities. In: 2016 Smart Cities Symposium Prague (SCSP), pp 1–6
40. Lopez J, Alcaraz C, Roman R (2013) Smart control of operational threats in control substations. Comput Secur 38:14–27
41. Lopez J, Rubio JE (2018) Access control for cyber-physical systems interconnected to the cloud. Comput Netw 134:46–54
42. Norris R, Miller J, Saint-Andre P (2017) XEP-0034: SASL integration. https://xmpp.org/extensions/xep-0034.html. Last retrieved in June 2018
43. OASIS (2012) OASIS advanced message queuing protocol (AMQP) version 1.0 Part 5: security. http://docs.oasis-open.org/amqp/core/v1.0/amqp-core-security-v1.0.html. Last retrieved in June 2018
44. OASIS (2014) MQTT and the NIST cybersecurity framework version 1.0. http://docs.oasis-open.org/mqtt/mqtt-nist-cybersecurity/v1.0/mqtt-nist-cybersecurity-v1.0.html. Last retrieved in June 2018
45. OMG (2015) Data distribution service specification v1.4. https://www.omg.org/spec/DDS/About-DDS/. Last retrieved in June 2018
46. Pearson S, Benameur A (2010) Privacy, security and trust issues arising from cloud computing. In: 2010 IEEE Second International Conference on Cloud Computing Technology and Science (CloudCom). IEEE, pp 693–702
47. Qiu Y, Ma M (2016) A mutual authentication and key establishment scheme for M2M communication in 6LoWPAN networks. IEEE Trans Ind Inf 12(6):2074–2085
48. Raza S, Shafagh H, Hewage K, Hummen R, Voigt T (2013) Lithe: lightweight secure CoAP for the Internet of Things. IEEE Sens J 13(10):3711–3720 (2013)
49. Rubio JE, Alcaraz C, Lopez J (2017) Preventing advanced persistent threats in complex control networks. In: 22nd European Symposium on Research in Computer Security (ESORICS 2017), vol 10493, pp 402–418

50. Rubio JE, Alcaraz C, Roman R, López J (2017) Analysis of intrusion detection systems in industrial ecosystems. In: Proceedings of the 14th International Joint Conference on E-Business and Telecommunications (ICETE 2017), vol 4, pp 116–128
51. Rubio JE, Roman R, Alcaraz C, Zhang Y (2018), Tracking advanced persistent threats in critical infrastructures through opinion dynamics. In: European Symposium on Research in Computer Security. Springer, Barcelona, vol 11098, pp. 555–574
52. Schwab D, Yang L (2013) Entity authentication in a mobile-cloud environment. In: Proceedings of the Eighth Annual Cyber Security and Information Intelligence Research Workshop (CSIIRW'13). ACM, New York, pp 42:1–42:4
53. Sharaf-Dabbagh Y, Saad W (2017) Cyber-physical fingerprinting for Internet of Things authentication: demo abstract. In: Proceedings of the Second International Conference on Internet-of-Things Design and Implementation (IoTDI'17). ACM, New York, pp 301–302
54. Shrouf F, Ordieres J, Miragliotta G (2014) Smart factories in industry 4.0: a review of the concept and of energy management approached in production based on the internet of things paradigm. In: 2014 IEEE International Conference on Industrial Engineering and Engineering Management (IEEM). IEEE, pp 301–302
55. Sicari S, Rizzardi A, Grieco LA, Coen-Porisini A (2015) Security, privacy and trust in Internet of Things: the road ahead. Comput Netw 76:146–164
56. Thamhain H (2013) Managing risks in complex projects. Proj Manag J 44(2):20–35
57. Thomas MV, Chandrasekaran K (2016) Identity and access management in the cloud computing environments, chap. 3. ISI Global, Hershey, Pennsylvania, USA, pp 61–89
58. Thota P, Kim Y (2016) Implementation and comparison of M2M protocols for Internet of Things. In: 2016 4th International Conference on Applied Computing and Information Technology/3rd International Conference on Computational Science/Intelligence and Applied Informatics/1st International Conference on Big Data, Cloud Computing, Data Science Engineering (ACIT-CSII-BCD), pp 43–48
59. University of Malaga (2018) DISS-IIoT: design and implementation of security services for the industrial internet of things. https://www.nics.uma.es/projects/diss-iiot. Last retrieved in June 2018
60. Vollbrecht J, Calhoun P, Farrell S, Gommans L, Gross G, de Bruijn B, de Laat C, Holdrege M, Spence D (2000) AAA authorization framework. RFC 2904
61. Wang H, Xiong D, Wang P, Liu Y (2017) A lightweight XMPP publish/subscribe scheme for resource-constrained IoT devices. IEEE Access 5:16393–16405
62. Wang KH, Chen CM, Fang W, Wu TY (2018) On the security of a new ultra-lightweight authentication protocol in IoT environment for RFID tags. J Supercomput 74(1):65–70
63. Wu XW, Yang EH, Wang J (2017) Lightweight security protocols for the Internet of Things. In: 2017 IEEE 28th Annual International Symposium on Personal, Indoor, and Mobile Radio Communications (PIMRC), pp 1–7
64. Yassein MB, Shatnawi MQ, Al-zoubi D (2016) Application layer protocols for the Internet of Things: a survey. In: 2016 International Conference on Engineering MIS (ICEMIS), pp 1–4
65. Zheng P, Wang H, Sang Z, Zhong RY, Liu Y, Liu C, Mubarok K, Yu S, Xu X (2018) Smart manufacturing systems for industry 4.0: conceptual framework, scenarios, and future perspectives. Front Mech Eng 13(2):137–150
66. Zhong RY, Xu X, Klotz E, Newman ST (2017) Intelligent manufacturing in the context of Industry 4.0: a review. Engineering 3(5):616–630
67. Zhou W, Zhang Y, Liu P (2018) The effect of IoT new features on security and privacy: new threats, existing solutions, and challenges yet to be solved. CoRR abs/1802.03110

Part IV
Cybersecurity

Analysis and Triage of Advanced Hacking Groups Targeting Western Countries Critical National Infrastructure: APT28, RED October, and Regin

Henry Mwiki, Tooska Dargahi, Ali Dehghantanha, and Kim-Kwang Raymond Choo

Abstract Many organizations still rely on traditional methods to protect themselves against various cyber threats. This is effective when they deal with traditional threats, but it is less effective when it comes to Advanced Persistent Threat (APT) actors. APT attacks are carried by highly skilled (possibly state-sponsored) cyber criminal groups who have potentially unlimited time and resources.

This paper analyzes three specific APT groups targeting critical national infrastructure of western countries, namely: APT28, Red October, and Regin. Cyber Kill Chain (CKC) was used as the reference model to analyze these APT groups activities. We create a Defense Triage Process (DTP) as a novel combination of the Diamond Model of Intrusion Analysis, CKC, and 7D Model, to triage the attack vectors and potential targets for these three APT groups.

A comparative summary of these APT groups is presented, based on their attack impact and deployed technical mechanism. This paper also highlights the type of organization and vulnerabilities that are attractive to these APT groups and proposes mitigation actions.

Keywords Critical national infrastructure · Advanced persistent attack · APT · APT28 · Red October · Regin

---

H. Mwiki (✉) · T. Dargahi
Department of Computer Science, University of Salford, Manchester, UK
e-mail: H.Mwiki@edu.salford.ac.uk; T.Dargahi@Salford.ac.uk

A. Dehghantanha
School of Computer Science, University of Guelph, Guelph, ON, Canada
e-mail: adehghan@uoguelph.ca

K.-K. R. Choo
Department of Information Systems and Cyber Security, University of Texas at San Antonio, San Antonio, TX, USA
e-mail: raymond.choo@fulbrightmail.org

D. Gritzalis et al. (eds.), *Critical Infrastructure Security and Resilience*, Advanced Sciences and Technologies for Security Applications,
https://doi.org/10.1007/978-3-030-00024-0_12

1 Introduction

Cyber security is increasingly been placed on one of the top priorities in organizations, particularly for those in the critical national infrastructure sector [1]. Advanced Persistent Threat (APT) has been the most challenging threat, if not one of the most challenging threats, to the security and safety of critical national infrastructures [2]. APT attacks have become hard to tackle because unlike many other cyber-attacks which depend on automated scanning and exploitation of known vulnerability(ies), APTs are sophisticated, and in many instances human-driven attacks, with specific targets [3]. What makes APT even more complicated is the fact that they are often under the radar, targeted and very focused cyber-attacks and when an individual or a group get unauthorized access to an IoT (Internet of Things) network they can stay undetected for a long period of time [4]. APT groups consist of experts that leverage open source intelligence and social engineering methods to compromise government and commercial entities in a systematic manner [5]. To be undetected APT groups exploit vulnerabilities that are not normally known to the public, and they use encrypted communication, abuse standard protocols, and find-day vulnerabilities [6]. APT groups pose a tremendous risk to many organizations' infrastructures due to the nature of their attacks [7]. APT groups tend to customize their Malware [8] to attack a specific target. These malware are not usually detectable using traditional security mechanisms [9] and require utilization of advanced techniques such as deep learning [10], machine learning [11], and pattern analysis [12]. All these bring major complications to what kind of protection an organization can use to defend itself against APT attacks, as different APT groups use different Tactics, Techniques, and Procedures (TTPs) [13].

The aim of this paper is to analyze three specific APT groups mainly targeting Western countries' critical national infrastructure, namely: APT28, Red October and Regin. We develop a Defense Triage Process (DTP) which has been tailored to these APTs' TTPs, TTPs by using a novel combination of the Diamond Model of Intrusion Analysis, the Cyber Kill Chain, and the 7D model. The purpose of DTP is to start a progression of notifying and understanding a set of APT security controls that will create the base of successful, complete analyses of the organization using established best practices. This will help organizations to find an ideal solution to protect themselves from these APT groups. We answer the following questions about the aforementioned APT groups in this study:

1. Which organizations are attractive to APT28, Regin, and Red October?
2. How can target organizations defend themselves from these APT groups attacks?
3. What existing routes can APT groups use to get inside these organizations' network?
4. What actions are taken by these APT groups to access and manipulate organizations network?
5. What does our incident response plan need to define and test to mitigate residual risks acknowledged in the analysis?

The rest of this paper is organized as follows. The next sections provides a general overview of APT groups (APT28, Regin, and Red October). Afterwards, we discuss the CKC, the Diamond Model and the 7D model. We provide a detailed discussion of three APT groups' tactics, techniques and procedures and offer relevant mitigation mechanisms. Finally, we conclude the paper and offer several future works.

2 An Overview of APT28, Regin and Red October APT Groups

2.1 APT 28

This is the Russian APT group which is also known by many other names like Fancy Bear, Strontium, Pawn Storm, Sofacy, Sednit, Tsar Team etc. [14]. This group appears to focus on collecting intelligence that would be most useful to the Russian government [15].

Since at least 2007, the APT28 espionage movement has mostly targeted national critical infrastrucutre in the USA, Europe and the countries of the former Soviet Union, including governments and militaries, security organizations, media entities, and dissidents and entities with conflict with the current Russian Government [15].

APT28 steals internal data after compromising the victim and sometimes publicize them [16]. Up to now, this group has been involved in the Syrian conflict, NATO-Ukraine relations, the European Union refugee and migrant crisis, the 2016 Olympics and Paralympics Russian athlete doping scandal, public accusations regarding Russian state-sponsored hacking, and the 2016 U.S. presidential election [17].

APT28 depends on four main tactics to compromise planned targets. These include sending spear-phishing emails that either deliver an armored document that deploys malware onto a user's systems or contain a malicious URL intended to harvest the recipients' email credentials and provide access to their accounts. APT28 has also compromised and positioned malware on legitimate websites aiming to infect site visitors, and has gained access to organizations by compromising their web-facing servers.

2.2 Regin

This is the APT group which used a unique malware called Regin. The group created an extremely complex malware that can be modified with an extensive range of various capabilities which can be deployed depending on the target [18]. It can withstand long-term intelligence-gathering processes without being detected. It can hide itself and its activities on compromised computers [19].

The key purpose of Regin is intelligence-gathering and to support other types of attacks [19]. It has been used to collect data from government organizations, financial institutions, infrastructure operators, businesses, academics, and private individuals [19].

The precise technique used for the primary compromise is still a mystery, although numerous theories exist, such as the use of man-in-the-middle attacks with browser zero-day exploits [18]. Some of the victim tools and modules were designed for lateral movement. The replication modules are copied to remote computers using Windows administrative shares and then executed. This technique needs administrative privileges inside the victim's network [18]. In numerous incidents, the infected machines were also Windows domain controllers [18]. Targeting system administrators via web-based exploits is an easy way of attaining instant administrative access to the entire network [19].

2.3 *Red October*

This is also a Russian group [20]. They mainly target embassies and diplomatic agencies, Universities and research firms, Commercial organizations, Nuclear energy labs, Oil and gas companies, Aerospace institutions and Military in different countries, mostly related to the region of Eastern Europe, former USSR members, and countries in Central Asia.

This group aims at stealing classified information, obtain geopolitical intelligence as well as selling classified information on black market [21].

They use spear phishing to create initial infection and direct to a specific target or organization based on known information. They make use of known vulnerabilities of different applications such as Microsoft Office, PDF and Java [20].

Malware dropper is used to infect victims machine after the malicious file has been opened then core component is installed and communication with command and control (C&C) server is established via a backdoor module after that encrypted communication is established between victim machine and C&C server [21]. More than 60 different domains hardcoded in malware code to communicate with C&C servers [21]. The malware contains components that infect machines on the same local network without the initial phishing attack [21].

3 An Overview of CKC, Diamond and 7-D Models

3.1 *Cyber Kill Chain*

Cyber Kill Chain (CKC) term originated from the military, whereby, a systematic procedure taken by the enemy to attack the target to make a desired effects is known as the kill chain [22]. The Cyber Kill Chain describes phases of intrusions which adversary take during an attack as follow (see Fig. 1):

Fig. 1 Lockheed Martin Cyber Kill Chain [22]

(i) *Reconnaissance* – This is the initial stage where adversary gathers information to identify and select targets. Adversary uses Open Source Intelligence (OSINT) and go through public websites, social media and use any publicly available information. This stage also can involve some technical strategies such as port scan to find vulnerabilities, service, and application to exploit.

(ii) *Weaponization* – Here adversary starts to analyze the data collected on their target to figure out the best attack approaches to apply. Specific operating system, firewalls, and other technologies may be targeted by the attacker. Specific people also may be targeted through phishing and drive-by download attacks onto the endpoints.

(iii) *Delivery* – This is the stage where the weapon is delivered. Lockheed Martin Computer Incident Response Team (LM-CIRT) for the years 2004–2010 observed that USB removable media, email attachments, and websites were the most three dominant delivery vectors for weaponized payloads by APT actors

(iv) *Exploitation* – This is when intruders' code is triggered after the weapon being delivered to the victim host. Vulnerabilities in an application or operating system are what exploitation targets most of the time, but sometimes can be just exploiting the users themselves or control certain feature of an operating system that auto-executes code.

(v) *Installation* – This always starts with one infected system and then spread quickly. Remote access trojan or backdoor can be installed on the victim machine and enable the adversary to keep persistence inside the environment.

Infections can use numerous tactics such as tempering with security processes to hide their existence.

(vi) *Command and Control (C2)* – To have a bidirectional communication, attacker set up command and control channels to function between themselves and infected devices. At this stage, an attacker has access and control of the target environment. To hide their tracks, attackers use encryption in command and control channel.

(vii) *Actions on Objectives* – This is the final stage in which attacker take actions to accomplish their original purposes. The aim here is data exfiltration which includes collecting, encrypting and extracting information from the victim environment. Confidentiality, Integrity, and Availability (CIA) of data is often violated at this stage. The attacker also may use the initial compromised system as a means of compromising more systems and move across inside the network.

All APT groups follow certain stages of Cyber Kill Chain, and each of these stages leaves behind a certain trace [23]. Therefore, Cyber Kill Chain can be used to block APT attacks by mapping adversary kill chain indicators to defender course of action, identify patterns that link individual intrusions into broader campaigns, and understand the iterative nature of intelligence gathering form the basis of intelligence-driven computer network defense (CND) [24].

3.2 Diamond Model of Intrusion Analysis

This model establishes the basic atomic element of any intrusion activity. It is composed of four fundamental features: adversary, infrastructure, capability, and the victim [25]. Adversary can be insider, outsider, individual, group, and even an organization. Capabilities are the tools and/or techniques adversary used in the event. Infrastructure can be physical and/or logical communication structures the adversary uses to deliver a capability, maintain control of capabilities such as command and control, and achieve results from the victim such as data exfiltrating. The victim is the target of the adversary and against whom vulnerabilities and exposures are exploited and capabilities used. Additional meta-features are also defined to support higher-level constructs such as linking events together into activity threads and further coalescing events and threads into activity.

As shown in Fig. 2, the event defines the four fundamental features existing in all malicious incident: that for each intrusion event there is an *adversary* making a step to a planned goal by means of a *capability* over *infrastructure* against a *victim* making a consequence. The name of the model: Diamond Model, is due to the way these features are edge-connected representing their unique relationship and arranged in the figure of a diamond.

Relationships between features are grounded on analytic pivoting and how possible it is to reach the other connected angle from any angle on the Diamond

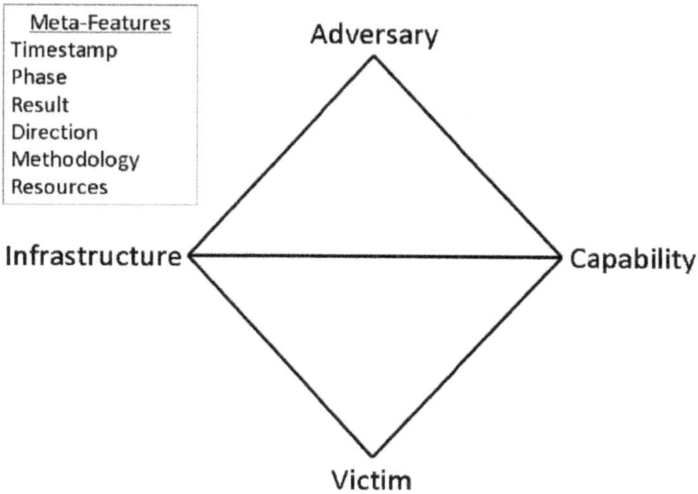

Fig. 2 The diamond event [25]

model. Example, from the victim angle, it is possible to see the capabilities which have been used against the victim over which infrastructure. From the infrastructure angle, it is possible to see capabilities which have been used over the infrastructure and to which victims. Also from the infrastructure angle, it is possible to potentially see the adversary controlling the infrastructure[25]. This whole movement from one component to another in a Diamond model is called *analytic pivoting*. Analytic pivoting is a fundamental analytic application of the model, it allows maximization of opportunities and clarification of intelligence gaps during analysis.

There is a further expansion of the Diamond model by two additional meta-features which defines the technology and social-political meta-features. Figure 3 shows the Extended Diamond model.

The Infrastructure and capability are connected by the technology meta-feature and defines the technology allowing the infrastructure and capabilities to interact efficiently. For example, Domain Name System (DNS) will be considered as a part of the technology meta-feature if malware uses it to determine its command and control point [26].

The relationship which always exists, and sometimes enduring between adversary and victim is described by the social-political meta-feature. This meta-feature also describes that there are primary needs, goals and intent exists behind all malicious events and the victim has a unique part in that connection. Concepts from criminology and victimology are allowed to be applied in Diamond model during intrusion analysis to enable one to realize the motive behind in choosing the victim, what value does adversary get from the victim and eventually how mitigation can be enhanced by influencing and manipulating that relationship.

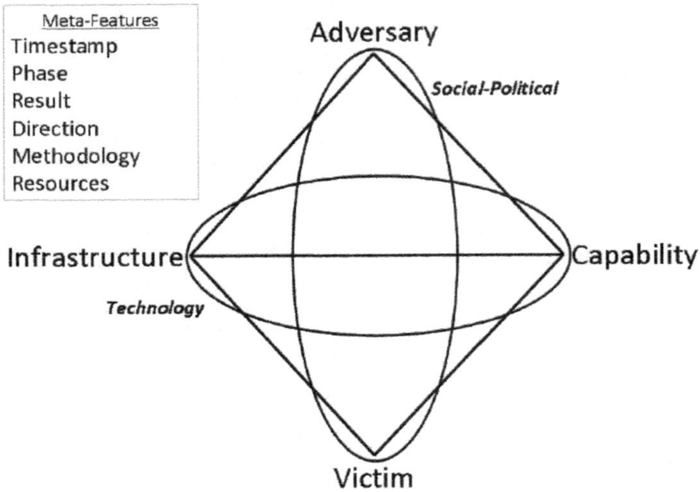

Fig. 3 Extended Diamond Model [25]

The existence of *Shared threat space* is highlighted also, whereby the needs of one or more of the same adversary are satisfied by two or more victims. The presence of this space shows that the sharing of threat intelligence is more profitable with those most likely to be affected by the same adversary as well as allow the associates of the space to potentially do prediction and forecasting of upcoming malicious movements.

The Cyber Kill Chain and the Diamond model are extremely complementary [25]. With the Cyber Kill Chain, it is possible to target and engage an adversary to make wanted effects [24]. The Diamond model assists in developing tradecraft and understanding to create and organize the knowledge which can be used to implement the Cyber Kill Chain analysis [25].

As shown in the Cyber Kill Chain, adversaries operate in multiple phases when they carry out an attack. A minimum of only two or more events is essential to make a malicious result. For example, the adversary must first prepare target selection and that's when malicious action can follow, hence those are two steps already [25].

An activity being found and events have been categorized and examined, they are well-ordered by the stages of malicious activity and connected by their causal relationship into threads. These threads are called *activity threads* [25]. Adversaries can take advantage of the knowledge and access gained in one operation and use it to enable other operations and hence being able to span horizontally and not only vertically along a single adversary-victim pair [27].

Figure 4 illustrates an activity thread whereby two victims are against an adversary's operations and a third unrelated victim against unknown an adversary's operations. Moreover, the dashed elements demonstrate the possibility of integrating hypotheses that can then be tested more or get a support of extra evidence collection.

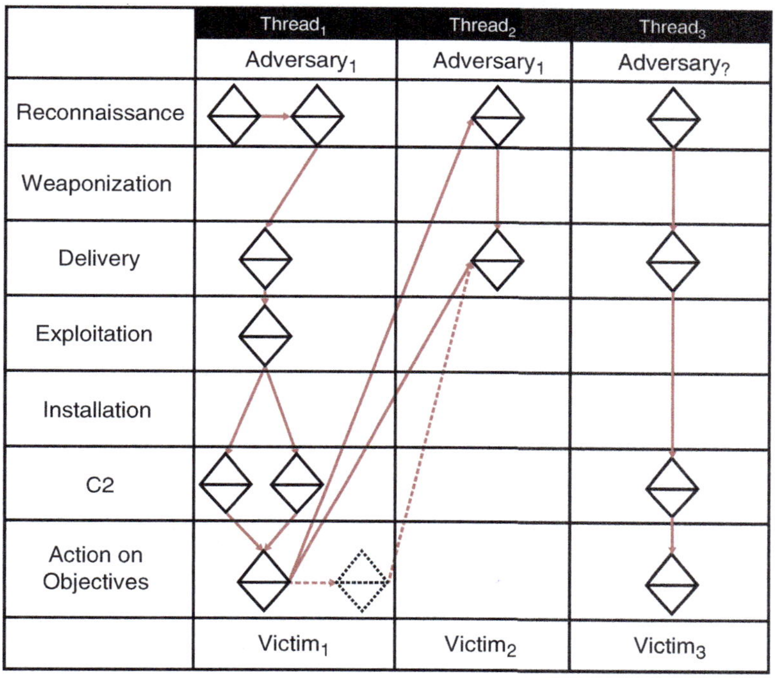

Fig. 4 Activity thread example [25]

Knowledge gaps and adversary campaigns can be identified with this organization of knowledge which can also be useful in generating hypothesis and documentation [28]. *Adversary processes* are the sub-graphs of these threads; they can be used to group and categorize activity based on the process rather than single indicators.

Even though activity threads organize the intelligence of adversary actions, attack graphs have often been applied by information security and assurance groups to hypothesize all exploit routes to an asset [29]. Decisions about defense have been planned using this methodology, and to make a defense decision there are things which have been considered such as the cost of a defensive action, how many exploit routes does it cover, and what the value of the particular asset. This has been addressed by the Diamond model and so integrates activity threads and attack graphs into a new plan known as the *activity-attack graph* as shown in Fig. 5.

This new graph assists in highlighting preferences of an attacker as well as possible alternative routes which can be taken by an attacker. Activity-attack graphs can help in making better strategies for mitigation by mitigating the existing threat while taking into consideration reactions or alternate tactics which can be taken or used by an adversary [27].

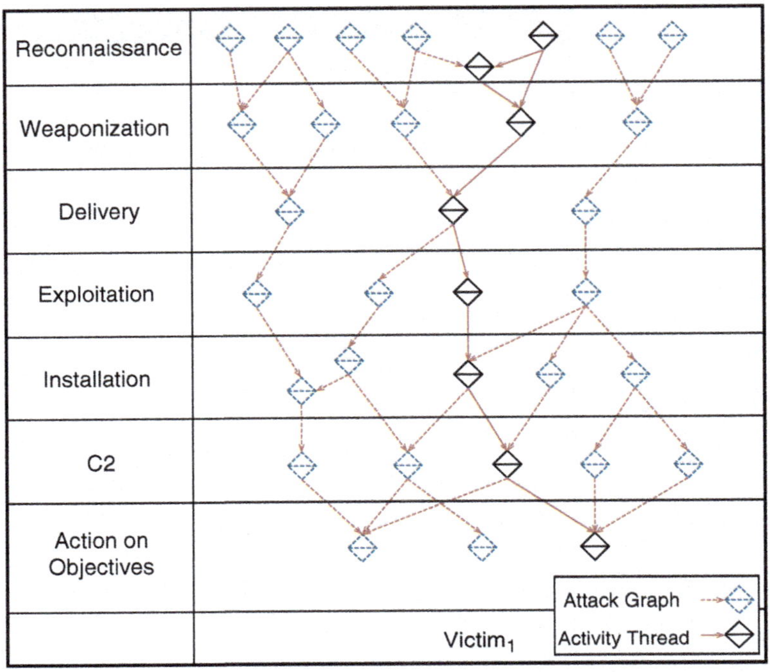

Fig. 5 Activity-attack graph [25]

3.3 7-D

7-D is a defence and mitigation modeling strategy that is well aligned with CKC. 7-D uses the Kill Chain Course of Action Matrix which determines how to detect, deny, disrupt, degrade, deceive and destroy the effectiveness of the adversary events along the kill chain phases as shown in Fig. 6.

Figure 6 shows 6 action categories: detect, deny, disrupt, degrade, deceive, and destroy (6D). In this paper, one action category has been added which is discover, hence there will be 7 action categories (7D). Discover is based on unknown bad activity while detect is based on known bad activity. To discover, regular threat hunting is required.

4 Analysis of APT28, Red October, and Regin Groups

In this part, APT groups named APT28, Red October and Regin have been studied to find out how they implement all the processes from the beginning to the end by analyzing their Cyber Kill Chain and Diamond model. Kill Chain Course of Action Matrix has been suggested using 7D model.

	Detect	Deny	Disrupt	Degrade	Deceive	Destroy
Reconnaissance	Web Analytics	Policy to Prevent Forum Use			Create fake postings	
Weaponization						
Delivery	NIDS, User Education	Email AV Scanning		Email Queuing	Filter but respond with out-of-office message	
Exploitation	HIDS	Patch	DEP			
Installation						
C2	NIDS	HTTP Whitelist	NIPS	HTTP Throttling		
Action on Objectives	Proxy Detection	Firewall ACL	NIPS	HTTP Throttling	Honeypot	

Fig. 6 Kill chain course of action matrix developed from threads 1 and 2 in Fig. 4 [25]

4.1 APT28

4.1.1 Cyber Kill Chain and Diamond Model

(i) *Reconnaissance*

According to Microsoft Security Intelligence report, this APT group takes their initial step by identifying and profiling potential victims. To identify their target, they rely mainly on OSINT [30]. APT28 scan websites with the aim of finding any web application vulnerabilities [31]. Before the 2016 U.S. election, APT28 is suspected to have performed a vulnerability scan trying to identify websites that are vulnerable to cross-site scripting (XSS) or Structured Query Language (SQL) injections attacks [31]. Between 10th of February and 14th of February 2015 in Ukraine, APT28 scanned 8,536,272 IPs to find possible vulnerabilities [32]. Microsoft OLE RCE CVE-2014-4114 also known as Sandworm, is an example of vulnerability exploited by APT28 to attack NATO, EU governments, the Ukrainian government, American academic organizations, and energy and telecommunications firms [33]. According to [32] the report, during reconnaissance APT28 pick victim organizations manually because they found the script with 11 IP classes hardcoded.

Legitimate domain name	Spoofed domain name controlled by STRONTIUM
accounts.google.com	accounts.g00qle.com
us-mg6.mail.yahoo.com	us-mg6mailyahoo.com
profile.live.com	privacy-live.com
mail.ukr.net	mail-ukr.net
www.nato.int	nato-news.com
www.bbc.com	bbc-press.org
www.osce.org	osce-press.com
www.eff.org	electronicfrontierfoundation.org

Fig. 7 Some of the domain names spoofed by APT28 in recent attacks

Diamond Model
The adversary is trying to find specific vulnerabilities which the victim has such as web application vulnerabilities, Microsoft Office vulnerabilities etc.

(ii) *Weaponization*

According to [31], this group analyses the information obtained from reconnaissance to get some intelligence which will enable them to find the best way to exploit identified vulnerabilities. APT28 can create a customized malware, or use known exploits and tools for the particular vulnerability as well as stolen certificates [34]. They weaponize their attacks which can use spear-phishing campaigns. They can create a fake link which has the domain name similar to a legitimate domain name which is used normally by the user [16, 30, 35], Fig. 7 shows some examples. Also, this means that they use social engineering as well so as to trick users to actually click the link [30].

Because they are well funded, this group has members who are highly skilled and who have enough time to find and weaponize identified vulnerabilities.

Sometimes they purchase the exploits from other sources [36].

Diamond Model
From the identified victim's vulnerabilities such as web application vulnerabilities the adversary is trying to use his capabilities such as malware, known exploits, and tools to weaponize the attack. This phase involves a pivot from the victim angle to the capability (v2c).

(iii) *Delivery*

This group uses email which may have a link to a fake website [37], or instead of sending a link, they can embed an exploit inside an attached malicious document [34]. According to a Google security team report, Sofacy is normally delivered by Microsoft word exploits as DOC, RTF and DOCX files [38]. For example, in 2017, Proofpoint researchers detected a Microsoft word attachment sent by email which malicious and was trying to exploit Adobe flash vulnerability, CVE-2017-11292.

The attack was attributed to APT28 [39]. In July 2017, APT28 spread out a malware with a spear phishing attack; through the email, a fake hotel reservation was delivered to the victims [40].

Also, APT28 uses a website to deliver the attack; they compromise websites which are most likely to be visited by the target [39].

USB removable media also have been used to deliver. In February 2015, Sofacy actors used USB removable media to deploy the attack [41].

Diamond Model
The adversary uses the capability they have to try to exploit the victim's vulnerability. Here the pivot is from the capability to the victim (c2v).

(iv) *Exploitation*

Most of the time APT28 use zero-day exploits. They tend to move quickly to take advantage of recently disclosed vulnerabilities [30]. They use exploits from a wide range of products of different vendors such as Adobe Flash Player, Microsoft Word, the Java Runtime Environment, Internet Explorer and some components of the Windows kernel. For example, the exploitation of a flash vulnerability CVE-2017-11292 can lead to arbitrary code execution across Windows, Mac OS, Linux and Chrome OS systems [39]. Sometimes one exploit can lead to further exploits; For example, a remote server may respond with a chain of exploits, zero-days, and privilege escalation that will infect the target's computer [42]. APT28 can exploit legitimate websites by injecting Browser Exploitation Framework (BeFF). To deploy a backdoor component, APT28 uses a dropper, CORE-SHELL, which eventually downloads other modules [30].

Diamond Model
This phase involved pivoting from the victim to the infrastructure such as web server, mail server, database etc. (*v2i*)

(v) *Installation*

APT28 installs various malicious tools to make sure that they maintain persistence. For example, they create AutoStart extensibility point (ASEP) registry entries and shortcuts to ensure that backdoor will always run immediately after the computer starts, and they tend to use multi-backdoor packages for extreme resilience [34]. Installation of APT28's signature GAMEFISH malware is the result of the execution of the macro within the malicious document [43]. If a malicious document is an office document, they use a mechanism called *Office test* to load a trojan each time a user opens Microsoft Office Applications [44]. Installation varies with the type of operating system the target is using. For example, when the target is running macOS, APT28 installs the backdoor trojan called XAgentOSX, and a tool named Komplex is used to download and install the XAgentOSX [45]. A Windows machine, persistence can be ensured by a kernel mode rootkit installed as a Windows service and also bootkit infecting the Master Boot Record (MBR) of the hard drive [36].

To evade detection, APT28 uses encrypted and compressed payloads which make things harder for antivirus and detection systems [38]. Each time Sofacy starts, it systematically disables crash reporting, logging and post-mortem debugging [38]. With the use of *NSFileManager:removeFileAtPath* method, they are able to delete a specified file, clear logs and reset timestamps. All these are an anti-forensic measure [35].

Diamond Model
This phase involved a pivot from the victim's infrastructure to the capability of the adversary (*i2c*).

(vi) *Command and Control*

Different network protocols such as SMTP, POP3, and HTTP, can be used by the backdoor to establish communication with its Command and Control (C2) servers [46]. The first action of the payload is to find a reliable means to communicate with its C2 server on the internet using a direct connection, through a proxy, or by injecting into a running browser. APT28's attack framework allows some exploits code to be loaded from C2 servers to carry on further exploits. For example, through an analysis of the dropped file which is an external C2 communication library, it was found that there were one primary C2 domain and three secondary C2 domains [47]. What happened is that the dropper made a contact with the primary C2 server and download components which enable communication with the secondary C2 servers for the second stage attack [32, 47]. FireEye Lab discovered a malware campaign targeting the hospitality sector in July 2017. In this attack there was a document holds a macro that after an enabled permit to finish the infection procedure. Macro was found to be a Visual Basic (VB) script which is able to extract the effective malware, which has to link to a C2 "mvtband.net" and "mvband.net" to download further malicious code to execute. For the time being these servers have already been blacklisted [40].

Diamond Model
This phase involved a pivot from the adversary's capability to the victim's infrastructure (*c2i*).

(vii) *Action on objectives*

This group has done so many damages on critical infrastructure networks [44]. Their intent most of the time is to manipulating a target, collecting information, and exfiltration [37]. For example, on April 2015 they attacked TV5monde, a global television network and almost destroy it by shutting down broadcasting [36]. This group can use the backdoor to harvest the entire contents of the USB device and then save it to the local computer to extract later [48]. They deploy a wide range of tools which can complete tasks like keylogging, email address and file harvesting [32]. In 2016 they gained access to an International Olympic Committee and were able to view and download athlete data [17]. They did cyber-attacks against the hospitality sector in July 2017 intended to collect information about hotel guests of interest [40].

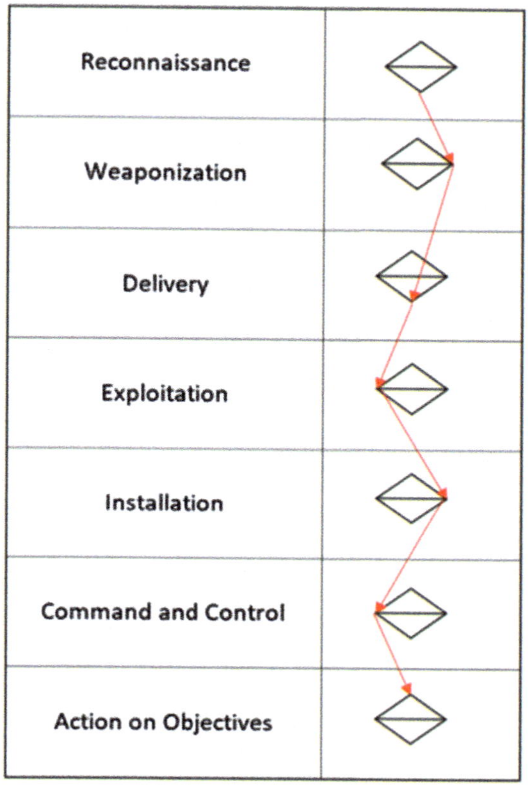

Fig. 8 Cyber Kill Chain and Diamond Model of APT28

Diamond Model
This phase involved a pivot from victim's infrastructure to the adversary's (*i2a*).

Figure 8 depicts Cyber Kill Chain and Diamond Model analysis for APT28 group.

4.2 Red October

4.2.1 Cyber Kill Chain and Diamond Model

(i) *Reconnaissance*

According to Kaspersky Labs report, the primary aim of this campaign was to target countries in Eastern Europe, former USSR Republics, and countries in Central Asia, even though victims were also found everywhere like Western Europe and North America [21]. These APT actors gathered information about vulnerabilities which can be found on their targets using applications such as Microsoft Word and Microsoft Excel [21].

During this reconnaissance phase, they used some modules which their main purpose was to collect overall information on the target so as to be able to focus and recognize the computers to infect, to evaluate the potentiality of the available informatics data and also to determine which other modules should be used later [49]. Other interesting information could also be gathered by these modules applications such as which websites are visited frequently, username and passwords which have been stored in cache memory and also FTP client settings [49]. For example, a Kaspersky report found out that these attackers deployed a module to actively scan the Local Area Network (LAN) to find hosts vulnerable for MS08-067 [21].

Diamond Model
The adversary is using his capability such as Reconnaissance modules trying to find specific vulnerabilities which the victim has in applications such as Microsoft Word and Microsoft Excel, and collect information of interest in the victim's infrastructure. (*c2i*),

(ii) Weaponization

Red October attackers designed their own malware which was called "Rocra". This malware had very unique modular architecture composed of malicious extensions, info-stealing modules and backdoor trojans [21]. After gathering information about the victim, they would do analyses and then determine which are the appropriate modules to use as a weapon to exploit the victim [50]. Some of the vulnerabilities which were used by these attackers are CVE-2009-3129 (MS Excel), CVE-2010-3333 (MS Word) and CVE-2012-0158 (MS Word) [21].

Diamond Model
Pivoting from the victims vulnerabilities, the adversary use his capabilities such as "Rocra" to take advantage of the victim's vulnerabilities (*v2c*).

(iii) *Delivery*

Attackers used spear-phishing campaigns to infect systems. They would send an email to a victim and include a customized Trojan dropper, or they used to bring infected USB drive in target premises and leave it there for someone to pick it and insert in a networked machine [21].

(iv) *Exploitation*

According to a Kaspersky Lab report, Red October attackers did not create the exploits from the documents sent via spear-phishing email; instead, they were created by other attackers and employed during other cyber-attacks such as Tibetan activists and also military and energy sector targets in Asia [21]. What these attackers did in the document used by Rocra was to change the embedded executable, so they replaced the existed one with their own code [21]. In particular, one of the commands in the Trojan dropper altered the default system codepage of the command prompt session to 1251, which is essential to render Cyrillic fonts [21].

Diamond Model
This phase involved pivoting from the adversary's capability to the victim's infrastructure such as web server, mail server, database, and mobile devices. (*v2i*).

(v) *Installation*

For malware installation to take place and infect the system, the malicious email encompassed exploits that were equipped with security vulnerabilities inside Microsoft Office and Microsoft Excel [49].

Diamond Model
This phase involved a pivot from the victim's infrastructure to the capability of the adversary (*i2c*).

(vi) *Command and Control*

Red October attackers created more than 60 domain names and numerous server-hosting sites in several countries, with the aim of controlling the network of infected machines [50]. Most of these sites were in Germany and Russia. Kaspersky Lab's investigation of Rocra's Command & Control (C2) infrastructure demonstrated that the chain of servers was working as proxies to hide the site of the 'mothership' control server [51].

Diamond Model
This phase involved a pivot from the adversary's capability to the victim's infrastructure (*c2i*).

(vii) *Action on Objectives*

According to a Kaspersky Lab report, Rocra had the following notable characteristics to achieve its objectives;

- *Advanced cryptographic spy-modules:* The key driver of the spying modules is to steal information. This contains files from various cryptographic systems, such as Acid Cryptofiler, which is identified to be used in organizations of NATO, the European Union, European Parliament and European Commission since the summer of 2011 to guard delicate information.
- *Mobile devices:* Apart from targeting traditional workstations, the malware has the ability to steal data from mobile devices, such as smartphones (iPhone, Nokia, and Windows Mobile). The malware is also able to steal configuration information from enterprise network equipment such as routers and switches, as well as deleted files from removable disk drives
- *"Resurrection" module*: A unique module that allows the attackers to "resurrect" infected machines. The module is embedded as a plug-in inside Adobe Reader and Microsoft Office installations and offers the attackers a foolproof means to regain access to a target system if the core malware part is exposed and detached, or if the system is patched. Once the C2s are working again the attackers direct a specific document file (PDF or Office document) to victims' machines through e-mail which will activate the malware again.

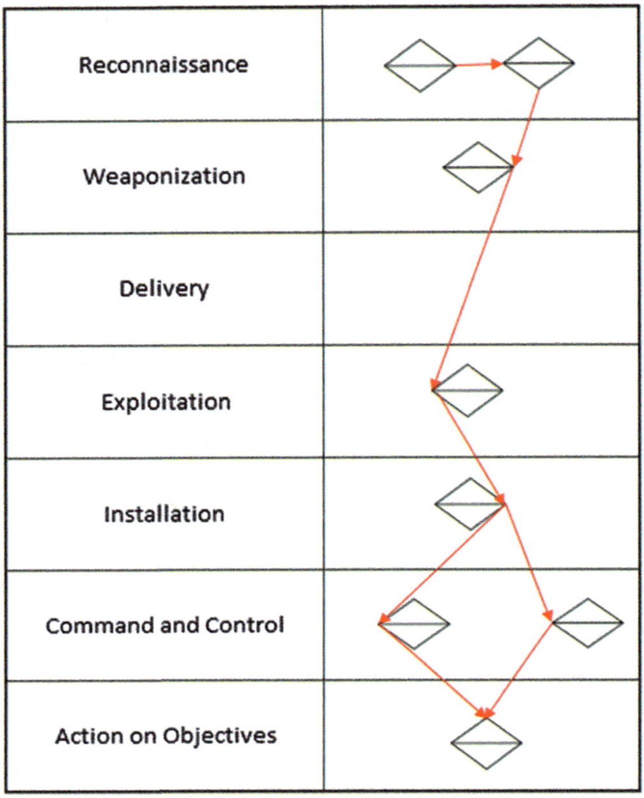

Fig. 9 Cyber Kill Chain and Diamond Model of Red October

Diamond Model
This phase involved a pivot from victim's infrastructure to the adversary (*i2a*).

Figure 9 depicts Cyber Kill Chain and Diamond model analysis of Red October.

4.3 Regin

4.3.1 Cyber Kill Chain and Diamond Model

(i) *Reconnaissance*

Regin targeted organizations such as government entities, business institutes, research institutes, infrastructure operators, and even private individuals who mostly were mathematicians and cryptography specialists mainly in ten different countries [18]. Having those targets means attackers had a good understanding of how they could be compromised, hence an in-depth reconnaissance phase was taken before the attacks. The attacker may have used OSINT or actively scan targets to gather the information.

Diamond Model
Adversaries used enough time to find information about the victim which can be used to conduct an initial compromise (*a2v*).

(ii) *Weaponization*

There are no clear data on how attackers built their payloads-delivery system or what exploits they used. What is known is that they used a backdoor-type Trojan called Regin and they customized it depending on the specific target [18]. Regin is a very complicated piece of malware whose assembly demonstrates a degree of technical capability hardly seen [52].

Diamond Model
From the victim's information, the adversary is using his capability to customize the attack (*v2c*).

(iii) *Delivery*

According to a Symantec report, it is believed that this malware was delivered to some targets by tricking them into visiting spoofed versions of famous and most visited websites (watering hole attack), and the threat may be installed via a web browser or by exploiting an application [18]. For example, in one case, it was found from log files that Yahoo! Instant Messenger is where Regin originated and it used an unconfirmed exploit [18]. It is also suspected that attackers used spear phishing attacks, and man-in-the-middle attacks to compromised targets, and Regin malware was discovered on a USB stick as well which belonged to one of the German Prime Minister Angela Merkel's staff [18].

Diamond Model
The adversary use his infrastructure to deliver the malicious payload to the victim's infrastructure (*i2i*).

(iv) *Exploitation*

Adversaries use customized payloads in order to generate a number of Regin payloads to exploit a target [18]. Attackers used standard capabilities such as Remote Access Trojan (RAT) to exploit targets and create a window of getting more information about vulnerabilities for further attacks, by downloading a number of other payloads to extend the functionality of the malware and exploit more [18]. More advanced payloads were identified such as Microsoft IIS web server traffic monitor and a traffic sniffer of the administration of mobile telephone base station controllers [18].

Diamond Model
From the adversary's capability, the victim's assets were compromised (*c2v*).

(v) *Installation*

Regin is a multi-staged Trojan and migrated to different processes by DLL injection. To avoid detection and maintain persistence in each stage, it is always hidden and encrypted, with the exclusion of the first stage [18]. Regin has a modular approach characteristic, meaning that it gives flexibility to the attackers and they can decide at each stage to load custom features tailored to the specific target when required [19]. During these stages, the malware can install itself and spread through the network by compromising system administrators and using their credentials to spread laterally across Windows administrative shares [53]. In a Kaspersky report, they found out that the malware was able to compromise a telco provider and spread through the network all the way to GSM base-stations where the malware could monitor calls [19].

Diamond Model
From the adversary's capability, the victim's infrastructure was compromised (*c2i*).

(vi) *Command and Control*

According to a Symantec report, C2 operations of Regin are extensive. The C2 communications can be transmitted over the network of Regin-infected computers [18]. The networking protocols are extensible and they are configurable between each pair of Regin-infected computers. Regin can connect with the attacker through ICMP/ping, embedding commands in HTTP cookies, and custom TCP and UDP protocols [18].

Moreover, compromised computers can serve as a proxy for further infections and command and control can also occur in a peer-to-peer style [18].

Diamond Model
The adversary's capability took control of the victim's infrastructure (*c2i*) and then a from communication link was created between the victim's infrastructure and the adversary's infrastructure (*i2i*).

(vii) *Action on Objectives*

This piece of malware offers its controllers a very powerful framework for mass surveillance, and has been used in spying acts against several government administrations, infrastructure operators, businesses, researchers, and even private individuals who mostly were mathematicians and cryptography specialists [54]. The capability of this malware provides the attackers with the ability to do things like passwords stealing, take screenshots or even take control of the mouse and keyboard, monitoring network traffic, and low-level forensics capabilities such as recovering deleted files [55].

Diamond Model
From the victim's infrastructure, information was sent to the adversary's infrastructure (*i2i*) and then finally information reached the adversary (*i2a*).

Figure 10 depicts Cyber Kill Chain and Diamond Model of Regin.

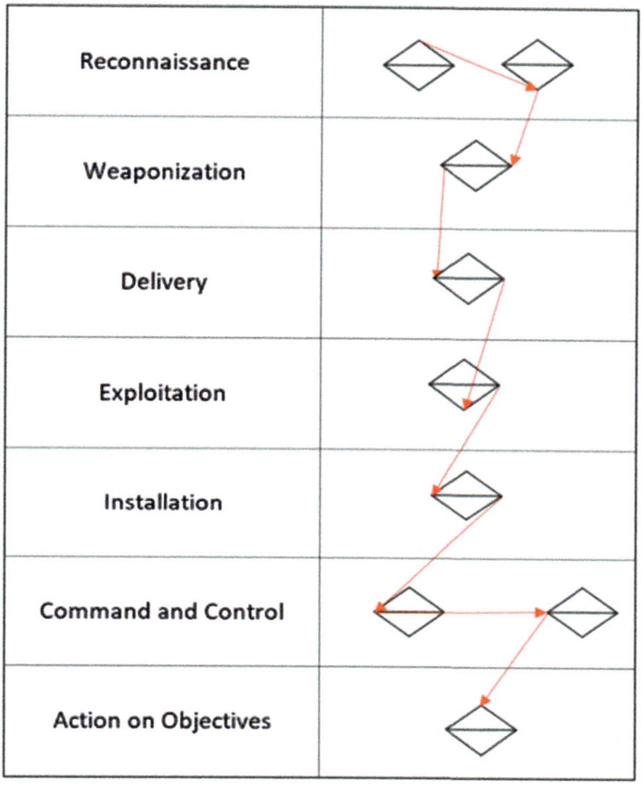

Fig. 10 Cyber Kill Chain and Diamond Model of regin

5 Mitigation Actions

We used Kill Chain Course of Action Matrix to determine how to detect, deny, disrupt, degrade, deceive and destroy the effectiveness of adversaries discussed in this paper. Table 1 shows seven mitigation actions for each category which were acknowledged to counter activities of APT28, Red October, and Regin.

6 Conclusion and Future Work

The key drivers for this study were (1) to analyze three specific APT groups – APT28, Red October and Regin – which mainly target critical national infrastructure of western countries, and (2) to develop a Defense Triage Process. To accomplish this, a novel combination of Diamond Model of Intrusion Analysis, Cyber Kill Chain, and the 7D model was used to make an effective triage of attack vectors and potential targets for a capable adversary.

Table 1 Kill Chain course of action matrix (7D Model) derived from 3 APT groups (APT28, Red October, and Regin)

	Discover	Detect	Deny	Disrupt	Degrade	Deceive	Destroy
Reconnaissance	Hunting	Web analytics	Policy to prevent forum use/traffic filtering			Create fake posting	
Weponization							
Delivery	Hunting	NIDS, user education	Email AV scanning		Email queuing	Filter but respond with out-of-office message	
Exploitation	Hunting	Hids	Patch	DEP			
Installation							
Command and control	Hunting	NIDS	HTTP whitelist	NIPS	HTTP throttling		
Action on objectives	Hunting	Proxy detection	Firewall ACL	NIPS	HTTP throttling	Honeypot	

By combining Cyber Kill Chain and Diamond Model of Intrusion Analysis, this study has identified methods used by these groups and how they attack and implement all the processes from reconnaissance to actions on objectives. This study has identified potential targets of studied APT groups and how those groups can use their capabilities to compromise targets infrastructure.

Defense Triage Process developed in this study will assist organizations to start a progression of notifying and understanding a set of security controls that will create the base of successful, complete analyses of the organization using established best practices. This will help organizations to find an ideal solution to protect themselves from these APT groups.

For future work, additional APT groups can be analyzed to get a broader understanding of malicious actors targeting national critical infrastructure, especially in terms of the way they implement their attacks and utilized TTPs. Moreover, real-world testing of suggested mitigation mechanisms of this study could be another interesting future work.

References

1. Walker-Roberts S, Hammoudeh M, Dehghantanha A (2018) A systematic review of the availability and efficacy of countermeasures to internal threats in healthcare critical infrastructure. IEEE Access 1–1
2. HaddadPajouh H, Dehghantanha A, Khayami R, Choo KKR (2017) A deep recurrent neural network based approach for internet of things Malware threat hunting, future generation computer system. Futur Gener Comput Syst 85:88–96

3. Ussath M, Jaeger D, Cheng F, Meinel C (2016) Advanced persistent threats: behind the scenes. In: 2016 Annual Conference on Information Science and Systems (CISS), pp 181–186
4. Azmoodeh A, Dehghantanha A, Choo K-KR (2018) Robust malware detection for internet of (Battlefield) things devices using deep Eigenspace learning. IEEE Trans Sustain Comput 1–1
5. Min M, Xiao L, Xie C, Hajimirsadeghi M, Mandayam NB (2017) Defense against advanced persistent threats: a Colonel Blotto game approach. In: 2017 IEEE international conference on communications (ICC), pp 1–6
6. Hopkins M, Dehghantanha A (2015) Exploit kits: the production line of the cybercrime economy? In: 2015 second international conference on Information Security and Cyber Forensics (InfoSec), pp 23–27
7. Conti M, Dehghantanha A, Franke K, Watson S (2017) Internet of things security and forensics: challenges and opportunities. Futur Gener Comput Syst 78:544–546
8. Pajouh HH, Dehghantanha A, Khayami R, Choo K-KR (2017) Intelligent OS X malware threat detection with code inspection. J Comput Virol Hacking Tech 14:213–223
9. Haughey H, Epiphaniou G, Al-Khateeb H, Dehghantanha A (2018) Adaptive traffic fingerprinting for darknet threat intelligence, vol 70
10. Homayoun S, Dehghantanha A, Ahmadzadeh M, Hashemi S, Khayami R (2017) Know abnormal, find evil: frequent pattern mining for Ransomware threat hunting and intelligence. In: IEEE transactions on emerging topics in computing
11. Azmoodeh A, Dehghantanha A, Conti M, Choo K-KR (2017) Detecting crypto-ransomware in IoT networks based on energy consumption footprint. J Ambient Intell Humaniz Comput 9:1–12
12. Kiwia D, Dehghantanha A, Choo K-KR, Slaughter J (2017) A cyber kill chain based taxonomy of banking Trojans for evolutionary computational intelligence. J Comput Sci 27:394–409
13. Conti M, Dargahi T, Dehghantanha A (2018) Cyber threat intelligence: challenges and opportunities. Springer, Cham, pp 1–6
14. Lemay A, Calvet J, Menet F, Fernandez JM (2018) Survey of publicly available reports on advanced persistent threat actors. Comput Secur 72:26–59
15. FireEye (2014) FireEye releases report on Cyber Espionage Group with possible ties to Russian Government
16. FireEye (2014) APT28: a window into Russia's cyber espionage operations?
17. FireEye (2017) APT28: at the center of the storm
18. Symantec (2015) Regin: top-tier espionage tool enables stealthy surveillance symantec security response
19. Kaspersky Lab (2014) The regin platform nation-state ownage of GSM networks
20. Chavez R, Kranich W, Casella A (2015) Red October and its reincarnation. Bost. Univ. | CS558 Netw. Secur
21. Kaspersky Lab (2013) Red October: an advanced cyber-espionage campaign targeting diplomatic and government institutions
22. Sager T (2014) Killing advanced threats in their tracks: an intelligent approach to attack prevention. SANS Institute InfoSec Reading. Room
23. Homayoun S, Ahmadzadeh M, Hashemi S, Dehghantanha A, Khayami R (2018) BoTShark: a deep learning approach for botnet traffic detection, vol 70
24. Hutchins EM, Cloppert MJ, Amin RM Intelligence-driven computer network defense informed by analysis of adversary campaigns and intrusion Kill Chains
25. Caltagirone S, Pendergast A, Org AP, Betz C, Org CB (2013) The diamond model of intrusion analysis
26. Shalaginov A, Banin S, Dehghantanha A, Franke K (2018) Machine learning aided static malware analysis: a survey and tutorial, vol 70
27. Pendergast A (2014) The diamond model for intrusion analysis
28. Caltagirone S (2013) The diamond model of intrusion analysis a summary why the diamond model matters
29. Christopher L, Choo K-KR, Dehghantanha A (2016) Honeypots for employee information security awareness and education training: a conceptual EASY training model

30. Microsoft (2015) Microsoft security intelligence report volume 19
31. FBI (2016) GRIZZLY STEPPE – Russian malicious cyber activity
32. Benchea R, Vatamanu C, Maximciuc A, Luncasu V (2015) APT28 under the scope: a journey into exfiltrating intelligence and government information
33. Weedon J, Fireeye JW (2015) Beyond 'Cyber War': Russia's use of strategic cyber espionage and information operations in Ukraine
34. Ostrowski M, Pietrzyk T (2014) APT28 Cybergroup activity
35. Crowdstrike (2016) Bears in the midst: intrusion into the democratic national committee
36. ESET (2016) En route with Sednit
37. Bitdefender TA, Botezatu B (2017) Dissecting the APT28
38. Mehta N, Leonard B, Huntley S (2014) Peering into the aquarium: analysis of a sophisticated multi-stage malware family
39. K. Pierre T (2017) APT28 racing to exploit CVE-2017-11292 flash vulnerability before patches are deployed
40. Pirozzi A, Farina A, Martire L (2017) Malware analysis report: APT28 – hospitality malware
41. Kaspersky Lab (2015) Sofacy APT hits high profile targets with updated toolset
42. T. Micro Incorporated (2017) Two years of pawn storm: examining an increasingly relevant threat
43. Smith L, Read B (2017) APT28 targets hospitality sector, presents threat to travelers
44. Falcone R (2016) Technical walkthrough: office test persistence method used in recent Sofacy attacks
45. Falcone R (2017) XAgentOSX: Sofacy's XAgent macOS tool
46. Hong K-F, Chen C-C, Chiu Y-T, Chou K-S (2015) Ctracer: uncover C&C in advanced persistent threats based on scalable framework for enterprise log data. In: 2015 IEEE international congress on big data, pp 551–558
47. Lee B, Falcone R (2016) New Sofacy attacks against US Government Agency
48. Kaspersky Lab (2015) APTs: a review and some likely prospects
49. Teto A (2014) Operation 'Red October': and it is cyber espionage
50. GReAT (2013) "Red October" diplomatic cyber attacks investigation
51. Kaspersky Lab (2013) Kaspersky lab identifies operation 'Red October,' an advanced cyber-espionage campaign targeting diplomatic and government institutions worldwide
52. Symantec (2015) Protect your IT infrastructure from zero-day attacks and new vulnerabilities
53. Kaspersky Lab (2014) Regin APT attacks among the most sophisticated ever analyzed
54. Schwartz MJ (2015) Regin espionage malware: a closer look
55. Winstanley A (2014) Is Israel behind the 'Regin' cyber-threat?

Aviation Cybersecurity and Cyber-Resilience: Assessing Risk in Air Traffic Management

Georgia Lykou, George Iakovakis, and Dimitris Gritzalis

Abstract Civil aviation is the safest transport mode in the world and probably also the most interconnected system of information and communication technology. Cyber-attacks are increasing in quantity and persistence, so the consequences of a successful malicious cyber-attack on civil aviation operations could be severe nowadays. New technologies, extension of connectivity and their integration in the aviation industry, especially in the field of Air Traffic Management (ATM), increase the risk to these critical assets. This chapter examines cyber security challenges and interoperability in ATM systems. We propose an extended threat model for analyzing possible targets and risks involved. We also introduce and analyze cyber resilience aspects in the aviation context and the need for holistic strategy of defense, prevention and response. Under the resilience umbrella, all actors should work on collaborative, risk-based framework to address security threats and increase the aviation systems resilience against future attacks.

Keywords Aviation cybersecurity · Aviation cyber-resilience · ATM cyber threats · Air navigation services cybersecurity

1 Introduction

Security threats to civil aviation operations have become more sophisticated and challenging. One that is emerging in the recent years and arguably even more advanced and complicated to manage is cyber-attack. Today, the global civil aviation community is relying on computer based and information technology (IT) systems

G. Lykou (✉) · G. Iakovakis
Information Security & Critical Infrastructure Protection (INFOSEC) Laboratory Department of Informatics, Athens University of Economics & Business (AUEB), Athens, Greece
e-mail: lykoug@aueb.gr; giakovakis@aueb.gr

D. Gritzalis
Department of Informatics, Athens University of Economics and Business, Athens, Greece
e-mail: dgrit@aueb.gr

D. Gritzalis et al. (eds.), *Critical Infrastructure Security and Resilience*, Advanced Sciences and Technologies for Security Applications,
https://doi.org/10.1007/978-3-030-00024-0_13

for their daily frontline and backroom operations. This reliance is expected to grow as new and modern airports are developed, new aircraft introduced into service and stakeholders seek to meet the growing demand of the more IT-savvy passengers with new passenger facilitation processes, using digital and IT-based systems [1].

Aviation is a key foundation for international trade, tourism, and investments crucial to the global economy development. The air transport industry supports 2.7 trillion dollars or 3.5% of the world's gross domestic product (GDP) providing 9.9 million direct jobs within the air transport industry [2]. According to the latest traffic forecasts, by 2034, both air passenger and air freight traffic are expected to double, compared to 2016. Passenger traffic is expected to overpass 14 trillion revenue passenger-kilometres (RPKs) with a growth of 4.5% per annum, and freight will expand by 4.2% annually over the same time period, reaching 466 billion freight tonne-kilometres (FTKs) [2].

The use of Information Technology in civil aviation has also increased exponentially in the last years. Digitalization, technological tools and systems often connected to the internet increase intelligence and interoperability on one hand, while on the other they may constitute serious risks for aviation cyber security. Therefore, it is necessary to keep a high level of attention and awareness on possible future developments of the cyber threat [3].

The overall aim is to reduce the vulnerability to cyber-related risks, to strengthen the air transportation systems resilience against cyber threats, which is seen as the capability of an organizational and technical system to protect itself from failures or losses, to mitigate impacts by adapting to changing conditions and to recover from degradation after the incident [4].

This chapter looks at some of the challenges and concerns about cyber security threats in the aviation sector. While in previous work [5] we have focused our concerns on the ground, analyzing cybersecurity measures and best practices to improve airports cyber resilience, in this research we present advanced services in surveillance systems of Air Traffic Control with the aim to address existing vulnerabilities and dependencies. Our purpose was to introduce and analyze resilience aspects in the aviation sector and then classify already proposed resilience recommendations, based on their technical, organizational, social, and economic dimensions.

The remainder of this article is organized as follows: section "Understanding ATM interoperability" examines ATM interoperability and recent advances in surveillance systems. Section "Aviation cyber threat agents" briefly presents related work on aviation cybersecurity and introduces an extended model with cyber-threat agents in the aviation sector. Security measures are presented in section "Security measures in ATM", while section "Cyber resilience in the aviation context" introduces resilience aspects within the aviation context and analyzes existing in literature resilience proposals on several dimensions. Finally, section "Cyber resilience in the aviation context" concludes resilience analysis and benefits for the aviation sector.

2 Understanding ATM Interoperability

In order for Air Traffic Control to safely manage airspace, each ground located air traffic controller needs to understand the status of each aircraft under their control. Traditionally, Primary and Secondary Surveillance Radar in various layouts have supported air traffic surveillance and management for decades. Both systems were designed at a time when radio transmission required a great financial investment and expertise. Hence, no security thought was given to these legacy systems, since it was presumed that they would remain out of reach. The rise of Software Defined Radio (SDR) voided this assumption and marked the shift from potential attackers being well resourced to those with much less resource and capability [6].

The ongoing move from traditional air traffic control systems, such as radar and voice, towards enhanced surveillance and communications systems using modern data networks, has caused a substantial shift in the security of the aviation environment. Implemented through aviation research programs like the Single European Sky ATM Research (SESAR) and the US American NextGen programs, several new air traffic control and communication protocols are currently being rolled out that have been in the works for decades [7].

In this section, we briefly describe the basic ATM systems serving surveillance and interoperability, used for air traffic control such as: Primary and Secondary Surveillance Radar, Automatic Dependent Surveillance-Broadcast, Traffic Collision and Avoidance System, and Wide Area Multilateration. All these systems interact with each other as graphically presented in Fig. 1. Then we discuss how recent advances in wireless technologies have changed the threat landscape in the aviation context.

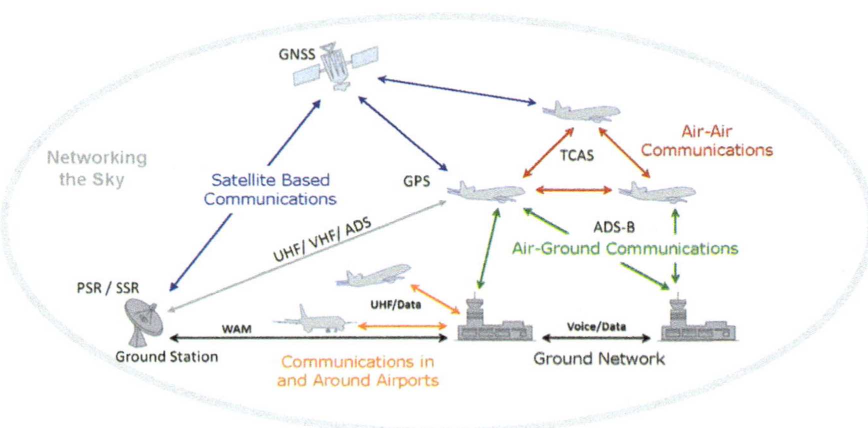

Fig. 1 ATM interoperabilities

Primary Surveillance Radar (PSR) describes non-cooperative radar localization systems. In civil aviation, these typically employ a rotating antenna radiating a pulse position-modulated and highly directional electromagnetic beam on a low GHz band. Potential targets in the airspace reflect the pulses and measurement of the bearing and round-trip time of these reflections provides the target's position. Whilst PSR is not data-rich, it is relatively hard to attack as it relies on physical properties [8].

Secondary Surveillance Radar (SSR) is a cooperative technology with modern communication versions, including the so-called transponder modes. SSR provides more target information on ATC radar screens compared to PSR. Ground stations interrogate aircraft transponders using digital messages on the 1030 MHz frequency, which reply with the desired information on the 1090 MHz channel. Commodity hardware can receive and transmit on these frequencies, making them accessible to attack [9]. Mode S is a particularly important for the current SSR system. It supports systems of increasing significance in modern aviation surveillance, in conjunction with multilateration techniques to provide redundancy. Being intentionally designed with lack of confidentiality, all SSR systems are subject to eavesdropping attacks by passive observers [7].

Automatic Dependent Surveillance-Broadcast (ADS-B) is a protocol in which aircrafts continually broadcast their own ID, position and velocity as well as further information such as intent or urgency codes. These broadcasts do not require interrogation but independently send the aircraft's position and velocity twice a second and unique identification every 5 s; ADS-B is currently in the roll-out phase and it is mandated for use by all aircraft from 2020 in all European and American airspace [7].

Traffic Collision and Avoidance System (TCAS) allows aircraft to interrogate nearby aircraft, in order to resolve airspace conflicts. For example, should another aircraft come within some predefined range, TCAS will initially produce a Traffic Advisory notifying the pilot of traffic nearby. Should the intruder enter the immediate airspace of the aircraft, an alarm will be produced which instructs one of the aircraft to change course. Since the first December 2015, TCAS is mandated for inclusion on civil aircrafts carrying more than 19 passengers or with a minimum take-off weight of 5700 kg [10].

Wide Area Multilateration (WAM) is particularly useful for ATM since it allows location estimation of an aircraft using 1090 MHz messages over large areas. WAM, combined with ADS-B, will form a key part of the next generation surveillance technologies [11] and can help to detect unusual ADS-B reports. Due to the number of sensors and data processing equipment required to cover large areas, the cost of installation is very high, which makes WAM quite hard to attack.

To aggregate information, all the above surveillance systems of Air Traffic Management with discussed characteristics, dependencies and vulnerabilities are presented in Table 1.

Table 1 Main characteristics, dependencies and vulnerabilities of ATM systems

System	Ground/air dependent	Deployment status	Technology	Dependency	Vulnerability
Primary Surveillance Radar (PSR)	Ground	In use	Measure the bearing and distance of targets using the detected reflections of radio signals	Airplane target independent	Not IT related
Secondary Surveillance Radar (SSR)	Ground	In use	Requests additional info from aircraft like identity, altitude, speed	Targets equipped with transponder	Eavesdropping
Traffic Collision and Avoidance System (TCAS)	Air	In use/mandatory since 2015	Target identity interrogation	Targets equipped with transponder	Eavesdropping, jamming, spoofing
Automatic Dependent Surveillance-Broadcast (ADS-B)	Air	Mandate by 2020	Targets broadcast information about identity, altitude, speed	Targets equipped with transponder	Eavesdropping, jamming, spoofing
Wide Area Multilateration (WAM)	Ground	In deployment	Combines ADS with PSR SSR Data for robustness	Central Processing IT based information	Data processing and IT related

3 Aviation Cyber Threat Agents

Although air transportation has a long history of risk management with a special focus on safety and physical security, the field of cyber risks has recently introduced a new landscape of threats. In 2016, at the 39th Assembly, International Civil Aviation Organization (ICAO) has announced preparation works on cybersecurity and cyber resilience. In this direction, Chapter 18 of the Aviation Security Manual which deals with cyber threats has been updated in September 2017. Moreover, Aviation Security Manual (Doc 8973) was enhanced to provide guidance, including minimum measures, to protect critical information systems against unauthorized access and use [11–12].

In addition, recent research studies revealed that cyber threat will most likely be one of the main security issues in aviation, since according to SESAR and NextGen programs the overall air transport system will massively migrate to an IP based infrastructure and operate in accordance with network centric operations concept, with real-time information sharing [4–9]. As a critical resource, information must be treated like any other critical asset which is essential to the efficiency and successful delivery of ATM systems.

In the area of aviation cybersecurity, research work has shown that complexity and criticality of information security and its governance demand the highest organizational security level. Civil Air Navigation Services Organization (CANSO) has issued in 2014 a guidance for Cyber Security [13] explaining how air navigation service providers should take into account cyber security risks in air traffic management, including cyber threats, vulnerabilities, motives of threat actors, as well as considerations about managing cyber risks and implementing a cyber-security program. Sathish et al. [14] proposed a vulnerability assessment framework for wireless threats in Aviation Cyber-Physical Systems (ACPS) by evaluating the tools and used them in their framework to assess the various threats associated with ACPS. Sampigethaya et al. [15] presented a comprehensive survey of security of the e-enabled airplane with applications such as electronic distribution of loadable software and data, as well as future directions such as wireless networked control and airborne ad hoc networks. Through his approach, Bernard Lim [16] looks at some of the challenges and possible ways to address the concern of cyber security threats confronting the global civil aviation community.

During their study, Stander and Ophoff [17] found that steps are taken by aircraft manufacturers and controlling bodies to prevent the occurrence of incidents as to compromise the information systems of an aircraft. D. Jeyakodi [18] justifies in her work how global aviation industry will remain a target for adversaries seeking to make a statement or cause substantial loss to life and financial bearing. The key to ensuring security would be to keep up with the developments thereby being in a position to confront the threats rather than evoking responsive action after its occurrence.

Strohmeier et al. [7] have presented a realistic threat model based on the up-to-date capabilities of different types of threat agents and their impact on a digitalized aviation communication system, where threat agents are classified based on their motivation and capabilities. We have extended this model by adding a new threat: "the insider". We strongly believe that this actor remains a considerable threat agent, not to be neglected from the scheme. We have also estimated risk exposure, taking into account implemented security controls and available security solutions and countermeasures, already proposed in literature. Table 2 presents this extended taxonomy applicable to wireless security in aviation ATM systems and we briefly describe each threat agent characteristics:

- *Passive observers* exploit the open nature of air traffic communication protocols. They use public and private websites and mobile applications, which display air traffic and its communications in real time, to gather information about private or secret air traffic movements. Alternatively, they can employ cheap SDR receivers to gather their own undistorted picture of all air traffic in their vicinity, in real time or stored for later analysis. The information collected can be exploited in many ways, ranging from privacy concerns to the detection of military operations. The risk exposure of ATM systems in such threat agents is rather low, due to no offensive capabilities in the aviation industry.
- *Activists and hobbyists* are the lowest active threat in our model, based on their abilities concerning both hardware and knowledge. Their aim is to exploit security holes with existing, easy-to-use attacks with typically low sophistication and they are able to monitor and interfere to aviation communication channels. Their motivation is regularly not rational, instead any identifiable impact is sought for publicity, thrill and recognition [19]. The risk exposure of ATM systems in such threat agents is considered low, since they can be detected and mitigated with the use of back-up surveillance systems.
- *Insiders* can be a serious threat and are often disgruntled employees, former employees, contractors, or even business associates. These users have inside information of the organization's security practices, data, and computer systems. Insiders can be greedy, malicious or unpredictable in their motivations. The fact that an insider has access to key applications and other critical systems makes him potentially even more dangerous than third-party cybercriminals who try to break in through malware and other mechanisms. Therefore, risk exposure of ATM systems is medium since it is really hard to promptly detect insider's malicious intent or actions.
- *Cyber-crime attackers* usually seek to attack systems for monetary gain, having a sufficient knowledge, using software-defined radios, and even small unmanned aerial vehicles (UAV), being able to inject new messages or modify existing ones in such ways that they are not flagged by current detection systems. They try to cause maximum damage and exert credible threats, as a pre-requisite for blackmail or to take advantage of inside knowledge. Consequently, they are seeking to exploit any possible and effective way to attack Air Traffic Control and aircraft systems. The risk exposure of ATM systems is medium and should be seriously taken into consideration in regular performed risk assessments.

Table 2 Threat agents in aviation systems

Threat	Resources	Goal motivation	Capabilities	Hardware cost	ATM target	Risk
Passive observers	Very Low	Information collection financial or personal interest	Eavesdropping, use of website & mobile apps	Internet access, SDR receiver stick ($10)	ADS-B	Low
Activists and lobbyists	Low	Any noticeable impact thrill and recognition	Eavesdropping, replay attacks, denial of service	COTS SDR transmitter ($300–$2.000)	ADS-B	Low
Insiders	Low – Medium	Disgruntlement, revenge, maximise financial gains selling proprietary information	Resources for specific impact on operations, based on proprietary knowledge	Low cost, enforced by inside use of tools and info on security gaps	SSR, RSR, ADS-B, TCAS	Medium
Cyber crime	Medium – High	Maximising impact financial gains using e.g. blackmail or valuable information	Resources for large-scale operations with sophisticated transponders	Directional antennas, small UAVs with SDR transmitters ($5.000)	SSR, PSR, ADS-B, TCAS	Medium
Cyber terrorism	Low – Medium	Political or religious motivation; Massive disruption and casualties	Resources for specific high-impact ops, though usually on a limited scale	As with cyber crime, potentially on a smaller, more targeted scale	SSR, PSR, ADS-B, TCAS	High
Nation state	Unlimited	Weapons targeting specific, potentially military objects	Anything physically and computationally possible	Military-grade radio equipment, capability for electronic warfare	SSR, PSR, ADS-B, TCAS, WAM	High

- *Cyber-terrorists* seek to threaten national security, cause mass casualties, weaken the economy, and damage public morale and confidence in aviation systems [19]. By exploiting the vulnerabilities in wireless aviation communications, terrorist groups, which traditionally hijack or crash planes using physical weapons, could mount attacks on planes from the ground and from safe distances. The risk exposure of ATM systems is high, due to the increased capacity of terrorists and extremists nowadays to use of IT and cyber technologies for their illegal purposes. This tendency is on the rise around the world, attributed to political and social instability in Middle East, North Africa and other conflict areas.
- *Nation state actors* can be part of the electronic warfare threat model, although traditionally this is outside the scope of securing civil aviation [7]. With sufficient knowledge of intrusion detection systems and near-unlimited resources, it is possible to bypass plausibility checks and redundancy-based defenses even in the ATM sector. The risk exposure of ATM systems is considered high and depends rather on specific political circumstances.

4 Security Measures in ATM

Since ATM Security is major component of Aviation Security, it plays a key role in the prevention and response to threats aimed at all parts of the aviation system including national and international high-value assets. In addition, ATM Security has an interface with Airspace Security revolving around national security and defense requirements, providing technological security and interoperability between civil and military systems [7]. Security threats may be directed at aircraft or through them to targets on the ground. The international dimension imposes the uniform and effective application of suitable measures. ATM has to support national security in respect of the identification of flights entering a State's national territory and Air Defense organizations have to be provided with all ATM information relevant to their task [3].

In general, security measures in aviation range across a number of security disciplines. It does not matter if the asset to protect is an aircraft, an airport, a control center or an information network, all security elements apply at a certain degree, as already referenced in Security Standards (ISO, NIST, ISA) and literature recommended practices [12–19]. In Table 3 we brief these basic security measures and disciplines.

While the above security principles can be implemented to a certain efficiency degree, there is a need for a 'holistic view' covering all challenges of aviation security for all phases of air transport, both on the ground and in the air, since the weakest link in the chain is the one likely to break.

Especially the last security element for operational continuity aims to handle degradations of the ATM system. Although it may encapsulate more managerial aspects, is an essential part of the overall aviation security cycle. It highlights the need for a holistic strategy of defense, prevention and response and introduces the

Table 3 Basic security measures and disciplines in ATM systems

Security discipline	Security measures
Physical security	Access control, perimeter protection, screening, control checks, assets responsibility, redundancies, environmental protection
Personnel security	User access management, security clearances, segregation of duties, recruitment policy, staff regulations, vetting, staff awareness and training
Information security	Protection of information CIA: Confidentiality, Availability, Integrity; cryptography, media handling, backups, software updates and patches
Communication security	Network segregation, security management, intrusion detection management, event logging, teleworking and mobile devices policies
Intelligence support	Security without intelligence is meaningless; intelligence support is a transverse requirement for threat assessments, threat watch and security alert levels declaration
Security information exchange	Information exchange between national authorities, security and intelligence organizations and ATM security managers, security warnings, threat and alert levels, incident identification and notification, reporting and incident resolution follow-up
Operational continuity	Emergency response, business continuity management and contingency plans

need for resilience management. The idea of resilience and its related aspects is introduced and analyzed in the next chapter section.

5 Cyber Resilience in the Aviation Context

The idea of cyber-resilience in ICT, in its most basic form, is the evaluation of what happens before, during and after a digitally networked system encounters a threat. Resilience is not event-specific: it accrues over the long term and should be included in overall business or organizational strategy. The different understandings of resilience are described in IMPROVER[1] project taking into account a combination of different properties [20–21]. Some definitions target on foresight, robustness, resourcefulness, redundancy, rapid recovery and adaptability. Others take prevention, preparedness, respond and recovery into consideration. According to IMPROVER, Resilience concepts encompass several dimensions, such as technical, organizational, social, and economic ones, as presented below:

- The *technological* dimension refers primarily to the physical properties of infrastructure components and systems and refers to the characteristics and behavior of these in the case of a change or incident.

[1] IMPROVER is a Horizon 2020 project focusing on how to improve European critical infrastructure resilience

- The *organizational* dimension, as it relates to the organizations and institutions that manage the physical components of the systems, i.e. CI operators or owners. It covers aspects such as culture, people, business continuity, risk and disaster management at the organizational level.
- The *social* dimension encompasses population and community characteristics that render social groups either more vulnerable or more adaptable to hazards and disasters.
- The *economic* dimension focus on reducing both direct and indirect economic losses resulting from disasters, in various levels.

In aviation context, Eurocontrol Research program uses for resilience the following definition: "*Resilience* is the ability to prevent disruptions, to prepare for and adapt to changing conditions and to respond and recover rapidly from disruptions to ensure the continuity of services at an acceptable performance level". The aim of this definition is to achieve the understanding that caring for resilience is more affiliated to the management of risks rather than to the elimination of them [22].

Being resilient implies minimizing reductions in performance (acceptable drop of performance) in the face of a successful attack. This means to be able to work properly also in several levels of degraded mode, while healing measures and repair works can be undertaken. It is therefore essential to provide methods and means to allow the solution to recover, as quick as possible from such degraded modes, achieving minimum recovery time.

As presented in Fig. 2, under the resilience umbrella, the whole set of measures, which are required for sufficient resilience against cyber-attacks, is a combination of different actions and proper behavior before, during and after the incident

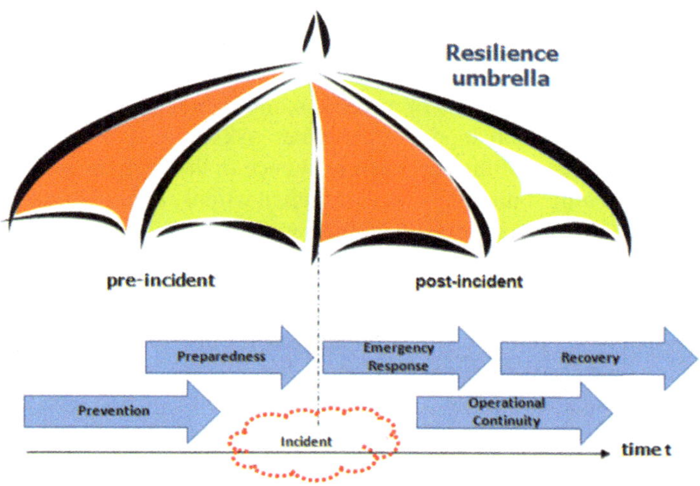

Fig. 2 The resilience umbrella. (Source: *EUROCONTROL*)

[23]. The flow of cyber resilience actions already starts when the services, tools or systems under concern are in the development phase, which is related to the "security by design". When taking this into account, the first step of "*Prevention*" is profound established. Controls may have to be put in place to address potential risks emanating from other parts of the system or "system of systems". Another pillar of resilience is "*Preparedness*" for possible attacks, which can be achieved by procedures and training of staff. Being prepared for any cyber-attack begins with thinking about daily activities and the way work is organized and conducted. This includes also the knowledge about the fastest and most secure ways of decoupling software tools from the system or network and safely/securely shutting down infected systems [23].

When being under attack the "*Emergency Response*" to the attack is also important. The first response focusses on identifying the problem, containing it, eradicating it. Responsive measures may also include the restriction of services or the unwinding of trained sequences. The focus shall be kept on the secure delivery of services and data whilst being aware of the attack in progress. This supports and enhances "*Operational Continuity*". The response phase needs to be continued until the cause and even the cascading effects of the attack have been eliminated, accounted for or phased out. When at any point in time this can surely be confirmed the phase of "Recovery" may be initiated. This phase again needs to be as short as possible in order to have all services, tools and systems in full operation after a cyber-attack.

Although resilience engineering has been introduced in the aviation mainly for enhancing safety sector, it has not been thoroughly expanded to the cybersecurity aspects and resilience in the air traffic management area. Resilience in aviation sector has been partially discussed in previous research [4, 5, 23, 24]. However, in our work we have studied a recent research, the ARIEL project for Air Traffic Resilience, which aimed to perform a holistic risk analysis and evaluation of critical infrastructures in aviation [4]. According to ARIEL, resilience is seen as the ability of a system to absorb or avoid damage without suffering complete failure and integrates the aspects of protection, mitigation and recovery. It proposes a continuous dynamic and model-based cyber risk analysis process, in order to establish persisting capabilities of cyber resilience in the air transportation system. Project report identified the following recommendations, as essential for resilience implementation in the aviation sector, which are listed and briefly explained below:

R-1) ***Develop the structural and procedural basis for continuous intra- and inter-organizational cyber resilience analysis***: Combining classical information security and newly developed cyber operational resilience approaches. Establishing an organizational structure that brings together the personnel of all relevant disciplines inside and across air traffic organizations to cope with the evolving cyber threat landscape in a holistic way. This has to be combined with suitable continuous processes aligned with the existing information security norms and standards.

R-2) ***Develop and manage interdisciplinary cyber risk analysis teams:*** To facilitate the establishment of interdisciplinary collaborating teams, there is a need to develop and apply the necessary methods and management approaches comprising elements of a common language; knowledge management and transfer; ignorance management for balanced evaluation of findings; widespread basic IT knowledge and security awareness by personnel of all disciplines including middle and top management.

R-3) ***Develop and maintain a portfolio of cyber threat scenarios:*** In contrast to the currently applied ad hoc way of threat scenario development and utilization, the introduction of a structured continuous process for the development and evolution of air-traffic cyber threat scenarios is recommended. This is to be combined with suitable methodology to develop scenarios and to apply them in the areas of knowledge development, training as well as verification and validation.

R-4) ***Ensure the interoperability of cyber-relevant models and data:*** Developing standardized meta-models for computer-based data exchange and collaboration of different models is needed. The integration and comparison of cyber-relevant results and findings in tool-based analysis and decision support is also essential. To enable interdisciplinary or even inter-domain collaboration based on a comprehensive approach, data sharing concepts are needed for a reuse of existing data, which include technical, methodological and organizational aspects.

R-5) ***Refine and Evolve Dynamic Risk Analysis Methods:*** Additional effort into the further evaluation and evolvement of the model-based dynamic risk analysis method should be developed. This semi-automated analysis method enables to dynamically model and analyze cyber risks in complex systems, large organizations or even in between several interconnected organizations. The high potential of this approach enables a comprehensive, dynamic cyber risk assessment in the aviation sector.

R-6) ***Safety & Security – Ensure consistency and enable synergies:*** Since cyber threats and potential cyber-attacks can have a direct impact on safety-critical system functions. Therefore, the development of a comprehensive risk management approach aligning the formerly separated considerations of safety and security under a common roof is requested.

R-7) ***Enhance design methodologies to ensure resilient system characteristics throughout a complete lifecycle:*** The restructuring of architectures of sociotechnical systems could support cyber resilience in addition to protective measures. Existing approaches of resilience engineering, which focus mainly on human factors in complex systems, have to be extended in a technical sense towards integration of cyber resilience capabilities. Some of the more important aspects to be considered are: the preparation of architectures for ongoing changes; the consideration of mitigation and recovery strategies in the system design; and the addition of system functions supporting the detection of cyber-attacks.

R-8) ***Exploit simulation methodologies to support cyber threat and risk analysis of complex systems:*** To achieve a holistic understanding of the effects of potential cyber-attacks in complex systems, simulation is a valuable method

Resilience Dimension	Technological	Organizational	Social	Economic
ARIEL Resilience Recommendation				
R - 1		✓	✓	
R - 2		✓	✓	
R - 3	✓	✓		
R - 4	✓	✓		
R - 5	✓	✓		
R - 6	✓	✓		
R - 7	✓	✓	✓	
R - 8	✓	✓	✓	✓

Fig. 3 Resilience dimension in ARIEL recommendations

to complement more traditional analysis methods. A widespread application of simulation models for processes and systems should be identified by cyber threat and risk analysis to be critical for system operation. Simulation increases the understanding of the impact of identified cyber threats and supports the validation of risk analysis results. Besides using existing simulation models as standalone tools, it is important to develop simulation models "from gate-to-gate" to support holistic analysis of aviation processes. Finally, using human-in-the-loop simulation is vital with operational staff to research the fundamentals of human factors in the face of potential cyber-attacks.

Based on the ARIEL recommendations, we have analyzed the resilience dimensions which are encompassing, according to IMPROVER Resilience concepts (technical, organizational, social, and economic) and the results of this analysis are presented in Fig. 3.

What we can comprehend from the above table analysis is that most recommendations cover at least two resilience dimensions with the technological and organizational ones to be the most common used. There is a core difference of resilience recommendations from cybersecurity disciplines, which usually handle a single dimension at a time. Resilience measures appear to be a synthesis of interactions, collaboration and evolution of current cybersecurity approaches.

The organizational dimension is common to all recommendations, since cyber resilience is really a matter of effective risk management, combined with collaborative working and interdisciplinary strategies to ensure contingency and efficient business continuity.

While there is a lack of recommendations that enforce the economic dimension of aviation resilience, the social dimension is also less developed. The only recommendation that covers all four resilience dimensions is the last one, about achieving a holistic understanding of the effects of potential cyber-attacks with simulation methodologies using human in the loop, which better support cyber risk analysis of complex systems.

For promoting an overall cyber-resilience approach in the aviation sector, long-term strategy should combine all resilience dimensions that is technological, organizational, societal, and economic. This cyber-resilience approach can ensure greater performance and readiness, making systems more efficient and effective.

6 Conclusions

Recent approaches to increase capacity and efficiency of the existing air traffic system have led to an enormous effort of transition towards digitalization and automation. As a result, formerly separated IT systems get connected via newly established networks for information and data exchange. Due to a growth of complexity, the attack surface of the overall aviation system is increasing and previously unknown interdependencies are being created. Limiting security risk management to "traditional" physical aspects like air terrorism is no longer sufficient to ensure a stable and robust operation of the air transportation system. The component of cyber-security has to be expanded from traditional risk mitigation approaches to more resilient focused approaches.

The domains of air transportation and cybersecurity are organized with a strong focus on protective mechanisms, in terms of their operational and technical implementation. To fulfil the requirements of continuous adaption to a rapidly changing threat environment, the architectures of operational and technical systems have to be restructured based on the results and dynamic simulation risk analysis. According to that, we strongly recommend to balance the cost and performance-driven development and prioritize a sustainable, comprehensive and continuous improvement in order to improve the overall system's cyber resilience.

As both safety and security are drivers for the determination of resilience requirements, it is sensible to take an integrated view on both subjects to foster the consistency of resilience concepts in aviation.

Since cyber resilience is really a matter of risk management, there isn't a single point at which it begins or ends. Instead, it comes from building strategy and working to ensure that the risk-transfer mechanisms that work for more traditional threats are also brought to bear on new cyber threats.

Being resilient requires those at the highest levels of a company, organization or government to recognize the importance of avoiding and mitigating risks. While it is everyone's responsibility to cooperate, in order to ensure greater cyber resilience, leaders who set the strategy for an organization are ultimately responsible and have increasingly been held accountable for including cyber resilience in organizational strategy.

References

1. Lim B (2014) Aviation security – emerging threats from cyber security in aviation – challenges and mitigations. J Aviat Manag
2. Industry High-Level Group (IHLG) (2017) Aviation benefits 2017 report
3. De Zan T, d'Amore F, Di Camillo F (2015) The Defence of civilian air traffic systems from cyber threats
4. Kreuzer M, Kiesling T (2017) Recommendations to strengthen the cyber resilience of the air traffic system, ARIEL, Air Traffic Resilience
5. Lykou G, Anagnostopoulou A, Gritzalis D (2018) Implementing cyber-security measures in airports to improve cyber-resilience, WIIoTS in the 2nd global IoT summit
6. Strohmeier M et al (2014) Realities and challenges of nextgen air traffic management: the case of ADS-B. IEEE Communations Magazine 52(5):111–118
7. Strohmeier M et al (2016) Assessing the impact of aviation security on cyber power. In the 8th international conference on cyber conflict cyber power
8. Strohmeier M et al (2016) On perception and reality inWireless air traffic communications security
9. Costin A, Francillon A (2012) Ghost is in the air(traffic): on insecurity of ADS-B protocol and practical attacks on ADS-B devices. In black hat USA
10. The European Commission (2011) Commission regulation laying down common airspace usage requirements and operating procedures for airborne collision avoidance, no. 1332. European Union
11. International Civil Aviation Organisation (2013) Initial capability for ground surveillance. In global air navigation plan 2013–20282
12. International Civil Aviation Organization (ICAO) (2017) Aviation security manual, 10th edition, https://www.icao.int/Security/SFP/Pages/SecurityManual.aspx
13. CANSO (2014) Cyber security and risk assessment. Civil Air Navigation Services Organization
14. Kumar S, Xu B (2017) Vulnerability assessment for security in aviation cyber-physical systems. IEEE 4th international conference on cyber security and cloud computing
15. Sampigethaya K, Poovendran R, Bushnell L (2008) Secure operation, control and maintenance of future E-enabled airplanes, Network Security Lab (NSL), EE Department, University of Washington, Seattle
16. Lim B (2014) Aviation security – emerging threats from cyber security in aviation – challenges and mitigations, J Aviat Manag
17. Stander A, Ophoff J (2016) Cyber security in civil aviation
18. Jeyakodi D (2015) Cyber security in civil aviation
19. Stouffer K, Falco J, Scarfone K (2007) Guide to industrial control systems (ICS) security. Recommed. NIST., no. SP 800–82, pp 1–157
20. Theocharidou M et al (2016) D1.3-final lexicon of definitions related to critical infrastructure resilience, IMPROVER, European Union's horizon 2020 research
21. Lange D et al (2017) Framework for implementation of resilience concepts to critical infrastructure, IMPROVER, European Union's horizon 2020 research
22. EUROCONTROL (2012) Manual for national ATM security oversight, Eurocontrol Publications
23. EUROCONTROL (2009) A white paper on resilience engineering for ATM. https://www.eurocontrol.int/sites/default/files/article/content/documents/nm/safety/safety-a-white-paper-resilience-engineering-for-atm.pdf
24. Delgado L, Cook A, Tanner G, Cristóbal S (2016) Quantifying resilience in ATM, in the proc. of 6th SESAR innovation days, Technical University of Delft, The Netherlands

Open Source Intelligence for Energy Sector Cyberattacks

Anastasis Keliris, Charalambos Konstantinou, Marios Sazos, and Michail Maniatakos

Abstract In March 2018, the U.S. DHS and the FBI issued a joint critical alert (TA18-074A) of an ongoing campaign by Russian threat actors targeting U.S. government entities and critical infrastructure sectors. The campaign targets critical infrastructure organizations mainly in the energy sector and uses, among other techniques, Open Source Intelligence (OSINT) to extract information. In an effort to understand the extent and quality of information that can be collected with OSINT, we shadow the threat actors and explore publicly available resources that can generate intelligence pertinent to power systems worldwide. We undertake a case study of a real, large-scale power system, where we leverage OSINT resources to construct the power system model, validate it, and finally process it for identifying its critical locations. Our goal is to demonstrate the feasibility of conducting elaborate studies leveraging public resources, and inform power system stakeholders in assessing the risks of releasing critical information to the public.

1 Introduction

Electric power systems have significantly evolved over the years and grew to become essential in our everyday life. Our expectation of uninterrupted power supply in everyday life is further exemplified by the far-reaching impact of power outages, also known as *blackouts*. Table 1 lists notable power outages of the twenty-

---

A. Keliris (✉)
New York University Tandon School of Engineering, Brooklyn, NY, USA
e-mail: anastasis.keliris@nyu.edu

C. Konstantinou
Center for Advanced Power Systems, Florida State University, Tallahassee, FL, USA
e-mail: ckonstantinou@fsu.edu

M. Sazos · M. Maniatakos
New York University Abu Dhabi, Saadiyat Island, Abu Dhabi, United Arab Emirates
e-mail: marios.sazos@nyu.edu; michail.maniatakos@nyu.edu

D. Gritzalis et al. (eds.), *Critical Infrastructure Security and Resilience*,
Advanced Sciences and Technologies for Security Applications,
https://doi.org/10.1007/978-3-030-00024-0_14

Table 1 Notable power outages of the twenty-first century

Year	Country	People affected	Cause
2012	India	620 million	Misoperation [21]
2015	Pakistan	140 million	Malicious destruction [15]
2014	Bangladesh	100 million	Equipment failure [1]
2009	Brazil & Paraguay	87 million	Adverse weather conditions [3]
2015	Turkey	70 million	Maintenance and oversupply [50]
2003	U.S. & Canada	55 million	Shortcircuit because of trees [19]

first century. The examples showcase the diversity of possible causes, and are sorted by their impact measured in millions of people affected.

Most blackouts observed to date are the result of equipment faults, natural phenomena, animals, or human errors. However, there is increased concern in the international community regarding *cyberattacks* that target power grids [38]. When it comes to cyberattacks against cyberphysical systems and critical infrastructure, Pandora's box was opened in 2010 with Stuxnet, a worm targeting equipment in a nuclear plant in Iran [41]. The first cyberattack targeting power systems is an incident in Ukraine reported in December 2015. The attack targeted computer systems of three energy distribution utilities and is believed to be the work of a nation-state actor [6]. The outage mainly affected the Ivano-Frankivsk region, leaving about 230,000 end consumers without power for hours [42]. A second, smaller scale cyberattack against the Ukrainian power grid hit Kiev a year later, this time targeting the transmission network [18].

A large scale, ongoing Advanced Persistent Threat (APT) campaign by Russian actors targeting U.S. critical infrastructure sectors was jointly reported in March 2018 by the U.S. Department of Homeland Security (DHS) and the Federal Bureau of Investigation (FBI) in US-CERT Technical Alert 18-074A [61]. According to Symantec, who has been closely monitoring the group behind this campaign, the energy sector is the main target of the campaign, and the attack focus is not limited to the U.S. [58]. For gathering information during the reconnaissance phase, the threat actors are believed to employ several techniques, including open source reconnaissance, also known as *Open Source Intelligence* (OSINT).[1]

In this chapter we undertake a study of publicly available resources that are pertinent to power systems across the globe, in an effort to understand the extent and quality of intelligence that can be generated with OSINT, as well as the feasibility of constructing exploitation vectors based on these resources. We showcase the practical applicability of OSINT-based studies by modeling a *real national power grid* using OSINT resources, cross-validating our model using secondary resources, and identifying the model's critical operational points through power security studies. Our contributions can be summarized as follows:

[1]OSINT refers to data and information passively collected and analyzed from publicly available sources. It is not related to open source software.

- We provide a corpus of publicly available resources pertaining to power systems. These can be leveraged using OSINT techniques to extract intelligence, identify critical operational points, and construct attack vectors against target systems.
- We demonstrate the significance of OSINT-based intelligence by undertaking an in depth case study of a real, large-scale power system. We build and successfully validate the model of this system from the ground up and analyze it to identify its critical operation points.

To the best of our knowledge this is the first broad study of publicly available resources regarding power systems. Moreover, this is the first work where a model of a real power system was built from the ground up, leveraging and fusing publicly available information using OSINT techniques. Our motivation for this study is the uncertainty currently observed among the various stakeholders of the power industry, including governments, vendors, and power utilities, regarding the real threat cyberattacks pose to power systems. Our study can assist stakeholders and regulators take informed decisions by raising awareness relating to the dangers of divulging more-than-required information to the public, and showcasing potential implications of public dissemination of sensitive information. Due to the sensitive nature of the extracted information, we anonymize certain critical parts of the study.

The remainder of the chapter is structured as follows: We present our threat model, target analyses, and methodology in Sect. 2. Publicly available resources for modeling a power system are presented in Sect. 3. Section 4 provides details on contingency analysis, a technique that can be used to derive the critical locations of a system. Section 5 outlines OSINT resources for constructing attack vectors. We evaluate the practicality of an open source campaign in Sect. 6 by constructing, validating, and analyzing the model of a real power system. A discussion on the significance of cyberattacks against the power grid and possible mitigation strategies are presented in Sect. 7, and we conclude the chapter in Sect. 8.

2 Target Analyses and Methodology

The threat model we assume considers adversaries with power systems expertise, whose objective is to cause large scale power outages. We assume that adversaries do not necessarily have footholds inside target organizations, nor physical access to the Supervisory Control And Data Acquisition (SCADA) center or power substations. Since we focus on publicly available resources for generating intelligence and formulating attack vectors, adversary groups represented in our threat model are not limited to nation-states or heavily funded private/government organizations.

2.1 Strategic Target Analysis

One of the first steps of a campaign is to study the target system at a high level of abstraction in order to understand and identify the strategic assets of interest [37]. For campaigns against the energy sector, this step concerns identifying which power systems stage or combination of stages are more suitable for achieving the required objectives. In general, power systems are comprised of four stages: generation, transmission, distribution, and consumption. The first stage is *generation*, where electricity is produced. It is then transferred near consumption centers in the *transmission* stage, and distributed to end consumers in the *distribution* stage. Finally, electricity is utilized in the *consumption* stage by industrial, commercial, or residential consumers.

For causing a large scale power outage, the consumption stage is not particularly attractive because a very large number of consumers would need to be compromised to achieve the required outcome. Employing a similar rationale, adversarial campaigns would likely not target the distribution network, because the attack would require the compromise of a large number of distribution substations, possibly controlled by several power utility companies. The generation stage, at which electricity is produced could be a promising target. However, power plants are manned 24/7 and typically employ a variety of protection mechanisms, including physical security, rendering attacks against them significantly more difficult. In addition, restoration of lost capacity from generator losses can be very quick, as demonstrated by the 2011 Cyprus explosion which destroyed 60% of the state-island's installed capacity [62]. Although the island's grid is not interconnected to other national grids, demand was met with distributed generation in a short period of time.

Considering attack difficulty and attack impact tradeoffs, the most attractive target is the *transmission network*. It fulfills the requirement of large scale impact, while at the same time reduces the difficulty of launching an attack. Several transmission substations are unmanned and situated in remote, not populated areas. This finding is also supported by the fact that the majority of impactful blackouts were caused by transmission stage failures.

2.2 Tactical Target Analysis

Following the strategic target analysis, tactical target analysis can identify *specific* targets in a power system (e.g., transmission lines and substations) that could fulfill the requirement of causing a large scale power outage, and how to attack these targets. For successful target selection we identify two prerequisites. First, adversaries need to create a *model* of the target system that enables power studies and can generate intelligence regarding the entire power system. Second, the constructed model must be processed and analyzed towards identifying the specific points of interest whose compromise could lead to a large scale power outage. To that end, well known tools and techniques from power system research, such as *contingency analysis*, can be used.

Once the specific critical points of interest are identified, *attack vectors* for exploitation of these points must be constructed. In general, Circuit Breakers (CBs) and their corresponding relay signals constitute attractive targets when attacking specific locations of a power system. The operation of CBs guarantees normal service of grid equipment due to system separation into protective zones, and the isolation of faulty zones as necessary to change load routing. In addition, the control of CBs via relay signals allows the control circuity to command the various CBs to open and interrupt or re-route the flow of electricity. Testifying to their criticality, 70% of the major disturbances in the U.S. are associated with faulty operation of relay controllers [45]. For constructing attack vectors, OSINT techniques can be leveraged towards forcing CBs to open/close connections in a target system.

2.3 Methodology

In fulfilling the steps against targets in the transmission stage as they are identified in the tactical target analysis, adversaries need to create a representative model of the power system, conduct power studies on the model to derive its critical operational points, and finally construct attack vectors against these specific points. In this work, we focus on OSINT-based intelligence that can be leveraged towards achieving these steps. We shadow threat actors seeking to cause a large scale blackout, investigating the feasibility of utilizing publicly available resources to achieve this objective.

We begin by carrying out extensive research on the sources of information on power systems that are available to the public. We provide a corpus of our findings in Sect. 3. Such sources can provide fragmented information that can be then combined towards creating models of power systems. Subsequently, we explore the types of power system studies that are necessary for identifying critical operational points of a target system. Contingency analyses are particularly relevant in achieving objectives such as large-scale power outages. An investigation of contingency analyses is provided in Sect. 4. We then examine OSINT resources that enable or can assist the construction of attack vectors against power systems. We consider a broad spectrum of possible attack-enabling resources, and provide a corpus for OSINT exploitation sources in Sect. 5.

Finally, in investigating the feasibility of leveraging OSINT intelligence and assessing the quality of analyses enabled by it, we undertake an in depth case-study of a real, large-scale power grid. We rely solely on OSINT intelligence, and fuse information from a variety of sources to first build the entire model, and then validate it using secondary OSINT sources. Using contingency analysis tools on the constructed model, we additionally identify the model's critical points.

3 OSINT Resources for Modeling a Power System

In this section we provide representative, but non-exhaustive publicly available resources, which can generate intelligence and provide sensitive information for power systems and their components. Leveraging OSINT, it is possible to obtain the information required to model a power system, enabling tactical target analyses through power studies on the constructed model.

Power system databases Several power system databases are publicly available, providing access to data relating to real systems across the globe. For example, the Open Power System Data platform provides data regarding power plants, generation capacities, and loads for several European power systems [14]. Another source of data is the European Network of Transmission System Operators for Electricity (ENTSO-E), which was established in an effort to ensure the optimal functioning of the EU internal energy market. Data include maps of transmission networks, grid interconnection details, real time cross border flows, historical and forecast loads and generation statistics, as well as development plans [5]. For the U.S., the Open Energy Information (OpenEI) is a government collaborative website that provides public access to energy data [13]. On a worldwide scale, the Enipedia semantic database features a plethora of load, generation, topology and line characteristics data for power systems across the globe [25].

Geographic Information Systems The topology of a power system can be constructed or validated by observing the physical components of the system and their interconnections. Instead of on-the-ground tracking of physical structures, it is possible to generate the network topology of a system using satellite imagery from Geographic Information Systems (GIS). This enables a bird's-eye-view analysis of a system that can be performed remotely. Examples of power system components as they appear in a GIS are presented in Fig. 1. The ability to map a power system with

Fig. 1 Images of power system components from GIS. Top left: transmission substation, top right: power plant, bottom: power lines

GIS can be likened to the analog loophole problem in digital rights management. Location information of physical power structures cannot be hidden from orbiting satellites or aerial photography aircraft, except in the case of underground cables. However, the vast majority of transmission level power lines are overhead [20]. Finding the topology of a power system with GIS can be possibly automated through machine learning and image processing techniques. In addition, tagging power structures and power networks can be crowd sourced. The Power Networks subproject of OpenStreetMap follows this exact approach [47].

Public reports Oftentimes, power utilities, Transmission System Operators (TSOs), or government agencies release reports to the public that contain operational details and information regarding their power system. These reports may be in the form of reports required by law, annual financial reports to shareholders, and statements that outline future requirements and how they will be met (e.g., [17, 28, 29]). Somewhat ironically, blackout reports may also contain sensitive information regarding a power system. Blackout reports are released to the public, typically for transparency reasons, and usually contain technical details in an effort to pinpoint and explain the source of the blackout (e.g., [50]). In addition, reports from initiatives that aim to enhance the resiliency of the power grid and accelerate grid modernization can contain information pertinent to specific power systems. For example, the North American SynchroPhasor Initiative (NASPI) website includes a report listing the geographical location of PMUs in the U.S. [10]. Information may be publicly available by design, as is the case with the Northern Regional Load Despatch Centre (NRLDC) of India. The NRLDC shares real-time data regarding frequency, scheduled and available capacity, and next-day load forecasts for operational reasons [7].

Press releases Adversaries can extract information regarding a power system from information released to the press. Such information can be in the form of newsletters, press releases by power system operators, success stories by the vendors who installed components of the system, corporate presentations, etc. Some examples include the media centers of vendors that include references to awarded, completed and ongoing projects [55], reports from turnkey solutions providers where information on voltage levels and transmission lines is listed [54] and vendor success stories that reveal communication protocols and system topology [51].

The resources presented here, as well as additional information that can be collected through other OSINT channels or through other more invasive techniques and non-public sources, can be fused to create a model of a power system. The information can be used to derive the topology of the system, as well as to extract, or estimate operational characteristics for performing power studies. Furthermore, information from different sources can be compared to evaluate the accuracy of the constructed model, a process we undertake in our experimental evaluation section.

4 Power System Studies: Contingency Analysis

Modeling a power system enables carrying out power studies on the system. Different power studies may be necessary for different campaign objectives. For campaigns that aim to disrupt the system and cause blackouts, power system security studies can provide valuable information to adversaries as to which specific targets are necessary and sufficient for destabilizing the entire system. Power system security is defined as the probability of the system's operating point to remain within acceptable ranges given the system constraints, the probabilities of changes (contingencies), and its environment [44].

Contingency analysis is a well known operation in modern Energy Management Systems (EMS), which provides necessary information to the system operator about the static security of the system. In contrast to state estimation, which is considered an online application, contingency constrained analysis is an offline application for power system planning and operation [49]. Abstractly, contingency analysis can be viewed as a "what if" scenario simulator that assesses, produces and ranks the impact of unscheduled events on a power system. For example, a contingency can be the failure, or loss of an element of the system (e.g., generator, transmission line, transformer), or the unplanned opening of a CB. These events form the contingency list, which is then used by contingency analysis algorithms to evaluate effects on the overall system.

In its basic form, contingency analysis generates a power flow solution for each event specified in the contingency list. The objective of the power flow analysis is to obtain a set of voltage magnitudes and angles for each bus in the power system corresponding to a specified load and generation condition. Subsequently, active and reactive power flows on each branch and generator are analytically determined. The loss or failure of each contingency event is simulated in the network model by removing that part from the simulated power system. The resulting network model is solved to compute the corresponding power flows, currents, and voltages for the remaining elements. The outcomes from each contingency test are then compared with the operational limits for every element (e.g., thermal ratings of transmission lines) to determine if a limit violation occurs.

Since contingency analysis relies on the execution of a power flow study, the first step is acquiring the required data to develop a power flow model of a power system. Specifically, for a power system model to be sufficient for contingency analyses, the following data are required [32]:

- System topology (Edges: Transmission lines/Transformers, Nodes: Buses).
- Transmission line parameters.
- Tie-line locations and ratings.
- Transformer and phase shifter parameters.
- Location, ratings, and limits of generators.
- Load location and load compensation.

In general, power systems must be able to sustain a single contingency condition ($N - 1$) to enable maintenance operations, where N is the number of components (typically the N branches of a network). North America Electric Reliability Corporation (NERC) and other regulatory agencies around the world enforce strict power security standards that require power system operators to satisfy the $N - 1$ security constraint [46]. NERC standards also necessitate that operators ensure sufficient system performance in the event of multiple outage contingencies. Nevertheless, the problem of contingency identification remains computationally challenging due to the total number of possible initiating events: it increases exponentially with k, where k is the number of outaged elements. The complexity is further exaggerated if outage scenarios are analyzed with a full AC power flow technique, which requires significant computational resources.

Instead of using full non-linear AC power flow analysis, approximate, but much faster techniques based on DC approximation can be used to estimate post-contingency values of interest [35]. In general, DC power flow analyses are commonly used in contingency studies where approximate real power flows are more important than voltage limits on buses [60]. The DC formulation is based on the same parameters as the AC problem, with additional simplifying assumptions: the voltage profile is flat, meaning that all bus voltage magnitudes are close to 1 p.u., line resistances and charging capacitances are considered negligible, and voltage angle differences between neighboring nodes are small enough such that $\sin(\theta_{ij}) \approx \theta_{ij}$.

Regarding algorithmic approaches that address the complexity of calculating $N - k$ contingencies, multiple techniques have been proposed in literature. Ranking and selection methods are traditional techniques that rank configurations of outages based on a heuristic index [56]. Advancements of such methods study contingencies based on Line Outage Distribution Factors (LODF), which are used to approximate the change in the flow on one line caused by the outage of a second line [24]. For $N - 2$ contingency screening in particular, recent work on LODF based approaches can mathematically guarantee identification of all the dangerous $N-2$ contingencies with low computational costs [60].

For every system, given its topology and flows, there *always* exist a number p of multiple contingencies, which cannot be sustained and will lead to cascading failures. Adversaries can leverage contingency analysis techniques for tactical target analyses, towards identifying these p contingencies. Having constructed a model of the target system, they can identify which *specific* p locations are critical, and target them explicitly to materialize an attack.

5 Constructing Attack Vectors with OSINT

With knowledge of the critical points of a power system from the modeling and analysis steps, adversaries need to find attack entry points and construct attack vectors against the system. More specifically, they need to devise means

Table 2 Internet connected power grid devices indexed by Shodan

Protocol	Port	Indexed devices
DNP3	19999/20000	341
Modbus	502	13575
IEC 104	2404	445
IEC 61850	102	161

towards disconnecting the critical transmission lines capable of a non-sustained contingency scenario, as they are identified using techniques outlined in Sect. 4. We provide representative, but non-exhaustive sources of public information that can be leveraged towards this objective through OSINT analysis.

Network One possible entry point is over the network. Direct network channels to industrial devices in the identified target locations over the public internet may be available. To this end, Shodan, a "search engine for internet-connected devices", can be employed [16]. Shodan uses crawlers that periodically index the web, searching for open ports and a wide variety of protocols including several industrial protocols. Table 2 is a snapshot of indexed devices by Shodan for the top four most commonly used protocols in the power industry taken at March 30, 2018. Moreover, to ensure non-stale results it is possible for attackers to launch their own crawlers. The release of efficient open source scanning tools such as ZMap, which can scan the entire IPv4 range in a matter of a minutes, have enabled large scale scans of the internet with limited resources [26]. Network telescope studies focusing on industrial protocols have shown that several scanning campaigns specifically target industrial protocols employing these tools [30]. The situation is exacerbated given the poor security the majority of industrial protocols employ, allowing unauthenticated access [33]. To extract the IPs of interest from the set of results, attackers can identify the organization in control of the target locations (e.g., power utility or governmental organization) and find IP addresses relating to the organization through reverse WHOIS searches. All IP addresses belonging to the organization are possible entry points; their compromise could enable lateral movement within the organization's network. The most promising results lie at the intersection between IPs owned by the company and industrial devices indexed by scanning campaigns. With the public IP address of a device known, attackers can employ remote fingerprinting techniques to identify the specifics of the industrial device [36].

Supply chain When specific make and model information for target devices are known, attackers can carry out device-specific studies. To that end, a possible option for an attacker is to obtain a physical copy of the target device for further hands-on experimentation. With access to physical copies of power system devices, attackers can validate known vulnerabilities and test their developed attack strategies to increase the success probability of the final attack. Besides official vendors, online marketplaces such as eBay, Amazon, Alibaba and other third party companies offer used or new industrial equipment for sale. The majority of listings concern surplus or decommissioned equipment, typically sold at a fraction of the original price.

Table 3 Microprocessor-enabled power grid devices from top vendors listed in eBay

Vendor	Listings
ASEA Brown Boveri (ABB)	216
General Electric (GE)	458
Schneider electric	373
Siemens	271

Table 3 contains eBay listing statistics regarding microprocessor-enabled power grid devices from the four vendors with the largest market share [59], gathered on March 30, 2018.

Vulnerability reports Publicly available vulnerability databases, such as the Industrial Control Systems Cyber Emergency Response Team (ICS-CERT) advisories and alerts, and the National Vulnerability Database (NVD) can provide a public source of vulnerabilities for target devices [8, 11]. Such databases are constantly updated with vulnerabilities discovered in industrial devices, including power system devices. If no vulnerabilities in the specific target device are published publicly, attackers can investigate vulnerabilities in the same family of products which will likely affect the target device because of intellectual property and code reuse between products in the same product line. Even if a patch was issued to address a known vulnerability, there is high probability that the target system is still vulnerable. Beyond reluctance and financial reasons, a major prohibitive factor for timely updates is that power systems must be available at all times and thus any modifications can only take place at prescheduled maintenance times [31].

Vulnerability development In case no known vulnerabilities exist, or the objectives of an attack cannot be fulfilled with known vulnerabilities, adversaries can develop their own zero-day vulnerabilities. This approach is more resource demanding but has a higher probability of success and lower probability of detection. To achieve this, attackers can follow several strategies. With access to a physical device, they can extract the firmware of the device and analyze it for vulnerabilities, monitor the network traffic exchanged looking for vulnerabilities in the network stack, and examine the configuration software for attack entry points. Several publicly available blog articles outlining step-by-step approaches and techniques for hacking embedded systems and Internet of Things (IoT) devices contain knowledge that is directly transferable to power system devices [4]. For example, copies of firmware images may be obtained from vendor websites. If that is not possible, or if the firmware is encrypted, it can be directly dumped from the physical device via debug ports, or extracted from the device's flash memory using chip-off forensics techniques [39]. Reverse engineering of firmware images can be accelerated with the use of open source tools such as binwalk [2]. Fuzzing, a black box technique for testing software, can also yield exploitable results. As regards to the network stack, information exchanges between the target device and the configuration software can be intercepted and analyzed using open source tools such as Wireshark [48].

Open source projects An observation that can be drawn from the analysis above is that there exists an abundance of open source software that can play an enabling role when designing an attack. The majority of this software was designed for benign uses (e.g., gathering statistics, education, penetration testing), but it can be misused by adversaries with a malicious agenda. This is the case for the ongoing campaign against U.S. critical infrastructure, where open source tools are employed [61]. In addition to the direct enabling role of open source projects, open source tools and libraries can become attack entry points. Adversaries can contribute to open source projects they know are used by the target organization (e.g., a widely-used open source project in the power industry is openSCADA [52]). Hidden within updates, they can inject malicious code and backdoors.

6 Experimental Evaluation of OSINT Techniques

For experimentally evaluating the impact and quality of information generated with OSINT techniques we use an OSINT approach to construct a model of a large, real system, and analyze it with contingency analysis techniques to identify its critical operation points. Compromise and adversary control of these points is sufficient to create a system wide blackout.

6.1 Modeling a Real Power System

In testing the feasibility of modeling an entire real complex power grid using publicly available information, we select a real system and set out to find the required information outlined in Sect. 4. Because of the sensitive nature of this study, we refrain from identifying the system and present only anonymized and non-identifying information. For the remainder of this section, the system under study is referred to as *Outage Land*.

We employ OSINT techniques to generate the model of Outage Land. We identify several sources of information for the power system including publicly available corporate presentations, expansion planning reports, and vendor success stories. From these public sources we initially extract all the high voltage buses and connections between them, as well as the location of transmission and generation substations. Through this analysis we create a topological map of the system. We cross-validate the constructed topology through GIS services, by manually tracing high voltage lines and transmission towers throughout Outage Land. Figure 2 shows an example of a generation plant, a transmission substation, and the incoming/outgoing transmission lines. Transmission towers are depicted as black dots and transmission lines as red lines. We undertake the laborious manual process of generating the topology of Outage Land's entire transmission network through GIS services and compare it with the topology extracted through other public resources,

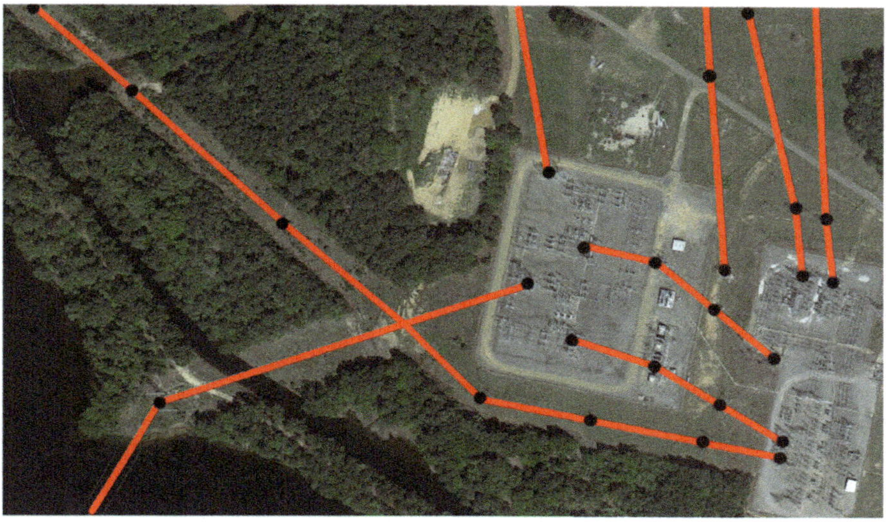

Fig. 2 Tracing transmission lines on GIS services

validating our findings. GIS mapping for the entire system required 40 man-hours. Although this process may be automated, we refrain from doing so because of the restrictive Terms and Conditions of commercial GIS software that do not allow automatic mining of satellite imagery.

Towards enabling contingency studies of Outage Land we again rely on OSINT techniques to extract operational characteristics of the power system. We fuse information from various sources to identify bus voltage levels, installed generation details, transmission line characteristics and load estimations. We again cross-validate the extracted information by comparing TSO reports, vendor success stories and news articles. The transmission network of Outage Land operates at three voltage levels: 132, 220 and 400 KV. The components and connections for each voltage level are presented in Table 4.

6.2 *Identifying Critical Operation Points*

The public information and the model of Outage Land is appropriately formatted to be used as input to MATPOWER, an open source MATLAB package, that solves power flow and optimal power flow problems [63]. The numerical testing we perform considers only the contingencies associated with tripped transmission lines, since these are the targets identified in Sect. 2.1. For each $N-1$ contingency scenario (tripping of a power line) we compare the power flows in the resulting tripped network with the corresponding thermal limits of the transmission lines. If

Table 4 Outage Land power system statistics

Type	Number
400 KV buses	36
220 KV buses	64
132 KV buses	75
Total buses	**175**
400 to 400 KV branches	84
220 to 200 KV branches	115
132 to 132 KV branches	155
400 to 220 KV branches	29
400 to 132 KV branches	34
220 to 132 KV branches	4
Total branches	**421**
Generation stations	12
Maximum load forecast	15 GW
Installed generation capacity	17 GW

the contingency does not violate these thermal constraints, then the line is marked as "safe", otherwise it is characterized as "dangerous". Similarly, in the $N-2$ contingency case a pair of different lines (i, j) are tripped simultaneously, and a set of constraints are used to identify the events that lead to thermal limits violations.

We rely on the DC approximation for the $N-k$ contingency problem. The power system is described by the vectors of voltage angles θ_i, where $i = 1, \dots, n$ and n the number of buses in the system. In this scenario, the DC power flow equations have the form of $B\theta = p$, where B is the $n \times n$ nodal DC susceptance matrix and p is the vector of real power injections at the buses of the system. The matrix B can be represented as $B = MYM^T$, where M is the $n \times N$ connection matrix with $1s$ representing the beginning bus of the branch and -1 its end. Y is the diagonal $N \times N$ matrix of branch susceptances. Therefore, the vector of power flows can be described as $v = YM^T\theta = YM^TB^{-1}p$.

In addition to the $N-1$ contingencies we also want to identify the dangerous $N-2$ contingencies to increase the impact and probability of a successful attack. To that end, we use the pruning algorithm proposed in [60]. The algorithm excludes islanding conditions as they do not cause cascading failure propagation. The effect of each tripped line is described with a LODF matrix L which relates the change of flow in a monitored line i that follows after the tripping of line j with original flow v_j, i.e., the matrix element $L_{ji} = (\hat{v_j} - v_j)/v_i$, relates the change of the flow through line j from v_j (before outage) to $\hat{v_j}$ (after outage) with the flow v_i through line i before the outage. In order to find the relation between single and two line contingency LODFs, the LODF matrix becomes [60]: $L = YM^TB^{-1}\widetilde{M}(1 - \widetilde{Y}\widetilde{M}^TB^{-1}\widetilde{M})^{-1}$, where $\widetilde{M}$ is the $n \times k$ submatrix of M corresponding to the outaged lines and similarly $\widetilde{Y}$ is the $k \times k$ outaged line submatrix of Y. This expression is applicable both to single and double line outage events. Direct comparison of these expressions allows us to relate the two LODF matrices. The double outage effect is:

$$\hat{v}_l - v_l = \frac{L_{li}(v_i + L_{ij}v_j)}{1 - L_{ji}L_{ij}} + \frac{L_{lj}(v_j + L_{ji}v_i)}{1 - L_{ji}L_{ij}} \tag{1}$$

In this relation we denote the outage lines by i, j and consider the change of the flow on some arbitrary line l. The contingency occurs whenever the absolute value of the flow at line l exceeds a critical value, i.e., $|\hat{v}_l| > v_l^{critical}$ that can for example be the thermal rating of a transmission line. We preprocess the model by converting it to a line-reduced network by aggregating radial branches into single nodes, a typical procedure for contingency analyses. We name the resulting nodes *transmission links*, as they are collections of transmission lines that connect the same edges (i.e., transmission substations).

From our analysis we identify 228 dangerous $N-1$ transmission link contingencies. They result from at least one violation of 72 transmission links in the 231-line reduced network, and are drawn as blue nodes in Fig. 3. All 228 dangerous $N-1$ contingencies include more than one transmission lines to form their transmission link, meaning that more than one transmission lines need to be compromised to realize the contingency scenario they describe.

To investigate $N-2$ transmission link contingencies, we increase limits on the lines that caused $N-1$ contingencies, and rerun the $N-1$ analysis in search of

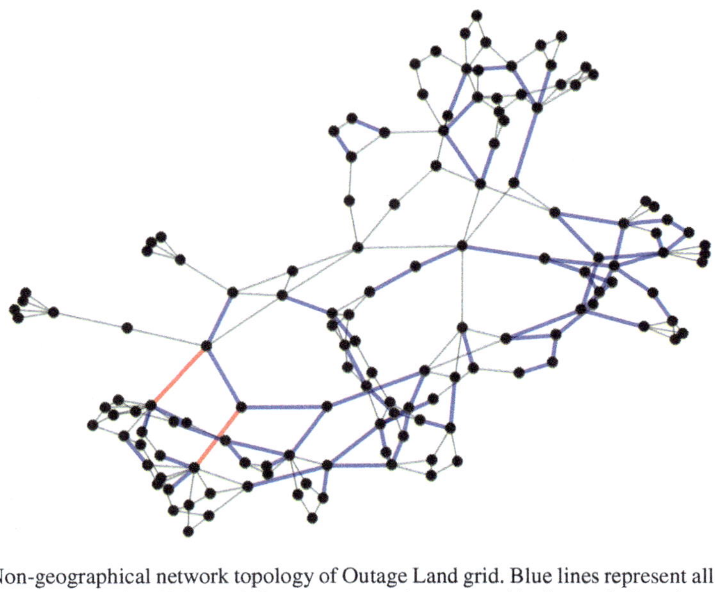

Fig. 3 Non-geographical network topology of Outage Land grid. Blue lines represent all the $N-1$ contingency transmission links, red lines represent the most critical set of $N-2$ contingency transmission links

$N-2$ contingencies. We identify 1174 such $N-2$ transmission link contingency scenarios in the reduced network, out of the possible 26,565 combinations. In these results, some lines appear more frequently than others. To identify the most critical links, we rank the results based on their frequency of appearance. The set of most critical $N-2$ contingency links are presented as red nodes in Fig. 3.

The identified transmission links from the above study pose the greatest risk for a wide area power outage, hence they would constitute a natural target for a malicious adversary. The significance of the identified links is intuitively corroborated by their physical proximity to densely populated areas in Outage Land where end consumers are mostly situated. In theory, *any* of the above identified contingencies will result in a blackout, allowing an adversary to select which links to attack based on accessibility, whether the transmission lines are overhead or underground, or other criteria.

7 Discussion and Related Work

Ostensibly, this study seems to suggest that the cyber threat against power systems is under-estimated, as it serves as a proof-of-concept that attacks are enabled by the plethora of public sources of information. However, before tripping the alarm bells, we must take into consideration the resiliency and robustness of power grids around the world, as well as the readiness and experience of power engineers in handling blackouts and power outages. In the U.S. alone, 3879 blackouts were reported just in 2016, lasting an average of 48 min and affecting almost 18 million people across the country, causing an annual monetary loss of a staggering $150 billion [27]. All across the globe, blackouts are a usual occurrence typically caused by weather phenomena, accidents, animals, equipment failures etc. As a consequence, power system operators have experience in handling power outage scenarios. Furthermore, the power industry and government stakeholders have direct financial incentives in addition to societal obligations to direct efforts and funding in blackout prediction and prevention mechanisms, and ensuring shorter recovery times [23]. The objective investigated in this chapter, per the threat model, is causing a wide area power outage. This is not the same as a *prolonged* power outage. For example prolonged outages may require the destruction or incapacitation of critical not-in-stock equipment (e.g., large transformers and generators), something that typically requires physical tampering.

In general, public dissemination of information and transparency are beneficial for progress. With this work, our aim is not to constraint the release of information, classifying all power grid information as confidential. Rather, we aim to highlight the sensitivity of certain pieces of information that could enable malicious adversaries to launch cyberattacks against a power system, and assist in better understanding public information regarding power systems.

7.1 Why Is Sensitive Information Publicly Available?

Throughout our study, we discovered a multitude of publicly available resources that contain sensitive details on power systems. Besides understanding what intelligence can be generated using such information, it is also important to understand *why* such information is part of the public domain. Regarding data that can be used for modeling a system, it is sometimes made publicly available by mistake, for example due to misconfiguration errors or improper access controls on public websites. In the case of identifying the location of power system assets through GIS, there is unintended leakage of information because of the nature of the satellite mapping techniques. Improper risk assessments from power system operators can also result in releasing sensitive information. Information can be leaked by third parties, such as power system devices vendors, when they release success stories that include operational details of real systems.

Alternatively, information can be intentionally released to the public. Given the strong dependence of nations on electric power supply, the power industry is often considered part of public utility infrastructure. Even in cases where power utilities are private corporations, they are regulated by public utilities commissions. For example, the U.S. Federal Energy Regulatory Commission (FERC) regulates all power utility companies. As public utilities, certain details regarding the operation of power systems must be transparent to citizens, in some cases mandated by law. For example, in some jurisdictions operational and procedural details must be provided in order to justify electricity pricing rates [9]. Furthermore, because of the threat of climate change, public interest groups and international organizations require publication of data to enable monitoring and regulating the environmental impact of electric utilities. Governments may also require the release of data to ensure transparency and fair competition between private companies [43], and data may be released towards promoting a more reliable and efficient power grid [53].

7.2 Prevention and Mitigation Strategies

In dealing with the cyber threat to power systems, efforts should be directed to efficient and effective prevention and mitigation strategies. These can start by following appropriate cybersecurity practices. Several best practices guides, such as the Guide to Industrial Control Systems Security by NIST [57] and standards, such as ANSI/ISA-62443 [34], offer practices that can thwart or impede attacks against power systems. Vendors of power system devices should harden the security of their products by revisiting their threat models, which might not consider malicious tampering as a serious threat. Given the direct impact of power system resiliency to the wider public, policymakers could accelerate action with regulatory policies that would incentivize power system operators and vendors to adopt good security practices.

From a technical perspective, field devices could be hardened at the different layers of the system, namely the hardware, firmware, software, network and operation layers [40]. When it comes to hardware, access through debug ports, such as JTAG and UART, should be disabled and the supply chain should be more strictly controlled to disallow adversaries to obtain physical copies of critical devices. Hardware support for cryptographic operations can enable the use of secure cryptographic primitives in power systems. For firmware, static images should be encrypted and firmware updates should be signed by the respective vendors. On the software layer, proper security mechanisms and risk assessments are necessary, along with the ability of secure and verifiable updates that do not require downtimes. Regarding the network layer, industrial devices deployed in power systems should never be directly connected to the internet, and field networks and business networks should be segmented to thwart attacks that rely on lateral movement. Industrial protocols with no security mechanisms should be redesigned or replaced with secure counterparts. On the operation layer, effective physical security mechanisms should be employed, and the overall operation of the system could inform anomaly detection schemes that aim to act as early indicators of attacks by gathering system-wide data.

As regards to information that can expose critical operational characteristics and/or enable attack vectors against a power system, more careful consideration must be taken when deciding its classification and targeted audience. Power system shareholders can carry out periodic reviews of publicly available information, gaining visibility into the different sources of information that concern their systems. In addition, requirements and methods for proper handling and control of potentially critical information can be established. For example, techniques that anonymize or randomize the information before its release could simultaneously provide transparency while protecting critical and sensitive data.

7.3 Related Work

In the power engineering academic community, there exist studies that utilize real data to model complex power systems. Several studies use the Polish grid model included in MATPOWER cases [63], which is based on data collected in 2000 and 2001 from the website of the Polish transmission system operator. However, information about the current state of the Polish grid is not available. In addition to the Polish grid, real data from the power grid in Great Britain (GB) in 2013 are available as a MATPOWER case. These data were assembled from National Grid public data and reports and used for optimal power flow studies [12]. Real U.S. power grid data obtained from the Platts GIS database are used in [22] to estimate the vulnerability of the U.S. power grid to geographically correlated failures. In contrast, in this chapter, we build a model of a real system from the ground up leveraging and fusing a variety of public resources, and validate the created model using GIS imagery.

8 Conclusions

Motivated by the observation that adversaries actively employ OSINT techniques for campaigns against the energy sector, we undertake a study of the amount and quality of intelligence that can be extracted using publicly available resources. We present a broad study on the sources of intelligence that can be leveraged to model a large power system, analyze the model and finally study how to exploit it. We experimentally evaluate the feasibility of an open source campaign by constructing and validating the model of a real system using only publicly available information, and analyzing the system to identify its critical points using contingency analysis. With this study we aim to provide insight into the threat cyberattacks based on publicly available resources pose to power systems. Our work can assist energy sector stakeholders and regulators take informed decisions, and more carefully handle information dissemination concerning sensitive characteristics of their power systems.

References

1. Bangladesh power cut plunges millions into darkness. http://reuters.com
2. Binwalk firmware analysis tool. https://github.com/ReFirmLabs
3. Blackout watch: Brazilian blackout 2009. http://pacw.org
4. Embedded device hacking. http://devttys0.com
5. European network of TSOs for electricity. http://entsoe.eu
6. How an entire nation became Russia's testlab for cyberwar. http://wired.com
7. India Northern Regional Load Despatch Centre. http://nrldc.in
8. Industrial Control Systems Cyber Emergency Response Team. http://ics-cert.us-cert.gov
9. International energy statistics (2017). http://eia.gov
10. Map of PMUs with synchrophasor data flows in North America. http://naspi.org
11. National Vulnerability Database. http://nvd.nist.gov
12. Network data of real transmission networks (2013). http://maths.ed.ac.uk
13. Open energy information. http://openei.org
14. Open power system data platform. http://open-power-system-data.org
15. Rebels tied to blackout across most of Pakistan. http://nytimes.com
16. Shodan search engine. http://shodan.io
17. U.K. Electricity Ten Year Statement 2016. http://nationalgrid.com
18. Ukraine's power outage was a cyber attack: Ukrenergo. http://reuters.com
19. Abraham S, Efford JR (2004) Final report on the August 14, 2003 blackout in the U.S. and Canada. Technical report, Power System Outage Task Force
20. Alonso F, Greenwell C (2016) Underground vs. Overhead: power line installation-cost comparison and mitigation. Electr. Light Power 22
21. Bakshi S (2012) Report of the enquiry committee on grid disturbance in Northern region on 30th July 2012 and in Northern, Eastern & North-Eastern region on 31st July 2012. Technical report, Indian Ministry of Power
22. Bernstein A, Bienstock D, Hay D, Uzunoglu M, Zussman G (2014) Power grid vulnerability to geographically correlated failures – analysis and control implications. In: IEEE INFOCOM 2014 – IEEE conference on computer communications, pp 2634–2642. https://doi.org/10.1109/INFOCOM.2014.6848211

23. Campbell RJ (2012) Weather-related power outages and electric system resiliency. Technical report, Congressional Research Service
24. Davis CM, Overbye TJ (2011) Multiple element contingency screening. Trans Power Syst 26(3):1294–1301
25. Davis C, Chmieliauskas A, Nikolic I (2015) Enipedia. Energy & Industry group, Faculty of Technology, Policy and Management, TU Delft
26. Durumeric Z, Wustrow E, Alex Halderman J (2013) ZMap: fast internet-wide scanning and its security applications. In: Presented as part of the 22nd USENIX security symposium (USENIX Security 13), Washington, DC. USENIX, pp 605–620. https://www.usenix.org/conference/usenixsecurity13/technical-sessions/paper/durumeric,ISBN:978-1-931971-03-4
27. Eaton: Blackout tracker (2017) United States annual report 2016
28. Eles: Slovenia's transmission network: annual report 2015. http://eles.si
29. Elia: Belgium electrical transmission network: annual report 2016. http://elia.be
30. Fachkha C, Bou-Harb E, Keliris A, Memon N, Ahamad M (2017) Internet-scale probing of CPS: inference, characterization and orchestration analysis. In: Proceedings of the 24th network and distributed system security symposium (NDSS'17), San Diego, Feb 2017
31. Gharavi H, Ghafurian R (2011) Smart grid: the electric energy system of the future. In: Proceedings of the IEEE. IEEE, Piscataway
32. Grainger J, Grainger W, Stevenson W (1994) Power system analysis. McGraw-Hill Education, New York
33. Igure VM, Laughter SA, Williams RD (2006) Security issues in SCADA networks. Comput Secur 25(7):498–506
34. International Society of Automation (2018) ANSI/ISA 62443 security for industrial automation and control systems. ISA
35. Kaplunovich P, Turitsyn K (2016) Fast and reliable screening of N-2 contingencies. Trans Power Syst 31(6):4243–4252
36. Keliris A, Maniatakos M (2016) Remote field device fingerprinting using device-specific modbus information. In: 2016 IEEE 59th international Midwest symposium on circuits and systems (MWSCAS), pp 1–4. https://doi.org/10.1109/MWSCAS.2016.7870006
37. Keliris A, Maniatakos M (2017) Demystifying advanced persistent threats for industrial control systems. ASME Mech Eng 139(03):S13–S17. https://doi.org/10.1115/1.2017-Mar-6
38. Knake R (2017) A cyberattack on the U.S. power grid. Contingency planning memorandum, vol 31. Council on Foreign Relations. https://www.cfr.org/report/cyberattack-us-power-grid. 3Apr2017
39. Konstantinou C, Maniatakos M (2015) Impact of firmware modification attacks on power systems field devices. In: International Conference on Smart Grid Communications. IEEE, pp 283–288. https://doi.org/10.1109/SmartGridComm.2015.7436314
40. Konstantinou C, Maniatakos M (2017) Security analysis of smart grid. Commun Control Secur Challenges Smart Grid 2:451
41. Langner R (2011) Stuxnet: dissecting a cyberwarfare weapon. Secur Priv 9(3):49–51
42. Lee RM, Assante MJ, Conway T (2016) Analysis of the cyber attack on the Ukrainian power grid. Technical report, SANS Industrial Control Systems
43. Momoh J, Mili L (2009) Economic market design and planning for electric power systems, vol 52. Wiley, Hoboken
44. Morison K, Wang L, Kundur P (2004) Power system security assessment. Power Energy Mag 2(5):30–39
45. NERC: Disturbance Reports 1992–2009
46. NERC (2009) FAC-011-2: system operating limits methodology for the operations horizon
47. OpenStreetMap: Power networks. http://openstreetmap.org
48. Orebaugh A, Ramirez G, Beale J (2006) Wireshark & Ethereal network protocol analyzer toolkit. Syngress, Rockland

49. Pajic S (2007) Power system state estimation and contingency constrained optimal power flow: a numerically robust implementation. Worcester Polytechnic Institute
50. Project Group Turkey (2015) Report on blackout in Turkey on 31st March 2015. Technical report, European Network of Transmission System Operators for Electricity
51. ProSoft Technology: Power success stories. http://prosoft-technology.com
52. Reimann J, Rose J (2015) Eclipse SCADA: the definite guide
53. Roland Berger (2014) Study regarding grid infrastructure development: European strategy for raising public acceptance. Technical report, European Commission Tender No. ENER/B1/2013/371
54. SCADA Innovations: Success stories. http://scadainnovations.com
55. Siemens: High voltage substation references. http://energy.siemens.com
56. Stott B, Alsac O, Alvarado F (1985) Analytical and computational improvements in performance-index ranking algorithms for networks. Int J Electr Power Energy Syst 7(3): 154–160
57. Stouffer K, Falco J, Scarfone K (2011) Guide to industrial control systems security. NIST Spec Publ 800(82):16
58. Symantec: Dragonfly: Western energy sector targeted by sophisticated attack group. http://symantec.com
59. Technavio: Global smart grid transmission and distribution equipment market 2016–2020. http://technavio.com
60. Turitsyn KS, Kaplunovich PA (2013) Fast algorithm for N-2 contingency problem. In: 46th Hawaii international conference on system sciences (HICSS). IEEE, pp 2161–2166. https://doi.org/10.1109/HICSS.2013.233
61. U.S. DHS and FBI: US-CERT: advanced persistent threat activity targeting energy and other critical infrastructure sectors. http://us-cert.gov/ncas/alerts/TA18-074A
62. Zachariadis T, Poullikkas A (2012) The costs of power outages: a case study from Cyprus. Energy Policy 51(Supplement C):630–641
63. Zimmerman RD, Murillo-Sánchez CE, Thomas RJ (2011) MATPOWER: steady-state operations, planning, and analysis tools for power systems research and education. Trans Power Syst 26(1):12–19

A Taxonomy of Side Channel Attacks on Critical Infrastructures and Relevant Systems

Nick Tsalis, Efstratios Vasilellis, Despina Mentzelioti, and Theodore Apostolopoulos

Abstract Information disclosure leads to serious exploits, disruption or damage of critical operations and privacy breaches, both in Critical Infrastructures (CIs) and Industrial Control Systems (ICS) and in traditional IT systems. Side channel attacks in computer security refer to attacks on data confidentiality through information gained from the physical implementation of a system, rather an attack on the algorithm or software itself. Depending on the source and the type of information leakage, certain general types of side channel attacks have been established: power, electromagnetic, cache, timing, sensor-based, acoustic and memory analysis attacks. Given the sensitive nature of ICS and the vast amount of information stored on IT systems, consequences of side channel attacks can be quite significant. In this paper, we present an extensive survey on side channel attacks that can be implemented either on ICS or traditional systems often used in Critical Infrastructure environments. Presented taxonomies try to take into consideration all major publications of the last decade and present them using three different classification systems to provide an objective form of multi-level taxonomy and a potentially profitable statistical approach. We conclude by discussing open issues and challenges in this context and outline possible future research directions.

General Terms Security, Privacy, Side channel attacks, ICS, Critical infrastructures, Timing, Electromagnetic, Sensor, IT, Cryptography, Cache

Keywords Side channel attack · Classification · Category · Targeted hardware · Targeted software · Critical infrastructure

---

N. Tsalis · E. Vasilellis (✉) · D. Mentzelioti · T. Apostolopoulos
Information Security & Critical Infrastructure Protection (INFOSEC) Laboratory Department of Informatics, Athens University of Economics & Business, Athens, Greece
e-mail: ntsalis@aueb.gr; vasilelliss@aueb.gr; dmentz@aueb.gr; tsa@aueb.gr

D. Gritzalis et al. (eds.), *Critical Infrastructure Security and Resilience*, Advanced Sciences and Technologies for Security Applications,
https://doi.org/10.1007/978-3-030-00024-0_15

1 Introduction

Side channel attacks (SCAs) are closely related to the existence of physically observable phenomena caused by the execution of computing tasks in electronic devices [1]. Processors consume time and power to perform assigned tasks, systems radiate electromagnetic fields, dissipate heat, and even produce noise. Through appropriate observations (with or without tampering of the device), malicious users can gather enough information to extract patterns and mount successful attacks; mostly against the confidentiality of a system. There exist plenty information sources that leak information and can consequently be exploited by malicious adversaries [1].

Critical infrastructures (CIs) include physical resources, services, and information technology facilities, networks, and infrastructure assets which, if disrupted or destroyed, would have a severe impact on the health, safety, public confidence, or economic well-being of citizens or the efficient functioning of governments. These categories comprise the sectors of water, gas, fuel, electricity, transportation, communication, national defense, financial services and food supply [2]. Contributions of this article can be summarized as follows:

1. We enhance taxonomies on side channel attack classification with reports and publications of known attacks along with their impact on critical infrastructures.
2. We aim to provide a single point of reference on novel, advanced side channel attacks published in the last decade, instead of providing a full bibliography of every potential attack; something that would possibly require a book on its own.
3. Given the large number of experiments using side channel analysis and the currently short number of proposed attack classification categories, we also extend existing side channel attack categories by proposing additional ones that could give a broader insight about side channel attack classification.
4. The presented classifications along with case studies can be used in a formative way and are thus useful during the preliminary stages of inquiry as a heuristic tool in the discovery, analysis, comparison and theorizing. A good classification connects concepts in a useful structure. There are many approaches to the process of classification and to the construction of the foundation of classification schemes. Each kind of classification process has different goals, and each type of classification scheme has different structural properties as well as different strengths and weaknesses. The goal of this paper is to provide a comprehensive taxonomy of a wide range of side channel attacks regarding the security and privacy, to analyze the methodology behind the categorization and
5. Interpret detected attacks and categories in a statistical way. Observations can be deduced by the proposed classification.

Section 2 describes the related work on side channel classification, presents current taxonomies and relates them with the studied attacks. In Sect. 3, we present the concepts with regard to the classification method used and provide our side channel attack classification taxonomies. In Sect. 4 we elaborate on results, while in Sect. 5 we summarize and pinpoint useful knowledge gained from this paper.

2 Related Work

The general classes of side channel attacks (SCAs) widely known [1] are summarized below, without being restrictive for additional categories, depending on the detail related to the source of the leaked information. We chose not to make any further description for every reference as it wouldn't serve the purpose of the classification method and statistical overview deployed in this paper.

Cache-based attack —CPU cache exists between the CPU and the main memory; a type of hierarchical structure that can speed program run-time. If the CPU accesses data that were not stored in the cache, a delay will be generated as the target data must be loaded from main memory into the cache. The measurement of this delay may enable attackers to determine the occurrence and frequency of cache misses [3–13].

Timing attack — A timing attack is, a way of obtaining user's private information by carefully measuring the time it takes for cryptographic operations to be carried out. The principle of this attack is to exploit the timing variance in the operation [14–19].

Power analysis attack — Power analysis attack can be divided into Simple and Differential Power Analysis (referred to as SPA and DPA, respectively). SPAs guess which particular instruction is being executed at a certain time, as well as the input and output values using power traces. The adversary needs an exact knowledge of the implementation to mount such an attack. DPA attacks need not have knowledge about the implementation details, because statistical methods are being deployed in the analysis process. DPAs can be mounted using very little resources [20–22].

Electromagnetic (EM) attacks —An adversary who observes electromagnetic emanations can understand their causal relationship with the underlying computation and infer a surprising amount of information. Electromagnetic Analysis (EMA) attacks are divided into: Simple Electromagnetic Analysis (SEMA) and Differential Electromagnetic Analysis (DEMA) [23–28].

Acoustic attack — One of the oldest eavesdropping channels, namely acoustic emanations, has received little attention recently, even though it has been demonstrated [29] that full 4096-bit RSA decryption keys can be extracted from laptop computers (of various models), within an hour, using the sound generated by the computer during the decryption of some chosen ciphertexts [29–31].

Memory attack — This type of attack is like the cache-based, with the difference that the attack vector is a memory module [32, 33].

Sensor-based attack — These attacks are based on two observations. First, smartphones are equipped with lots of sensors—both motion sensors as well as ambient sensors—that can be accessed without any permission, and second, these devices operate with fingers while being held in the users' hands [34–40].

Optical — Secrets and sensitive data can be read by visual recording using a high-resolution camera, or other devices that have such capabilities. It has also been proved that optical radiation emitted from computer LED (light-emitting diodes) status indicators can be analyzed to infer the data being processed by a device.

Fault attack — Hardware faults and errors occurring during the operation of a cryptographic module, seriously affect security. Fault attacks present practical and effective attacking against the cryptographic hardware devices, such as smart cards.

Template attack — In a theoretical sense this is the strongest kind of side channel attack. This attack requires that an adversary has access to an identical experimental device.

Combination of side channel attacks — Such a combination could be used to circumvent some countermeasures against specific side channel attacks. A simple example is the measurement of the time between significant features in the power trace. More recently, researchers have also examined the potential for multi-channel attacks which utilize multiple side-channels simultaneously, like power and EM.

Combination of SCA and mathematical attacks — Traditional cryptanalysis techniques can be combined with side channel attacks to uncover the secret key and/or break the implementation details of the ciphers. Consequently, even a small amount of side-channel information is sufficient to break common ciphers.

Another categorization system proposed in [41] classifies side channel attacks along three axes:

(i) Passive vs active: Attackers are distinguished into those who passively observe leaking side channel information, and those who actively influence the target via any side channel.
(ii) Physical vs logical properties: Side channel attacks are classified according to the exploited information, i.e., depending on whether the attack exploits physical properties (hardware) or logical properties (software features).
(iii) Local vs vicinity vs remote attackers: Side-channel attacks are classified depending on whether the attacker must be in physical proximity/vicinity with the target.

Attacks are sorted into the following classes: invasive attacks, semi-invasive attacks and non-invasive attacks. An invasive attack involves direct access to the internal components of cryptographic modules or devices. The semi-invasive attack involves access to the device, but without making electrical contact other than with the authorized surface. A non-invasive attack involves close observation or manipulation of the device's operation. This attack only exploits externally available information that is often unintentionally leaked.

The aim of our work, parallel to the suggested categorization process described in the methodology section, is to convey the knowledge that side channel attacks are notable threats for the pivotal facilities and critical infrastructures. The rest of this part is dedicated not only to certain broadly recognized attacks that have received significant attention from security experts possibly due to their destructive capability and robustness, but also to prominent experiments using side channels to assist in anomaly detection.

2.1 Acoustic Side Channel

Trippel et al. investigated how analogue acoustic injection attacks can damage the digital integrity of the capacitive MEMS accelerometer. Spoofing such sensors with intentional acoustic interference enables an out-of-spec pathway for attackers to deliver chosen digital values to microprocessors and embedded systems. For example, they showed how to inject fake steps into a Fitbit fitness tracker to earn financial rewards, and also, they accomplished a self-stimulating attack whereby they play malicious music files from a smartphone's speaker to control an app that drives an RC car [43].

Asonov et al. [44] proved that keylogging a computer keyboard input is possible with just the sound emanated by different keys. Their approach was based on differentiating the sound emanated by the keys with the use of a neural network to assist in recognizing which key was being pressed. On the other hand, Zhuang et al. [45] used a combination of standard machine learning and speech recognition techniques, like spectrum features, Hidden Markov Models, linear classification, along with feedback-based incremental learning.

In 2010 Backes et al. [46] recovered the content of medical prescriptions processed by a dot-matrix printer through analysis of the emanated sounds. Their experiments showed that 72% to 95% of printed words could be recovered, if contextual knowledge about the text was assumed.

Al Faruque's [31] demonstrated that sound emanated from additive manufacturing systems, like 3D printers, carry process information which can be used to indirectly reconstruct the printed objects, without having access to the original design. Chhetri [47] successfully managed to reconstruct various test objects with an average prediction accuracy of 86% and average prediction error of 11.11%. His attack model consisted of digital signal processing, machine-learning algorithms as well as context-based post processing to steal the geometry details. In another work, Chhetri used the relation between the cyber domain data and the physical domain analog emissions from the acoustic side-channel, in order to detect attacks on the firmware of cyber-physical additive manufacturing systems with an accuracy of 77.45% [48].

Krishnamurthy et al. [49] demonstrated that an attacker could exfiltrate sensitive information from a malware compromised PLC. In their testbed, the malicious program used the acoustic emissions of a motor controlling a valve in a feedback control loop as a covert channel. This kind of secret transmission could take place without affecting the performance, stability or the signal characteristics of the closed-loop process.

2.2 Cache Side Channel

In 2009 Ristenpart [50] showed that a potential attacker could co-locate another instance in public IaaS clouds, by examining the cache usage. Moreover, he demonstrated that having information about the computational load of an instance, could assist in detecting co-residency with a victim.

Zhang et al. in their work showed that a tenant could detect co-location in the same core through monitoring the L2 cache [12]. Furthermore, they demonstrated that de-duplication enables co-location detection from co-located VM's in PaaS clouds [10].

Finally, in 2012 Zhang et al. [9] managed to extract the cryptographic private ElGamal keys from incautious hosts using the most recent version of the libgcrypt cryptographic library.

2.3 Electroacoustic Side Channel

Although this type is not considered among the side channel categories, researchers were inspired by side-channel schemes used to detect Trojans in integrated circuits through the use of structural health monitoring (SHM) techniques [51]. Vincent et al. in their work [51] proposed that SHM techniques, and more specifically PZT augmented impedance based SHM, could be used in side-channel detection of changes to manufactured parts. During the manufacturing process the antenna/PZT assembly is joined to the manufactured parts, then the PZT is excited and the resulting impedance signature is acquired. Deviations in impedance provide a measure that allows the detection of damage or, in this case, the incipient intrusion and part modification. Critical manufacturing is crucial to the economic prosperity and continuity of a nation. Disruption of certain elements of manufacturing industry could affect essential functions across multiple critical infrastructure sectors. Thus, such attacks should not be underestimated or overlooked.

2.4 Electromagnetic Side Channel

Israeli researchers developed a new palm-sized device that can wirelessly steal data from a nearby laptop based on the radio waves leaked by its processor's power use. This spy bug, built for less than $300 and capable of fitting inside a piece of pita bread, is designed to allow anyone to "listen" to accidental radio emanations of a computer's electronics from 19 inches away and derive the user's secret decryption keys [23].

Grzesiak in his work proved that data being processed by a laser printer could be intercepted from a distance [52]. Lee et al. [53] using an H-Field probe and a digital oscilloscope, measured the electromagnetic emanations from VGA cables and, through analysis of display mode data, managed to successfully reconstruct the display.

Islam et al. [54] extracted information from the thermal side channel of multi-tenant data centers contain information about the tenant's runtime power usage, using a state-augmented Kalman filter. Their experiment revealed that an attacker could capture 54% of all attack opportunities to deploy power attacks and compromise the availability of the data center.

Al Faruque et al. [55] in their work introduced a novel methodology to reverse engineer the thermal images taken from additive manufacturing systems. They succeeded in extracting specific information available in the cyber-domain, like speed, axis movement, temperature, etc. Mowery et al. [56] studied the effectiveness of thermal attacks on ATMs that used plastic keypads and found that such attacks are feasible even after the user has been authenticated. Wodo et al. [57] successfully obtained the password or code entered on a variety of keyboard devices through thermal imaging attacks. Andriotis et al. [58] observed the resulted heat traces from a pattern entered for authentication and succeeded in retrieving parts of the pattern. Abdelrahman et al. [59] proved that thermal attacks are viable on touch screens, as they managed to infer PINs and patterns on mobile devices with an average success rate of 72% for PINs even with duplicate digits. As thermal cameras become ubiquitous and affordable, a new form of threat to privacy on touch screen rises.

In 2012 Stone [60] described a methodology for detecting anomalous operations of PLCs utilizing information extracted from radio-frequency (RF) features. Malicious actions or even system failures may result in changes in operating characteristics of PLCs. Specifically, a single collected waveform response provides sufficient separability to enable differentiation between normal and anomalous operating conditions. In 2015 Stone [61] expanded the previous work [60] to include Hilbert transformed sequence of unintentional PLC time domain emissions, a refined methodology to establish a more robust normal condition reference sequence, as well as a demonstration that involved multiple AllenBradley SLC-500 05/02 CPU programmable logic controllers. Van Aubel et al. [62] proposed a system that leverages electromagnetic side-channel measurements to detect behavioral changes of the software running on industrial control systems, utilizing methods from cryptographic side-channel analysis. Yi Han et al. [63] presented a contactless, passive, as well as non-intrusive control flow integrity monitoring solution for PLCs. ZEUS can identify malicious code execution through analysis of the electromagnetic emanation signals with an accuracy of 98.9% and with zero runtime overhead by its design. Boggs et al. [64] monitored involuntarily electromagnetic (EM) emissions from embedded devices to detect malicious code execution. In their work they used commercial off-the-shelf (COTS) software defined radio (SDR) hardware to detect code execution on an industrial control system (the Allen-Bradley 1756-EWEB module). Experimental results presented a prototype capable of detecting unknown (attack) code execution with 98% accuracy at 100% detection rate.

2.5 Optical Side Channel

LED status indicators on data communication equipment, under certain conditions, are shown to carry a modulated optical signal that is significantly correlated with information being processed by the device [65]. Physical access is not required; the attacker gains access, from a considerable distance, to all data going through the device, including plaintext in the case of data encryption systems. Modems and Internet Protocol routers, were found to be vulnerable. Authors describe design changes that can successfully block this kind of "Optical TEMPEST" attack [66].

Backes et al. [67] presented a novel eavesdropping technique capable of spying data being displayed at a distance on an arbitrary computer screen, including LCD monitors. Their approach can exploit reflections of the screen's optical emanations in a variety of objects that can be found in close proximity (up to 10 meters) and use these reflections in order to recover the original screen content.

Classen et al. [65] studied the feasibility of a VLC eavesdropper to intercept as well as decode a transmission even while being outside the direct beam. Such an attempt is not only feasible, but also implementable in many ways, like using a door gap, a window, or even a keyhole. Nearby attackers can often intercept VLC signals, potentially revealing information on personal habits in smart-home applications as well as sensitive health data. Blocking windows with a privacy film offers almost zero protection.

In 2017 Chakraborty et al. [68] demonstrated the feasibility of using the light sensor of a mobile device to recover information about the content being shown on a nearby flat-panel display (FPD). Although single-pixel light sensors have limited power, a judicious choice of features, that capture information that refers to changes in light intensity over time, assist in inferring sensitive information about the content type.

The aforementioned findings suggest that access to raw light-sensor readings, which can currently be done without special access controls, may carry nontrivial security ramifications. Consequently, possible data leakages from light-sensors in critical infrastructures should not be underestimated or overlooked.

2.6 Power Side Channel

Smartcard is widely used to restrict access to corporate and government buildings, and to process payments in public transit systems. Researchers at Germany's Ruhr University have circumvented the encryption used to protect a smartcard, a feat that makes it possible to clone perfect replicas of the digital keys and steal or modify their contents. The attack takes about seven hours to recover the secret key protecting the Mifare DESFire MF3ICD40. The hack leaves no trace that the card has been compromised, and requires equipment costing \$3000. The contactless card was adopted by NASA in 2004, although it's not clear if the agency has since upgraded [42].

Wei et al. [69] presented the first attack against the implementation of deep learning models. They performed an attack on a convolutional neural network accelerator based on FPGA and succeeded in recovering the input image by analyzing the collected power traces, without having any prior knowledge about the detailed parameters of the neural network. Wei's power-based side channel attack can achieve a recognition accuracy up to 89%.

Electrical network frequency (ENF) signals share the same patterns which could be used to identify not only the recorder time, but also the location of videos and sounds. Youngbae et al. [70] created a reference map of ENF signals that represent hundreds of locations worldwide and deployed a novel side channel attack which could identify the physical location of video or sound that was recorded or even streamed. Results show that their attack can infer the intra-grid location of the recorded audio files with an accuracy of 76% for those files that were 5 minutes or longer. Such an attack is feasible even when audio and video data are processed within a certain distortion range with audio codecs used in real VoIP applications. As a result, since critical infrastructure sectors are becoming increasingly dependent on VoIP for their telephony [71], such attacks should be taken seriously under consideration when selecting appropriate security countermeasures.

Meulenaer et al. [72] proved the feasibility of stealthy of power analysis attacks on AES as well as ECC implementations on MICAz and TelosB nodes. They designed a measurement setup which lets the attacker acquire power traces from the node without disturbing its normal operation or removing it from the network. The experiment showed that these attacks could not be detected by surveillance-based node capture defenses.

Hively et al. [73] proposed a novel approach to detect cyber anomalies through analysis of power information samples from a variety of computer components, like motherboard, internal aggregate DC power, external aggregate AC power, disk drive, CPU, graphic cards and network interface cards. They deployed phase-space analysis that used measurements of electrical power consumed by each active component to determine anomalous patterns, once affected by certain malware. This approach could be expanded to other cyber components in industrial control systems, apart from computers.

The urgency to produce a robust detection framework for rootkits has increased as depicted with recent examples of APT attacks utilized to disrupt critical infrastructure in Ukraine in 2015. In that direction Dawson et al. [74] presented such a detection framework that uses the voltage measurements of power supply and extracts time-serial system dynamics through a non-linear phase-space algorithm. Their result indicates that the algorithm is capable of detecting rootkit infection through power measurement analysis with an accuracy rate that meets or exceed the performance of similar machine learning algorithms.

Gunti et al. [75] proposed a method to drastically increase detection sensitivity of hardware trojans in integrated circuits, with little overhead in the design. The introduced security module can easily detect any tamper by the adversary without affecting at the same time the functionality of the system. In their model, once a block is activated, its static power consumption is measured and then compared with the circuit under test. Shende et al. [76] in their work followed a similar approach by comparing the mean of power traces of trojan infected and non-infected integrated circuits. They performed statistical data analysis to calculate statistical parameters of power, which are used as feature vectors. Principal Component Analysis (PCA) is used to reduce these feature vectors and then they are classified through Linear Discriminant Analysis (LDA). The proposed work is capable of detecting trojan infected IC with 100% accuracy.

Industrial Control Systems (ICS) rely in legacy systems that often maintain the same hardware for decades. Moreover, it is very important to be able to identify ongoing attacks on ES without interfering with real-time constraints. Moore et al. [77], in order to overcome the aforementioned problems, studied whether a buffer overflow attack generates distinct power consumption signatures once executed on a vulnerable ES. Clark et al. [78] proposed a monitoring system that analyses power consumption to detect malware during run-time. They tested their system on pharmaceutical compounder and a similar industrial control (SCADA) system with a detection accuracy of 94% for known malware and 85% for unknown one. Abbas et al. [79] proposed a novel approach of real-time anomaly detection utilizing the power profile of an application. Their approach showed it is sensitive enough to distinguish not only two different applications, but also two functionally equivalent applications that have different implementations. The existing anomaly detection methodologies rely on energy consumption and are capable of detecting anomalies only after an application has finished its execution, while Abba's technique can detect anomalies at an early stage, during execution.

Gonzalez et al. [80] utilized power fingerprinting to monitor the execution of systems with constrained resources like PLCs, without loading third party software on the platforms, as well as detect malicious software. Due to its negligible overhead along with zero-day detection capability, power analysis could potentially transform cyber security through enabling malware detection. Xiao et al. [81] proposed a non-invasive power-based anomaly detection scheme to detect attacks on PLCs, based on power consumption analysis. They implemented a real-time monitoring system for anomaly detection equipped with a read-time data acquisition module. The abnormal sample is identified through comparison of the actual sample with the predicted one. Xiao's detection scheme has an accuracy as high as 99.83%.

2.7 Timing Side Channel

Gong et al. [82, 83] studied the information leakage through the timing side channel that arises from a first come first serve (FCFS) scheduler that server two users. Such a timing channel gives room for one user to learn the traffic pattern of the other by analyzing the queueing delays that occur. Gong proved that an attack strategy exists, for example a sequence of jobs that the attacker issues, that can lead to learning the other's users traffic patter without ambiguity.

Goethem et al. [18] showed that modern browsers expose new side channels which could be used to obtain accurate timing measurements, regardless of network conditions. They introduced four novel web-based timing attacks against modern browsers and described the ways an attacker could acquire personal information based upon a user's state on cross-origin website.

Hoyos et al. [84] performed a timing attack on the IEC-61850-8-1 authentication mechanism which revealed that such attacks are feasible since the computational capacity of embedded processors that run authentication algorithms currently exceeds the needed 4 ms response time. A successful attack can create an automation breakdown, including damaging not only power transformers, but also circuit breakers as well.

Zhong et al. [85, 86] proposed a method, based upon the side channel of timing delays, to distinguish the packets that were generated by different phasor measurement units (PMUs) and were sent through an encrypted VPN tunnel. This timing delay side channel can be used by attackers to selectively drop one PMU from traffic without interrupting other packets. Islam et al. [87] studied the security of PMUs in smart grid communication infrastructures and suggested HMAC-SHA1 as an authentication algorithm for signing the measured values. Moreover, they analyzed the execution time of the authentication module, which is the suggested authentication algorithm in IEC 62351, applied to the communications of substations, in order to correlate it with the secret-related information of the algorithm.

Johnstone et al. [88] provided a novel solution to one proof of concept attacks against BACnet devices, using an ANN classifier for the time differences between frames of the same type. They suggested that state aware machine learning methodologies could be used to discover threats that comprise a collection of legitimate commands and may cause system failure. Dunlap et al. [89] presented a novel approach that leverages timing-based side channel analysis to establish a unique fingerprint capable of assisting in the detection of unauthorized modifications of embedded devices. Their approach is applied to an Allen Bradley ControlLogix PLC, where execution time measurements are collected and analyzed by a custom anomalous behavior detection system. The detection rate of the system reaches as high as 97.8%, confirming that it is feasible to use timing side channel analysis to detect anomalous behavior in PLCs.

Kocher et al. describe practical attacks that combine methodology from side channel attacks, fault attacks, and return-oriented programming that can read arbitrary memory from the victim's process. More broadly, the paper shows that speculative execution implementations violate the security assumptions underpinning numerous software security mechanisms, including operating system process separation, static analysis, containerization, just-in-time (JIT) compilation, and countermeasures to cache timing/side-channel attacks. These attacks represent a serious threat to actual systems, since vulnerable speculative execution capabilities are found in microprocessors from Intel, AMD, and ARM, which are used in billions of devices [90].

Meltdown [11] is a novel attack that exploits side effects of out-of-order execution on modern processors to read arbitrary kernel-memory locations including personal data and passwords. The attack is independent of the operating system and does not rely on any software vulnerabilities. Meltdown breaks all security assumptions given by address space isolation as well as paravirtualized environments and, thus, every security mechanism building upon this foundation. On affected systems, Meltdown enables an adversary to read memory of other processes or virtual machines in the cloud without any permissions or privileges affecting millions of customers who use personal computers [11].

2.8 *Traffic Analysis*

Various side channel attacks take advantage of network traffic analysis to infer user's web browsing activities. Hintz et al. [91] proved that transferred file sized could be used as a reliable fingerprint for websites. Lu et al. [92] successfully exploited not only packet size, but also packet ordering information in order to improve the success rate of webpage identification.

Chen et al. [93] managed to infer browsing activity through packet analysis on traffic, which was encrypted using not only WPA, but HTTPS as well, proving that encrypted channels are not fully protected against traffic analysis.

Tsalis et al. [94] showed the feasibility of information disclosure of functionality over encrypted TCP/IP running MODBUS RTU protocol, through targeted side channel attacks on encrypted packets. They proved that any web interface that implements unpadded encryption with specific block cipher modes or most stream ciphers to send MODBUS functions over TCP/IP is subject to differential packet size attacks. This is due to the fact that MODBUS has very small number of available commands and differences in packet sizes, that result to traffic distinctions.

2.9 *Vibration Side Channel*

Gerson de Souza Faria [95, 96] described an attack capable of identifying the sequence of keystrokes through analysis of mechanical vibrations generated by the pressed keys. Accelerometers are used as vibration sensors and the key recognition rates achieved are: 98.4% in ATM keypad, 76.7% in PIN-pad testing on a hard surface, as well as 82.1% in PIN-pad held in hand. Findings indicate that engineers must design human-machine interfaces in a more secure way in the future, as well as a new attack vector that processes certification must be addressed hereafter.

2.10 Combination of Side Channels

Chen [97] in his work presented a novel attack model and algorithm capable of extracting the exact schedules of real-time systems utilizing fixed priority algorithms. Real-time embedded systems can be found in a variety of domains, like space vehicles, medical devices, aircrafts, automobiles, industrial control systems as well as nuclear power plants. The cache-timing attack over a camera task that shows how the transition of the memory usage can be recovered by the observation of the cache usage, was demonstrated on an ARM-based development board, Zedboard, with a high success rate. The leaked schedules could be utilized to launch a side channel attack against a specific victim task. SheduLeak algorithm is robust not only in the presence of schedule randomization defenses, but also jitters.

Michael Weiβ et al. [98, 99] conducted a cache-timing attack in an implemented virtualization scenario using the PikeOS microkernel virtualization framework under different scheduler configurations. PikeOS is a microkernel-based real-time operating system which has been developed for both safety and security-critical applications with certification needs in the fields of Aerospace & Defense, Automotive & Transportation, Industrial Automation & Medical, Network Infrastructures as well as Consumer Electronics. Weiβ's approach was based on Bernstein's time-driven cache-based attack against AES and proved that dedicated cores for the crypto routine provide the highest amount of timing leakage.

3 Side Channel Attack Classification Taxonomies

3.1 Confidentiality Attacks Through Side Channels

The initial idea behind this paper is a suggestion of a target-based and multi-level classification that relies on already known and established taxonomies. Therefore, we use known categories of side channel attacks as a first level of classification, depending on the type of measurement that reveals the sensitive information, e.g. power, electromagnetic analysis, timing and more described previously in Sect. 2.

After extensive search in most recognized scientific associations, conferences and publishers, the current survey references 91 papers and articles from 48 different conferences, journals, technical reports or relevant published material.

Every side channel attack is performed on a computational or electronic device that incorporates a software program. The aim is to gather useful data, form patterns and extract sensitive information. This simple remark indicates that every side channel attack requires a device and a program leading to a certain effect. In general, a successful classification must include categories which can be applied to all types of side channel attacks. Considering the above requirements, the categories for our target-based classification method is expanded on three basic axes:

Hardware device target	Software device target	Result
3D Printer	Application	Break Heap ASLR
Computer monitor (PC, laptop)	Cryptographic library	Compromise Data Centre's Availability
Data center hardware	Cryptographic software	Equipment Control
Dot-matrix printer	Data center software	Intellectual Property (IP) theft
Embedded board	Development environment for programming controller applications	Key extraction
Integrated circuits	Enclave program	Leak of process-specific information
Keyboard (PC, notebook, telephone, ATM pads)	Hosting platform	Leak of sensitive information
Laser Printer	Modelling software	PIN inference
Mobile device (Smartphone or Tablet)	Operating system	Recovery of cryptographic secrets
PC (laptop or desktop)	Virtualization software	Schedule extraction
PC (laptop or desktop), Remote Terminal Unit (RTU)	Web browser	Unveil the source of encrypted packets
PC (laptop or desktop), Router, Switch, Intelligent Electronic Device (IED)		
PLC		
PMU		
POS terminal		
Router		
Smartcard		
Smartwatch		
Touch Screen (smartphone, smartwatch, computer monitor)		
WSN hardware		

For instance, Irazoqui et al. [37] demonstrated how AES keys were recovered by using a cross-core cache attack targeting an OpenSSL implementation of AES in a dual core machine. In this example, the attack falls into the cache-based category, the targeted hardware is the PC, the targeted software is the Cryptographic library and the result is the Key extraction.

To help the reader get a clearer idea of how this categorization works, it is practical to refer to an extra example. Liang Cai and Hao Chen [35] showed that they can infer user inputs on a touchscreen by observing the readings of the accelerometer and the orientation of the sensors on smartphones. It is easily induced that the authors have presented a Sensor-based attack, the targeted hardware was a Mobile device, the targeted software was an Application and the result was Information leakage. Same methodology has been implemented to the rest of the papers collected and studied for this purpose.

In this paper, 74 side channel attacks all published from 2010 and onwards are used for creating our taxonomy. Following tables demonstrate the frequency of the attacks per category, per targeted hardware, per targeted software and per result.

Table 1 assembles all levels of taxonomy starting from the category of side channel attack and ending with the result. The information extracted by observing this table is:

- Cache attack is implemented only on PC and to various software categories. This essentially means that a cache attack is mostly possible to be executed on a PC rather than any other device.
- Electromagnetic attack is used to extract a cryptographic key.
- Timing attack is used for information leakage.
- Sensor-based attack can be executed to various devices.

Table 2 sorts each class of the SCA to the targeted hardware category indicating the corresponding numbers of SCA. The contents of this table denote that PC is the hardware category that receives the greatest amount and essentially all types of side channel attacks.

Table 3 sorts each class of the SCA to the targeted software category indicating the corresponding numbers of SCA. The contents of this table denote that the timing attack is popular in an application program, whereas the electromagnetic is in operating systems.

Finally, Table 4 shows how the different types of SCA are distributed among the result categories remarking that information leakage and key extraction are the most often outcomes when a SCA is performed.

3.2 *Anomaly Detection Through Side-Channels*

During the last decade, scientists have begun to experiment with different uses of side-channel information; apart from malicious ones. Physical side channels (like indirect measurement of electromagnetic emissions, power consumption, etc., of program execution) can be utilized for anomaly detection, i.e. the execution status of digital circuit or a processor to assess using monitors. Unauthorized attempts to disrupt the normal operation of a target system, like a computer or an industrial control system, can be detected with extreme accuracy [51, 60–64, 73–81, 88, 89].

The aforementioned surveys describe methodologies capable of learning a baseline side-channel normal activity, processing it in order to extract key features, comparing both subsequent collected data and processed data for anomalous behavior, in order to identify such behavior. Although not mentioned in those researches, similar to regular IDPS systems, reports could be centrally managed and predefined actions can be executed once anomalous behavior is observed.

Table 1 Main table of all side-channel attack categories

	Side channel attack category	Targeted hardware category	Targeted software category	Result	References
1	Acoustic	3D printer	Modelling software	Intellectual Property (IP) theft	[30, 31, 47, 48]
		Dot-matrix printer	Operating system	Leak of sensitive information (e.g. user, content of medical prescriptions, etc.)	[46]
		Keyboard (PC, notebook, telephone, ATM pads)		Leak of sensitive information (e.g. user input, passwords, etc.)	[44, 45]
		PC (laptop or desktop)	Cryptographic library	Key extraction	[29]
		PLC	Development environment for programming controller applications	Leak of sensitive information (e.g. passwords)	[49]
				Leak of process-specific information (e.g. gains, process stability limits, mode switching parameters, etc.)	
2	Cache	PC (laptop or desktop)	Application	Key extraction	[5]
			Cryptographic library		[4, 7–9, 13]
			Cryptographic software		[3]
			Enclave program		[6]
			Virtualization software	Leak of sensitive information (e.g. user input, passwords, etc.)	[10, 12, 50]
				Leak of process-Specific Information (e.g. cache wiping, random delay insertion, adjusting each machine's perception of time, etc.)	
3	Electromagnetic	3D printer	Modelling software	Intellectual Property (IP) theft	[55]
		Computer monitor (PC, laptop)	Operating system	Leak of sensitive information (e.g. displayed content, passwords, etc.)	[53]
		Data center hardware	Data center software	Compromise data centre's availability	[54]

		Keyboard (PC, notebook, telephone, ATM pads)	Operating system	PIN inference	[56, 57]
		Laser Printer		Leak of sensitive information (e.g. user input, etc.)	[52]
		Mobile device (Smartphone or Tablet)	Cryptographic library	Key extraction	[25, 27]
			Operating system	PIN inference	[59]
		PC (laptop or desktop)	Cryptographic library	Key extraction	[24]
			Cryptographic software		[23, 26]
		Smartcard	Application		[28]
		Touch Screen (smartphone, smartwatch, computer monitor)	Operating system	PIN inference	[58]
4	Memory	PC (laptop or desktop)	Enclave program	Leak of process-specific information (e.g. timing information, session-key information, page faults inside enclaves, etc.)	[33]
			Web browser	Leak of sensitive information (e.g. HTTP password hashes)	[32]
				Break heap ASLR	
5	Optical	Computer monitor (PC, laptop)	Operating system	Leak of sensitive information (e.g. displayed content, passwords, etc.)	[67]
		PC (laptop or desktop)			[68]
		WSN hardware		Leak of sensitive information (e.g. personal habits in smart-home applications, health data, etc.)	[65]

(continued)

Table 1 (continued)

	Side channel attack category	Targeted hardware category	Targeted software category	Result	References
6	Power	Integrated circuits	Application	Leak of sensitive information (e.g. health data, etc.)	[69]
		PC (laptop or desktop)		Leak of sensitive information (e.g. geolocation, etc.)	[70]
				Key extraction	[22]
			Cryptographic software		[21]
		Smartcard	Application		[20]
		WSN hardware	Operating system	Recovery of cryptographic secrets	[72]
7	Sensor-based	3D printer	Modelling software	Intellectual Property (IP) theft	[36]
		Mobile device (Smartphone or Tablet)	Application	Leak of sensitive information (e.g. user input, passwords, etc.)	[34, 35]
				PIN inference	[39, 40]
		POS terminal	Operating system		[38]
		Smartwatch	Application	Leak of sensitive information (e.g. user input, passwords, etc.)	[37]
8	Timing	Mobile device (Smartphone or Tablet)	Application	Leak of process-specific information (e.g. statistical information of process interrupts, application running status, etc.)	[15]
				Leak of sensitive information (e.g. user data, user actions, unlock patterns, etc.)	
		PC (laptop or desktop)	Cryptographic library	Key extraction	[19]
			Hosting platform	Leak of sensitive information (e.g. user's documents, etc.)	[16]

			Operating system	Leak of process-specific information (e.g. deduce information about the privileged address space layout, etc.)	[14]
			Web browser	Leak of sensitive information (e.g. user's behavior in a website, passwords, etc.)	[17, 18]
		PC (laptop or desktop), Router, Switch, Intelligent Electronic Device (IED)	Operating system	Equipment control	[84]
		PMU	Application	Key extraction	[87]
				Unveil the source of encrypted packets	[85, 86]
		Router		Leak of sensitive information (e.g. user's behavior in networks, etc.)	[82, 83]
9	Traffic analysis	PC (laptop or desktop)	Web browser	Leak of sensitive information (e.g. health data, economic data, etc.)	[91–93]
		PC (laptop or desktop), Remote Terminal Unit (RTU)	Development environment for programming controller applications	Leak of process-specific information (e.g. underlying functionality of a SCADA system, etc.)	[94]
10	Vibration	Keyboard (PC, notebook, telephone, ATM pads)	Operating system	Leak of sensitive information (e.g. user input, passwords, etc.)	[95, 96]
11	Combination	Embedded Board	Operating system	Schedule extraction	[97]
			Virtualization Software	Key extraction	[98, 99]

Table 2 SCA attacks per targeted hardware

Targeted hardware category	Side channel attack category	# of side channel attacks per category	References
3D Printer	Acoustic	4	[30, 31, 47, 48]
	Electromagnetic	1	[55]
	Sensor-based	1	[36]
Computer monitor (PC, laptop)	Electromagnetic	1	[53]
	Optical	1	[67]
Data center hardware	Electromagnetic	1	[54]
Dot-matrix printer	Acoustic	1	[46]
Embedded board	Combination	3	[97–99]
Integrated circuits	Power	1	[69]
Keyboard (PC, notebook, telephone, ATM pads)	Acoustic	2	[44, 45]
	Electromagnetic	2	[56, 57]
	Vibration	2	[95, 96]
Laser printer	Electromagnetic	1	[52]
Mobile device (Smartphone or Tablet)	Electromagnetic	3	[25, 27, 59]
	Optical	1	[68]
	Sensor-based	4	[34, 35, 39, 40]
	Timing	1	[15]
PC (laptop or desktop)	Acoustic	1	[29]
	Cache	11	[3–10, 12, 13, 50]
	Electromagnetic	3	[23, 24, 26]
	Memory	2	[32, 33]
	Optical	1	[68]
	Power	3	[21, 22, 70]
	Timing	5	[14, 16, 17–19]
	Traffic analysis	3	[91–93]
PC (laptop or desktop), Remote Terminal Unit (RTU)	Traffic analysis	1	[94]
PC (laptop or desktop), Router, Switch, Intelligent Electronic Device (IED)	Timing	1	[84]
PLC	Acoustic	1	[49]
PMU	Timing	3	[85–87]
POS terminal	Sensor-based	1	[38]
Router	Timing	2	[82, 83]
Smartcard	Electromagnetic	1	[28]
	Power	1	[20]
Smartwatch	Sensor-based	1	[37]
Touch Screen (smartphone, smartwatch, computer monitor)	Electromagnetic	1	[58]
WSN hardware	Optical	1	[65]
	Power	1	[72]

Table 3 SCA attacks per targeted software

Targeted software category	Side channel attack category	# of side channel attacks per category	References
Application	Cache	1	[5]
	Electromagnetic	1	[28]
	Optical	1	[68]
	Power	4	[20, 22, 69, 70]
	Sensor-based	5	[34, 35, 37, 39, 40]
	Timing	6	[15, 82, 83, 85–87]
Cryptographic library	Acoustic	1	[29]
	Cache	5	[4, 7–9, 13]
	Electromagnetic	3	[24, 25, 27]
	Timing	1	[19]
Cryptographic software	Cache	1	[3]
	Electromagnetic	2	[23, 26]
	Power	1	[21]
Data center software	Electromagnetic	1	[54]
Development environment for programming controller applications	Acoustic	1	[49]
	Traffic analysis	1	[94]
Enclave program	Cache	1	[6]
	Memory	1	[33]
Hosting platform	Timing	1	[16]
Modelling software	Acoustic	4	[30, 31, 47, 48]
	Electromagnetic	1	[55]
	Sensor-based	1	[36]
Operating system	Acoustic	3	[44–46]
	Combination	1	[97]
	Electromagnetic	6	[52, 53, 56–59]
	Optical	2	[65, 67]
	Power	1	[72]
	Sensor-based	1	[38]
	Timing	2	[14, 84]
	Vibration	2	[95, 96]
Virtualization software	Cache	3	[10, 12, 50]
	Combination	2	[98, 99]
Web browser	Memory	1	[32]
	Timing	2	[17, 18]
	Traffic analysis	3	[91–93]

Table 4 SCA attacks per result

Result	Side channel attack category	# of side channel attacks per category	References
Break heap ASLR	Memory	1	[32]
Compromise data centre's availability	Electromagnetic	1	[54]
Equipment control	Timing	1	[84]
Intellectual Property (IP) theft	Acoustic	4	[30, 31, 47, 48]
	Electromagnetic	1	[55]
	Sensor-based	1	[36]
Key extraction	Acoustic	1	[29]
	Cache	8	[3–9, 13]
	Combination	2	[98, 99]
	Electromagnetic	6	[23–28]
	Power	3	[20–22]
	Timing	2	[19, 87]
Leak of process-specific information	Acoustic	1	[49]
	Cache	3	[12, 50]
	Memory	1	[33]
	Timing	2	[14, 15]
	Traffic analysis	1	[94]
Leak of sensitive information	Acoustic	4	[44–46, 49]
	Cache	3	[10, 12, 50]
	Electromagnetic	2	[52, 53]
	Memory	1	[32]
	Optical	3	[65, 67, 68]
	Power	2	[69, 70]
	Sensor-based	3	[34, 35, 37]
	Timing	6	[15–18, 82, 83]
	Traffic analysis	3	[91–93]
	Vibration	2	[95, 96]
PIN inference	Electromagnetic	4	[56–59]
	Sensor-based	3	[38–40]
Recovery of cryptographic secrets	Power	1	[72]
Schedule extraction	Combination	1	[97]
Unveil the source of encrypted packets	Timing	2	[85, 86]

To the best of our knowledge, our research is the first attempt that brings to surface this new point of view about side-channels, through gathering as much as possible of the scattered studies. Table 5 summarizes previous researches concerning the feasibility of using side-channels to detect anomalous behaviors not only on regular computer systems, but also in industrial control systems.

Table 5 Main table of all side-channels used for anomaly detection

	Side channel attack category	Targeted hardware category	Targeted software category	Result	References
1	Electroacoustic	Manufactured Part	Modelling software	Trojan detection	[51]
2	Electromagnetic	PLC	Development environment for Programming Controller Applications	Anomaly detection	[60, 61]
				Intrusion detection	[62]
				Malicious code detection	[63]
				Unknown code detection	[64]
3	Power	Compounder	Operating system	Malware discovery	[78]
		Embedded board	Application	Buffer overflow attack detection	[77]
			Operating system	Anomaly detection	[79]
		Integrated circuits	Operating system	Hardware trojan detection	[75, 76]
		Mechanical components	Operating system	Anomaly detection	[73]
		PC (laptop or desktop)		Rootkit detection	[74]
		PLC	Development environment for programming controller applications	Anomaly detection	[81]
				Malicious code detection	[80]
				Malware discovery	[78]
4	Timing	PC (laptop or desktop)	Application	Timing attack detection	[88]
		PLC	Development environment for programming controller applications	Anomaly detection	[89]

4 Discussion

A SCA classification method is considered effective and appropriate when it is applied to the plurality of side channel attacks in a uniform, generic and yet informative way. The process we followed to decide on a method included the collection of an indicative sample of related papers and a list of the SCA characteristics [1, 41].

The criteria by which we chose included papers were in descending order: Conference/Journal/Book impact factor, types of side channel attack presented, novelty of attack (since we focus on the last decade, novel publication plays an important role) and quality of text. All papers were evaluated by the aforementioned criteria, gathered through an extensive search in security and privacy conferences, as well as filtered by year range and publisher. A different approach would include a category-oriented research, i.e. to collect papers from every SCA category from late 1990s. Although this method would capture all publications and thus the categories of SCAs, nevertheless it would eliminate the actual needs of today's security experts. It would also eliminate our capability to withdraw useful conclusions on current trends of modern SCAs, along with categories which are currently more susceptible to attacks. It is our firm belief that this information is more useful to modern researchers by directing security experts to focus on modern machines that are vulnerable to side channel attacks. Gathering some statistics (specifically, the type of attacks and the amount of systems affected) generated interesting observations:

- The most frequent type of attack threatening security and privacy is *cache-based* and *electromagnetic-based.*
- A little less frequent compared to the cache is the *timing attack, acoustic, sensor-based and power analysis* attacks.
- The trend shows that *electromagnetic-based and timing attacks* are gaining more ground in the recent years.

Presented categories of side channel attacks are essentially a list of SCA characteristics. Power, electromagnetic, timing, etc. portray the attack vector that delivers the desired results. Each category of SCA targeted either the hardware or the software of the experiment's testbed, which resulted in the aforementioned outcomes. The categories outlined below can act as criteria for evaluating the importance of published SCAs. To be more specific and to promote further research, we provide below certain attributes that were found useful in terms of classifying the impact of SCAs. Each criterion has a 1-to-3 quantitative scale to describe each attack.

- **Invasiveness** (invasive-3, semi-invasive-2, non-invasive-1). Invasiveness is a property which defines the amount of tampering that needs to be done on a device in order for the attack to be successful. For example, if an attack can be performed simply by observing the device, then this attack is considered non-invasive, whereas opening the device and connecting on-board sensors is classified as a fully invasive attack.
- **Adaptability** (Flexible-3, Relatively flexible-2, No flexibility-1). Inspired by the classification criteria of Risk Assessment methods proposed by D. Gritzalis et al. [100], this category can also be applied to SCA and show to what extend can this attack be adjusted to other devices or software programs.
- **Performance** (Fast application speed −3, Medium application speed-2, Time consuming-1). The performance criterion describes how fast an SCA can be performed, given that all required equipment is present and connected. E.g. an attack that needs thousands of samples from multiple encryptions can take hours, whereas other attacks can be instant (single observation).
- **Complexity** (Very complicated-3, slightly complicated-2, Easy-1). This criterion indicates how difficult is of a given attack to be successful. Takes into consideration algorithm complexity, the amount of information that needs to be gathered and the mathematical complexity of the attack (e.g. differential electromagnetic attacks usually score high on this scale).

- **Cost** (Very Expensive (more than 1000$), Expensive (500–1000$), Inexpensive (0–500$)). This attribute describes the cost of equipment needed to perform an attack.

Quantitative scale (1 lowest – 3 highest)					
Type of attack	Invasiveness	Adaptability	Performance	Complexity	Cost
Acoustic	1	2	2	2	Inexpensive
Cache	3	3	2	3	Expensive
Electromagnetic	3	3	1	3	Very expensive
Memory	3	1	2	2	Expensive
Optical	2	1	2	2	Inexpensive
Power	2	2	2	2	Very expensive
Sensor-based	1	2	2	2	Expensive
Timing	2	2	2	3	Expensive
Traffic analysis	1	1	1	2	Expensive
Vibration	1	1	2	2	Inexpensive

In general, the selection of sample experiments and the discovery of suitable classification categories is a quite subjective procedure and depends on the criteria the authors choose to examine.

Attacks that monitor side channel information have recently been getting much attention by experts confirming that SCA can be quite powerful and need to be addressed. Side channel attacks are relatively easy to implement whilst powerful attacks against cryptographic implementations and their targets range from primitives, protocols, modules, and devices to even systems. These attacks pose a serious threat to the security of cryptographic modules. In consequence, cryptographic implementations have to be evaluated for their resilience against such attacks and the incorporation of different countermeasures has to be considered. The confidentiality of critical information such as passwords, encryption keys, patterns, detailed system layout map, etc., shall rank high when it comes to security concerns in the industry, hence applicable reinforcement should be imposed in this aspect. Adversaries will obviously choose attacks that maximize side channel information, so it is very important that the strongest attacks be considered when designing defensive strategies.

5 Conclusion

This work examined and presented a set of categories for the classification of side channel attacks (SCAs) in the area of security and privacy and for the recent years. The methodology proposed was based on the prerequisite criteria of uniformity, overall generality and entire applicability. The categories of our method are extended to three target-based axes, which are, to the best of our knowledge, applied to every SCA. The tables emanating from our classification method possess a statistical value as they can be used to extract useful information regarding risk assessment and help security experts deploy more targeted countermeasures.

Critical information stored in computing devices is a possible target of a security attack, let alone if this constitutes an asset of a critical infrastructure. A plausible leak can have devastating implications such as far-reaching damage to public interests, national security and economic sustainability. Any protection put into place to safeguard critical infrastructures should focus on preserving not only the physical elements of the infrastructure, but also and most importantly, its virtual elements, as a disruption of these assets may trigger the same damage as the disruption of physical components, putting the security and safety of these interconnected systems at risk. For this reason, the amplitude of side channel attacks cannot be contemplated as minor nor the effects derived from them. Thus, studying every detail related to SCAs including this classification research can help us gain a deeper understanding of feasible impacts and thereupon secure critical infrastructure in a more productive manner.

For further work, we plan to evaluate the criteria of selection, do more research about the limitations cited and verify that the indicated method applies to a larger collection of attacks.

References

1. Department of Homeland Security (2017) Office of infrastructure protection. [online] Available at: https://www.dhs.gov/office-infrastructure-protection. Accessed 5 June 2018
2. Zhou Y, Feng D (2005) Side-channel attacks: ten years after its publication and the impacts on cryptographic module security testing. IACR Cryptol ePrint Arch 2005:388
3. Liu F, Yarom Y, Ge Q, Heiser G, Lee RB (2015). Last-level cache side-channel attacks are practical. In: Security and privacy (SP), 2015 IEEE Symposium on. IEEE, pp 605–622
4. Gullasch D, Bangerter E, Krenn S (2011) Cache games–bringing access-based cache attacks on AES to practice. In: Security and Privacy (SP), 2011 IEEE Symposium on. IEEE, pp 490–505
5. Guanciale R, Nemati H, Baumann C, Dam M (2016) Cache storage channels: Alias-driven attacks and verified countermeasures. In: 2016 IEEE Symposium on Security and Privacy (SP). IEEE, pp 38–55
6. Moghimi A, Irazoqui G, Eisenbarth T (2017) CacheZoom: how SGX amplifies the power of cache attacks. In: International conference on cryptographic hardware and embedded systems. Springer, Cham, pp 69–90
7. Benger N, Van de Pol J, Smart NP, Yarom Y (2014) "Ooh Aah... Just a Little Bit": a small amount of side channel can go a long way. In: International workshop on cryptographic hardware and embedded systems. Springer, Berlin/Heidelberg, pp 75–92
8. Genkin D, Valenta L, Yarom Y (2017) May the fourth be with you: a microarchitectural side channel attack on several real-world applications of Curve25519. In: Proceedings of the 2017 ACM SIGSAC conference on computer and communications security. ACM, New York, pp 845–858
9. Zhang Y, Juels A, Reiter MK, Ristenpart T (2012) Cross-VM side channels and their use to extract private keys. In: Proceedings of the 2012 ACM conference on computer and communications security. ACM, New York, pp 305–316

10. Zhang Y, Juels A, Reiter MK, Ristenpart T (2014) Cross-tenant side-channel attacks in PaaS clouds. In: Proceedings of the 2014 ACM SIGSAC conference on computer and communications security. ACM, New York, pp 990–1003
11. Lipp M, Schwarz M, Gruss D, Prescher T, Haas W, Mangard S, ..., Hamburg M (2018) Meltdown. arXiv preprint arXiv:1801.01207
12. Zhang Y, Juels A, Oprea A, Reiter MK (2011) Homealone: co-residency detection in the cloud via side-channel analysis. In: 2011 IEEE symposium on security and privacy. IEEE, Piscataway, pp 313–328
13. Irazoqui G, Eisenbarth T, Sunar B (2015) S $ A: a shared cache attack that works across cores and defies VM sandboxing–and its application to AES. In: Security and privacy (SP), 2015 IEEE symposium on. IEEE, Piscataway, pp 591–604
14. Hund R, Willems C, Holz T (2013) Practical timing side channel attacks against kernel space ASLR. In: 2013 IEEE symposium on security and privacy. IEEE, Piscataway, pp 191–205
15. Diao W, Liu X, Li Z, Zhang K (2016) No pardon for the interruption: new inference attacks on android through interrupt timing analysis. In: Security and privacy (SP), 2016 IEEE symposium on. IEEE, Piscataway, pp 414–432
16. Wang L, Grubbs P, Lu J, Bindschaedler V, Cash D, Ristenpart T (2017) Side-channel attacks on shared search indexes. In: 2017 38th IEEE Symposium on Security and Privacy (SP). IEEE, pp 673–692
17. Vila P, Köpf B (2017) Loophole: timing attacks on shared event loops in chrome. In USENIX security symposium
18. Van Goethem T, Joosen W, Nikiforakis N (2015) The clock is still ticking: timing attacks in the modern web. In: Proceedings of the 22nd ACM SIGSAC conference on computer and communications security. ACM, New York, pp 1382–1393
19. Meyer C, Somorovsky J, Weiss E, Schwenk J, Schinzel S, Tews E (2014) Revisiting SSL/TLS implementations: new Bleichenbacher side channels and attacks. In: USENIX security symposium, pp 733–748
20. Kim TW, Kim TH, Hong S (2017) Breaking Korea transit card with side-channel analysis attack unauthorized recharging. In Black Hat Asia
21. Genkin D, Pipman I, Tromer E (2015) Get your hands off my laptop: physical side-channel key-extraction attacks on PCs. J Cryptogr Eng 5(2):95–112
22. Clavier C, Marion D, Wurcker A (2014) Simple power analysis on AES key expansion revisited. In: International workshop on cryptographic hardware and embedded systems. Springer, Berlin/Heidelberg, pp 279–297
23. Genkin D, Pachmanov L, Pipman I, Tromer E (2015) Stealing keys from PCs using a radio: cheap electromagnetic attacks on windowed exponentiation. In: International workshop on cryptographic hardware and embedded systems. Springer, Berlin/Heidelberg, pp 207–228
24. Genkin D, Pachmanov L, Pipman I, Tromer E (2016) ECDH key-extraction via low-bandwidth electromagnetic attacks on PCs. In: Cryptographers' track at the RSA conference. Springer, Cham, pp 219–235
25. Belgarric P, Fouque PA, Macario-Rat G, Tibouchi M (2016) Side-channel analysis of Weierstrass and Koblitz curve ECDSA on Android smartphones. In: Cryptographers' track at the RSA conference. Springer, pp 236–252, Cham
26. Espitau T, Fouque PA, Gérard B, Tibouchi M (2017) Side-channel attacks on BLISS lattice-based signatures: exploiting branch tracing against strongswan and electromagnetic emanations in microcontrollers. In: Proceedings of the 2017 ACM SIGSAC conference on computer and communications security. ACM, New York, pp 1857–1874
27. Genkin D, Pachmanov L, Pipman I, Tromer E, Yarom Y (2016) ECDSA key extraction from mobile devices via nonintrusive physical side channels. In: Proceedings of the 2016 ACM SIGSAC conference on computer and communications security. ACM, New York, pp 1626–1638
28. Bauer A, Jaulmes E, Lomné V, Prouff E, Roche T (2014) Side-channel attack against RSA key generation algorithms. In: International workshop on cryptographic hardware and embedded systems. Springer, Berlin/Heidelberg, pp 223–241

29. Genkin D, Shamir A, Tromer E (2014) RSA key extraction via low-bandwidth acoustic cryptanalysis. In: International cryptology conference. Springer, Berlin/Heidelberg, pp 444–461
30. Hojjati A, Adhikari A, Struckmann K, Chou E, Tho Nguyen TN, Madan K et al (2016) Leave your phone at the door: side channels that reveal factory floor secrets. In: Proceedings of the 2016 ACM SIGSAC conference on computer and communications security. ACM, New York, pp 883–894
31. Faruque A, Abdullah M, Chhetri SR, Canedo A, Wan J (2016) Acoustic side-channel attacks on additive manufacturing systems. In: Proceedings of the 7th international conference on cyber-physical systems. IEEE Press, New York, p 19
32. Bosman E, Razavi K, Bos H, Giuffrida C (2016) Dedup est machina: memory deduplication as an advanced exploitation vector. In: 2016 IEEE symposium on security and privacy (SP). IEEE, Los Alamitos, pp 987–1004
33. Wang W, Chen G, Pan X, Zhang Y, Wang X, Bindschaedler V et al (2017) Leaky cauldron on the dark land: understanding memory side-channel hazards in SGX. In: Proceedings of the 2017 ACM SIGSAC conference on computer and communications security. ACM, New York, pp 2421–2434
34. Xu Z, Bai K, Zhu S (2012) Taplogger: inferring user inputs on smartphone touchscreens using on-board motion sensors. In: Proceedings of the fifth ACM conference on security and privacy in wireless and mobile network. ACM, New York, pp 113–124
35. Cai L, Chen H (2011) TouchLogger: inferring keystrokes on touch screen from smartphone motion. HotSec 11:9–9
36. Song C, Lin F, Ba Z, Ren K, Zhou C, Xu W (2016) My smartphone knows what you print: exploring smartphone-based side-channel attacks against 3d printers. In: Proceedings of the 2016 ACM SIGSAC conference on computer and communications security. ACM, New York, pp 895–907
37. Maiti A, Armbruster O, Jadliwala M, He J (2016) Smartwatch-based keystroke inference attacks and context-aware protection mechanisms. In: Proceedings of the 11th ACM on Asia conference on computer and communications security. ACM, New York, pp 795–806
38. Liu X, Zhou Z, Diao W, Li Z, Zhang K (2015) When good becomes evil: keystroke inference with smartwatch. In: Proceedings of the 22nd ACM SIGSAC conference on computer and communications security. ACM, New York, pp 1273–1285
39. Simon L, Anderson R (2013) Pin skimmer: inferring pins through the camera and microphone. In: Proceedings of the third ACM workshop on security and privacy in smartphones & mobile devices. ACM, New York, pp 67–78
40. Maiti A, Jadliwala M, He J, Bilogrevic I (2015) (Smart) watch your taps: side-channel keystroke inference attacks using smartwatches. In: Proceedings of the 2015 ACM International Symposium on Wearable Computers. ACM, New York, pp 27–30
41. Spreitzer R, Moonsamy V, Korak T, Mangard S (2018) Systematic classification of side-channel attacks: a case study for mobile devices
42. Goodin D (2018) Scientists break card that secures homes, offices, transit. Retrieved from https://www.theregister.co.uk/2011/10/10/mifare_desfire_smartcard_broken/. Accessed 6 June 2018
43. Trippel T, Weisse O, Xu W, Honeyman P, Fu K (2017) WALNUT: waging doubt on the integrity of MEMS accelerometers with acoustic injection attacks. In: Security and privacy (EuroS&P), 2017 IEEE European symposium on. IEEE, pp 3–18
44. Asonov D, Agrawal R (2004) Keyboard acoustic emanations. In: Null. IEEE, p 3
45. Zhuang L, Zhou F, Tygar JD (2009) Keyboard acoustic emanations revisited. ACM Transactions on Information and System Security (TISSEC) 13(1):3
46. Backes M, Dürmuth M, Gerling S, Pinkal M, Sporleder C (2010). Acoustic side-channel attacks on printers. In: USENIX Security symposium, pp 307–322
47. Chhetri SR, Canedo A, Faruque MAA (2018) Confidentiality breach through acoustic side-channel in cyber-physical additive manufacturing systems. ACM Trans Cyber-Phys Sys 2(1):3

48. Chhetri SR, Canedo A, Faruque MAA (2016) Kcad: kinetic cyber-attack detection method for cyber-physical additive manufacturing systems. In: Proceedings of the 35th international conference on computer-aided design. ACM, New York, p 74
49. Krishnamurthy P, Khorrami F, Karri R, Paul-Pena D, Salehghaffari H (2018) Process-aware covert channels using physical instrumentation in cyber-physical systems. IEEE Trans Inf Forensics Secur 13(11):2761–2771
50. Ristenpart T, Tromer E, Shacham H, Savage S (2009) Hey, you, get off of my cloud: exploring information leakage in third-party compute clouds. In: Proceedings of the 16th ACM conference on computer and communications security. ACM, New York, pp 199–212
51. Vincent H, Wells L, Tarazaga P, Camelio J (2015) Trojan detection and side-channel analyses for cyber-security in cyber-physical manufacturing systems. Proced Manuf 1:77–85
52. Grzesiak K, Przybysz A (2010) Emission security of laser printers. In: Military communications and information systems conference, Wrocław, pp 353–363
53. Lee HS, Sim K, Yook JG (2015) Measurement and analysis of the electromagnetic emanations from video display interface. In: Electrical design of advanced packaging and systems symposium (EDAPS), 2015 IEEE. IEEE, pp 71–73
54. Islam MA, Ren S, Wierman A (2017) Exploiting a thermal side channel for power attacks in multi-tenant data centers. In: Proceedings of the 2017 ACM SIGSAC Conference on Computer and Communications Security. ACM, New York, pp 1079–1094
55. Mowery K, Meiklejohn S, Savage S (2011) Heat of the moment: characterizing the efficacy of thermal camera-based attacks. In: Proceedings of the 5th USENIX conference on offensive technologies. USENIX Association, pp 6–6
56. Wodo W, Hanzlik L (2016) Thermal imaging attacks on keypad security systems. In: SECRYPT, pp 458–464
57. Andriotis P, Tryfonas T, Oikonomou G, Yildiz C (2013) A pilot study on the security of pattern screen-lock methods and soft side channel attacks. In: Proceedings of the sixth ACM conference on Security and privacy in wireless and mobile networks. ACM, New York, pp 1–6
58. Abdelrahman Y, Khamis M, Schneegass S, Alt F (2017) Stay cool! understanding thermal attacks on mobile-based user authentication. In: Proceedings of the 2017 CHI conference on human factors in computing systems. ACM, New York, pp 3751–3763
59. Al Faruque MA, Chhetri SR, Canedo A, Wan J (2016) Forensics of thermal side-channel in additive manufacturing systems. In: CECS technical report# 16–01. University of California, Irvine
60. Stone S, Temple M (2012) Radio-frequency-based anomaly detection for programmable logic controllers in the critical infrastructure. Int J Crit Infrastruct Prot 5(2):66–73
61. Stone SJ, Temple MA, Baldwin RO (2015) Detecting anomalous programmable logic controller behavior using RF-based Hilbert transform features and a correlation-based verification process. Int J Crit Infrastruct Prot 9:41–51
62. Van Aubel P, Papagiannopoulos K, Chmielewski Ł, Doerr C (2017) Side-channel based intrusion detection for industrial control systems. arXiv preprint arXiv:1712.05745
63. Han Y, Etigowni S, Liu H, Zonouz S, Petropulu A (2017) Watch me, but don't touch me! contactless control flow monitoring via electromagnetic emanations. In: Proceedings of the 2017 ACM SIGSAC conference on computer and communications security. ACM, New York, pp 1095–1108
64. Boggs N, Chau JC, Cui A (2018) Utilizing electromagnetic emanations for out-of-band detection of unknown attack code in a programmable logic controller. In: Cyber sensing 2018, vol 10630, p 106300D. International Society for Optics and Photonics
65. Classen J, Chen J, Steinmetzer D, Hollick M, Knightly E (2015) The spy next door: eavesdropping on high throughput visible light communications. In: Proceedings of the 2nd international workshop on visible light communications systems. ACM, New York, pp 9–14
66. Loughry J, Umphress DA (2002) Information leakage from optical emanations. ACM Trans Inf Sys Secur (TISSEC) 5(3):262–289

67. Backes M, Dürmuth M, Unruh D (2008) Compromising reflections-or-how to read LCD monitors around the corner. In: Security and privacy, 2008. SP 2008. IEEE symposium on. IEEE, Piscataway, pp 158–169
68. Chakraborty S, Ouyang W, Srivastava M (2017) LightSpy: optical eavesdropping on displays using light sensors on mobile devices. In: Big Data (Big Data), 2017 IEEE international conference on. IEEE, pp 2980–2989
69. Wei L, Liu Y, Luo B, Li Y, Xu Q (2018) I know what you see: power side-channel attack on convolutional neural network accelerators. arXiv preprint arXiv:1803.05847
70. Jeon Y, Kim M, Kim H, Kim H, Huh JH, Yoon JW (2018) I'm listening to your location! Inferring user location with acoustic side channels. In: Proceedings of the 2018 World Wide web conference on world wide web. International World Wide Web Conferences Steering Committee, pp 339–348
71. Cao F, Malik S (2006) Vulnerability analysis and best practices for adopting IP telephony in critical infrastructure sectors. IEEE Commun Mag 44(4):138–145
72. De Meulenaer G, Standaert FX (2010) Stealthy compromise of wireless sensor nodes with power analysis attacks. In: International conference on mobile lightweight wireless systems. Springer, Berlin/Heidelberg, pp 229–242
73. Hively LM, McDonald JT (2013) Theorem-based, data-driven, cyber event detection. In: Proceedings of the eighth annual cyber security and information intelligence research workshop. ACM, New York, p 58
74. Dawson JA, McDonald JT, Shropshire J, Andel TR, Luckett P, Hively L (2017) Rootkit detection through phase-space analysis of power voltage measurements. In: 2017 12th international conference on malicious and unwanted software (MALWARE). IEEE, Piscataway, pp 19–27
75. Gunti N B, Lingasubramanian K (2015) Efficient static power based side channel analysis for hardware trojan detection using controllable sleep transistors. In: SoutheastCon 2015. IEEE, pp 1–6
76. Shende R, Ambawade DD (2016) A side channel based power analysis technique for hardware trojan detection using statistical learning approach. In: Wireless and optical communications networks (WOCN), 2016 thirteenth international conference on. IEEE, Piscataway, pp 1–4
77. Moore S, Yampolskiy M, Gatlin J, McDonald JT, Andel TR (2016) Buffer overflow attack's power consumption signatures. In: Proceedings of the 6th workshop on software security, protection, and reverse engineering. ACM, New York, p 6
78. Clark SS, Ransford B, Rahmati A, Guineau S, Sorber J, Xu W, ..., Holcomb D (2013) WattsUpDoc: power side channels to nonintrusively discover untargeted malware on embedded medical devices. In: HealthTech
79. Abbas M, Prakash A, Srikanthan T (2017) Power profile based runtime anomaly detection. In: TRON symposium (TRONSHOW). IEEE, Tokyo
80. Gonzalez CA, Hinton A (2014) Detecting malicious software execution in programmable logic controllers using power fingerprinting. In: International conference on critical infrastructure protection. Springer, Berlin/Heidelberg, pp 15–27
81. Xiao YJ, Xu WY, Jia ZH, Ma ZR, Qi DL (2017) NIPAD: a non-invasive power-based anomaly detection scheme for programmable logic controllers. Front Inf Technol Electron Eng 18(4):519–534
82. Gong X, Kiyavash N (2013) Timing side channels for traffic analysis. In: Acoustics, speech and signal processing (ICASSP), 2013 IEEE international conference on. IEEE, Piscataway, pp 8697–8701
83. Gong X, Kiyavash N (2016) Quantifying the information leakage in timing side channels in deterministic work-conserving schedulers. IEEE/ACM Trans Networking 24(3):1841–1852
84. Hoyos J, Dehus M, Brown TX (2012) Exploiting the GOOSE protocol: a practical attack on cyber-infrastructure. In: Globecom Workshops (GC Wkshps), 2012 IEEE. IEEE, Piscataway, pp 1508–1513

85. Zhong X, Ahmadi A, Brooks R, Venayagamoorthy GK, Yu L, Fu Y (2015) Side channel analysis of multiple pmu data in electric power systems. In: Power systems conference (PSC), 2015 Clemson University. IEEE, Piscataway, pp 1–6
86. Zhong X, Arunagirinathan P, Ahmadi A, Brooks R, Venayagamoorthy GK (2015) Side-channels in electric power synchrophasor network data traffic. In: Proceedings of the 10th annual cyber and information security research conference. ACM, New York, p 3
87. Islam CS, Mollah MSH (2015) Timing SCA against HMAC to investigate from the execution time of algorithm viewpoint. In: Informatics, electronics & vision (ICIEV), 2015 international conference on. IEEE, Piscataway, pp 1–6
88. Johnstone MN, Peacock M, den Hartog JI (2015) Timing attack detection on bacnet via a machine learning approach
89. Dunlap S, Butts J, Lopez J, Rice M, Mullins B (2016) Using timing-based side channels for anomaly detection in industrial control systems. Int J Crit Infrastruct Prot 15:12–26
90. Kocher P, Genkin D, Gruss D, Haas W, Hamburg M, Lipp M, ..., Yarom Y (2018) Spectre attacks: exploiting speculative execution. arXiv preprint arXiv:1801.01203
91. Hintz A (2002) Fingerprinting websites using traffic analysis. In: International workshop on privacy enhancing technologies. Springer, Berlin/Heidelberg, pp 171–178
92. Lu L, Chang EC, Chan MC (2010) Website fingerprinting and identification using ordered feature sequences. In: European symposium on research in computer security. Springer, Berlin/Heidelberg, pp 199–214
93. Chen S, Wang R, Wang X, Zhang K (2010) Side-channel leaks in web applications: a reality today, a challenge tomorrow. In: 2010 IEEE symposium on security and privacy. IEEE, Los Alamitos, pp 191–206
94. Tsalis N, Stergiopoulos G, Bitsikas E, Gritzalis D, Apostolopoulos T (2018) Side channel attacks over encrypted TCP/IP Modbus reveal functionality leaks. In: Proceeding. of the 15th International Conference on Security and Cryptography (SECRYPT-2018), Portugal
95. de Souza Faria G, Kim HY (2013) Identification of pressed keys from mechanical vibrations. IEEE Transactions on Information Forensics and Security 8(7):1221–1229
96. de Souza Faria G, Kim HY (2016) Identification of pressed keys by time difference of arrivals of mechanical vibrations. Comput Secur 57:93–105
97. Chen CY, Ghassami A, Nagy S, Yoon MK, Mohan S, Kiyavash N, ..., Pellizzoni R (2015) Schedule-based side-channel attack in fixed-priority real-time systems
98. Weiß M, Weggenmann B, August M, Sigl G (2014) On cache timing attacks considering multi-core aspects in virtualized embedded systems. In: International conference on trusted systems. Springer, Cham, pp 151–167
99. August M (2014) IDP: an analysis of a cache-based timing side channel attack and a countermeasure on PikeOS
100. Gritzalis D, Iseppi G, Mylonas A, Stavrou V (2018) Exiting the risk assessment maze: a meta-survey. ACM Comput Surv (CSUR) 51(1):11

The manufacturer's authorised representative in the EU is Springer Nature Customer Service Centre GmbH, Europaplatz 3, 69115 Heidelberg, Germany. If you have any concerns regarding our products, please contact ProductSafety@springernature.com

Printed and bound by CPI Group (UK) Ltd, Croydon, CR0 4YY
28/09/2026
02224085-0006